Number	Page	Formula
8.4	146	$P(X \text{ and } Y) = P(X) \cdot P(Y)$ (for statistically independent events)
8.5	147	$P(X \text{ and } Y) = P(X) \cdot P(Y|X)$ or $P(Y) \cdot P(X|Y)$ (for statistically nonindependent events)
8.7	160	$\sigma_{\bar{X}} = \sigma/\sqrt{n}$
9.2	184	$s_{\bar{X}} = s/\sqrt{n}$
9.3	188	$t = \dfrac{\bar{X} - \mu}{s_{\bar{X}}}$
10.1	198	$\text{CI}_{95} = \bar{X} \pm (t_{cv})(s_{\bar{X}})$
11.3	213	$s_{\bar{X}_1 - \bar{X}_2} = \sqrt{s^2\left(\dfrac{1}{n_1} + \dfrac{1}{n_2}\right)}$
11.4	213	$s^2 = \dfrac{(n_1 - 1)s_1^2 + (n_2 - 1)s_2^2}{n_1 + n_2 - 2}$
11.5	213	$s_{\bar{X}_1 - \bar{X}_2} = \sqrt{\dfrac{(n_1 - 1)s_1^2 + (n_2 - 1)s_2^2}{n_1 + n_2 - 2}\left(\dfrac{1}{n_1} + \dfrac{1}{n_2}\right)}$
11.6	214	$t = \dfrac{(\bar{X}_1 - \bar{X}_2) - (\mu_1 - \mu_2)}{\sqrt{\dfrac{(n_1 - 1)s_1^2 + (n_2 - 1)s_2^2}{n_1 + n_2 - 2}\left(\dfrac{1}{n_1} + \dfrac{1}{n_2}\right)}}$
11.8	218	$s_{\bar{d}} = \sqrt{\dfrac{s_d^2}{n}}$
11.9	218	$s_d^2 = \dfrac{\Sigma(d - \bar{d})^2}{n - 1} = \dfrac{\Sigma d^2 - \dfrac{(\Sigma d)^2}{n}}{n - 1}$
12.1	229	$s_p = \sqrt{\dfrac{pq}{n}}$

Continued inside back cover

Basic Behavioral Statistics

Basic Behavioral Statistics

Dennis E. Hinkle
Virginia Polytechnic Institute
and State University

William Wiersma
The University of Toledo

Stephen G. Jurs
The University of Toledo

Houghton Mifflin Company   Boston

Dallas   Geneva, Illinois Hopewell, New Jersey Palo Alto London

To
Mary, Chas, and Anya
and
Joan, Lisa, and Susan
and
Sara, Peter, and Andrew

Ceramic art on cover and chapter openers by Kenneth Goldstrom.

Library of Congress Catalog Card Number: 81-81700

ISBN: 0-395-31729-0

Printed in the U.S.A.

Contents

2

Frequency Distributions 23

3

Describing Distributions 47

4

Describing Individual Scores 69

5

The Normal Curve 85

6

Correlation 103

7

Regression and Prediction 123

8

Probability 141

9

Hypothesis Testing:
One-Sample Case for the Mean   169

10

Estimation: One-Sample Case for the Mean 197

11

Hypothesis Testing and Estimation: Two-Sample Case for the Mean 209

12

Hypothesis Testing for Other Statistics 227

13

Analysis of Variance   251

14

Two-Way Analysis of Variance 281

15

Selected Nonparametric Tests 303

Appendixes

Foreword to the Student

By Sheila Tobias
author of *Overcoming Math Anxiety* (Houghton Mifflin, 1980)

For some very unusual people mathematics is a gift. For the rest of us it is a collection of learnable (and easily forgettable) skills. Hence it may be wise, before you undertake to learn the material covered in this text, to do some systematic review of the algebra and precalculus courses you may have taken, along with a group of supportive and like-minded friends. Such re-exposure will remind you of mathematical notation and of some of the basic algebra you once knew. And it may help you "hit the ground running" when you start statistics.

A more problematical aspect of mathematics, somewhat harder to deal with, is the built-in prejudice some of us bring to mathematics and math-related fields of study. We call this phenomenon "math anxiety" or "math avoidance," and in numerous intervention programs around the country, students are being urged—before they confront their required statistics course—to undertake some "math anxiety reduction." Such a process involves self-examination of your own, personal math autobiography, some self-monitoring of the ways you may be defeating yourself when doing mathematics, and some study of mathematics in a tension-free and pressure-free environment.

Another technique that is used for reducing math anxiety is assertiveness training. Some of us feel embarrassed when we don't "get" a mathematical idea the first time it is presented, and consequently we don't ask questions in class. Assertiveness training teaches students how to ask their instructors for the help they need. Sometimes the text—even one as carefully prepared as this one—may not answer a question floating around in your mind. Note such questions and ask for clarification. If necessary, write your questions down and give them to your instructor. Being cumulative, your statistics course will build continuously on everything that has gone before. A missed concept or a misunderstanding is like a dropped stitch; it is all the harder to correct later on.

A third helpful technique is to work with other people. Your instructor may not permit you to take group tests (though some teachers are beginning to allow this in order to reduce tension and competitiveness among students), but group study is often worthwhile. Explaining concepts to one another enhances and deepens your own learning, and the group experience relieves some of the uncertainty of doing mathematics in isolation.

Above all, don't be demoralized by your mistakes. In a session held for math-anxious adults, a graduate student in mathematics once agreed to answer general questions from the group. An intrepid woman ventured this question: "How do you feel when you make a mistake?" After a short pause, the graduate student answered, "I find my mistakes interesting and my confusions even more so." The group gasped. They had come to believe that mistakes were like blemishes, emblems of their stupidity. The graduate student had pointed out, correctly, that mistakes are like windows into our minds.

Examine your mistakes. Are they the result of careless error—something you wouldn't do again—or do they indicate a missing skill? Did you mislearn an operation that worked for awhile on simpler data but will not work on the new material? Is the problem in the mathematics? In your comprehension of the verbal information presented? In the notation? (You will be learning more new notation in your statistics course than you learned in your algebra courses in high school. Some of this notation will be different from and contradictory to what you learned before.) Or is the problem in visualizing the operation when it is expressed on a graph?

Obviously, the better you analyze your errors (alone or with the help of your tutor or instructor), the less likely it is that you will make them in the future.

Finally, be flexible. Even mathematicians may modify a problem they are working on to get a better handle on it. They substitute smaller numbers or replace a long series with a shorter one, just to get an idea of the parameters of the problem and to test their methods. Guessing is O.K., so long as you find a good way to test your hunches. (Sometimes the way you test your hunch will give you a good clue to the formula you are seeking.) And use your intuition. Math-anxious people trust their intuition far less than successful learners of math and statistics.

This textbook has been skillfully prepared and carefully reviewed by many experienced teachers of statistics, with the aim of making the presentation as clear as possible. But this does not mean you will learn everything you need to know the first time around. "Reading" mathematics or statistics is very different from reading social science or general nonfiction. Read with a paper and a pencil; each statement is made once and you must master it before going on. Imagine examples, imagine counterexamples. Note any questions that arise and, if these are not answered by the time you are finished studying, give them to your instructor. Try to slow down your reading speed. Think of reading statistics as rather like "reading" a crossword puzzle. You don't just look at the cues for Down and Across; you work them out. And the same is true for this textbook. Don't just glance at the words; work out the ideas and apply them.

If you have anxiety to deal with, try to make yourself relax. Writing down your "self-defeating self-talk" is one way to relax. Try to get more time to complete your tests if your instructor will allow it. And you may want to consult other books on math anxiety.

Consider ways to make your first course in statistics easier for you. You might audit the course for one semester before taking it for credit. This will give you an idea of the scope of the material and some advance warning about what the more difficult areas will be. You might even discover that the subject of statistics is both fascinating and empowering: The opportunity to take control of data, to find patterns in ostensibly unrelated phenomena, and to deal critically with other people's research will give you thinking strengths that no other course can offer.

Preface

This book was written primarily as a text for a one-term or semester course in statistics. However, it can serve as a supplement for related courses with a quantitative orientation, such as courses in measurement and research methodology. It is an introductory text and is designed to teach concepts and procedures that are fundamental for further study in statistics. The book can be used for either undergraduate or graduate courses found in behavioral sciences curricula such as psychology, education, and sociology.

Conceptual Approach The approach of the text is conceptual and nonmathematical in nature. Any study of statistics must include some consideration of formulas and computation, but we have included formulas only when necessary and then with an emphasis on understanding the formula, not a ''cookbook'' use of it. We have selected the simplest version of formulas in most cases, and we do not confuse the student with three or four equivalent formulas. Basic concepts of statistical reasoning and the underlying assumptions are presented. For undergraduate students and beginning graduate students (especially those who may take additional statistics courses) such concepts are necessary, even though they require only a limited mathematical background. Computation is illustrated through the examples that are introduced to develop understanding of the concepts.

Learning and Teaching Aids The book contains several features designed to enhance its use as a teaching tool and a learning device. Statements summarizing the important points being made are displayed prominently throughout the chapters. Key concepts listed at the end of each chapter help the reader review the major concepts introduced in the chapter. Exercises are included for each chapter, and solutions to the exercises are provided in an appendix. Thus students can receive immediate feedback about their solutions. For students who use an electronic calculator to work problems, the key strokes for selected exercises have been included at the end of each chapter in a section identified by a calculator symbol, as on page 21. The use of pocket calculators in statistical analysis is described in Appendix A.

A *Student Workbook* is available and can be used effectively as a supplement to the text. It includes definitions of key concepts, exercises to reinforce these concepts, and additional exercises for further review.

The first chapter offers an optional review of elementary mathematics for those students with a very limited mathematics background. This review covers the basic operations of addition, subtraction, multiplication, and division, along with the real number system and elementary algebra. Notation is explained, and the Σ operator, so important in statistics, is introduced. But essentially, no mathematical background beyond elementary algebra is required for using the text effectively.

Coverage
The book includes both descriptive and inferential statistics, the latter through analysis of variance. The basic descriptive statistics are presented in Chapters 2 through 7; included are graphs, measures of location, measures of central tendency, measures of variability, standard scores, correlation, and regression. The learning process is enhanced through the use of realistic examples and exercises. It should be noted that the concept of regression is considered in the descriptive sense—that is, in terms of the development of the regression equation, the use of this equation in the prediction process, and the definition of the standard error of prediction.

Inferential statistics are introduced via a discussion of the concept of probability and its relationship to sampling and sampling distributions. These basic concepts are then used in developing the logic underlying hypothesis testing and parameter estimation. Hypothesis testing and estimation are discussed for the one-sample case for the mean in Chapters 9 and 10 and then expanded to the two-sample case in Chapter 11. One-sample and two-sample cases for other parameters are considered in Chapter 12.

Chapter 13 is devoted to one-way analysis of variance (ANOVA). Conceptual development of the partitioning of the sum of squares and *post hoc* multiple comparison procedures receive special consideration. Repeated measures analysis of variance is also included. Two-factor analysis of variance is treated in Chapter 14 as the logical extension of one-way ANOVA. Selected nonparametric statistics are presented in Chapter 15; these procedures are the analogues to the procedures discussed in Chapters 9 through 13.

Acknowledgments
We are grateful to the Literary Executor of the late Sir Ronald A. Fisher, F.R.S., to Dr. Frank Yates, F.R.S., and to Longman Group Ltd., London for permission to reprint Tables III, IV, VII, and XXIII from their book *Statistical Tables for Biological, Agricultural and Medical Research* (6th edition, 1974).

We would like to acknowledge several people who have contributed to this effort. First of all we would like to thank the reviewers who gave us most constructive criticisms as well as suggestions that have enhanced the manuscript: William A. Frederickson, Central State University; John Harsh, University of Southern Mississippi; Donald V. Huard, Phoenix College; Charles R. Kessler, Los Angeles Pierce College; and Alice L. Palubinskas, Tufts University. Special thanks goes to J. R. Cox for his review of the manuscript and to L. E. Ewing for his assistance in developing the exercises and their solutions. Our most special acknowledgment is to our families, to whom we dedicate the book.

Dennis E. Hinkle
William Wiersma
Stephen G. Jurs

1

Introduction

A common misconception in these days of advertising claims about the superiority of one product over another is that statistics can be used to prove almost anything. Such claims remind us of Benjamin Disraeli's statement that "there are three kinds of lies—lies, damned lies, and statistics." There is no doubt that statistics have been and are used and misused to demonstrate the superiority of a commercial product, the relationship between smoking and lung cancer, and so on. Without the statistics, there would be no basis for such claims. Thus it would seem that a basic understanding of statistics would enhance our ability to discriminate between sound scientific facts and exaggerated advertising claims.

The Meaning of Statistics

The meaning of the word "statistics" is often confusing, because the lay user of statistics and the professional/mathematical statistician use the term to mean different things. For example, lay users of statistics include the weather forecaster and the sports reporter. The weather forecaster reports such statistics as the high and low temperatures, the amount of rainfall, and the air quality index. The sportscaster reports the statistics for individual players as well as team statistics. On the other hand, the professional/mathematical statistician uses the term "statistics" to refer to the statistical techniques that have been developed to assist the research scientist in analyzing and understanding the data yielded by a research study.

The word will be used in both ways in this book, but we will be most concerned with understanding the basic statistical methods available to the research scientist. These procedures involve mathematical calculations that summarize the data collected in research studies. So, in this context, we will use *statistics* to refer to the entire body of mathematical theory and procedures that are used to analyze data.

Statistics is a word with various meanings. In some situations it simply means bits of information. In its broadest sense, it refers to the theory and procedures used for the purpose of understanding data.

Basic Mathematics Needed for Statistics

Whatever the reason for using statistics, the researcher should have an understanding of what the statistics are doing, the information they are providing, and the conclusions that can be drawn from them. Statistics do involve everything from basic to complex mathematics. In this introductory text, no complex mathematics is required, but it is necessary to be familiar with some elementary mathematical concepts and operations. Some arithmetic and elementary algebraic operations, along with basic statistical symbolism, will suffice. A brief review of mathematical concepts and operations follows. You must master them to understand the statistical concepts and procedures that you will work with in this text.

Four Basic Arithmetic Operations

Addition, subtraction, multiplication, and division are the four *basic arithmetic operations* performed with numbers. The algebraic sign for addition is the plus sign (+), and the addition of two numbers, say X and Y, is indicated as $X + Y$. For example, if $X = 15$ and $Y = 5$, then $X + Y = 15 + 5 = 20$. The result of the addition operation is called the *sum*. The algebraic symbol for subtraction is the minus sign $(-)$. Subtracting Y from X, we find that $X - Y = 15 - 5 = 10$. The result of the subtraction operation is called the *difference*.

Multiplication of two numbers can be algebraically denoted in three ways: $X \cdot Y$ or $(X)(Y)$ or just simply XY. All three ways will be used in this book.

$$X \cdot Y = 15 \cdot 5 = 75$$

$$(X)(Y) = (15)(5) = 75$$

$$XY = (15)(5) = 75$$

The result of the multiplication operation is called the *product;* the individual numbers that are multiplied are called *factors*.

Division of two numbers can also be algebraically denoted in three ways: $X \div Y$ or X/Y or $\dfrac{X}{Y}$. In this latter form, the number on the top is called the *numerator* and the number on the bottom is called the *denominator*.

$$X \div Y = 15 \div 5 = 3$$

$$X/Y = 15/5 = 3$$

$$\frac{X}{Y} = \frac{15}{5} = 3$$

The result of the division operation is called the *quotient*.

It is important to note the relationship between the multiplication operation and the division operation. The quotient X/Y can also be expressed as the product $(X)(1/Y)$. For example,

$$X/Y = 15/5 = 3$$

$$(X)(1/Y) = (15)(1/5) = 3$$

Arithmetic Operations with Real Numbers

In elementary arithmetic, we are most familiar with positive numbers—that is, all numbers zero (0) or greater. In the study of statistics, however, we must deal with what mathematicians refer to as real numbers. *Real numbers* are all positive and negative numbers from negative infinity $(-\infty)$ to positive infinity $(+\infty)$. Thus real numbers include not only the positive and negative integers (whole numbers) but also all fractions and decimals.

Consider the real number line, which is illustrated in Figure 1.1. It represents the positioning of the real numbers along the line. For convenience, the fractions are shown above the line and the integers below it. Note that, though only fractions in multiples of one-half have been included, any fraction could have been used.

When both positive and negative numbers are used in any arithmetic operation, we often refer to them as *signed numbers*. The algebraic sign in front of the number affects the results of these arithmetic operations. Sometimes we want to ignore the

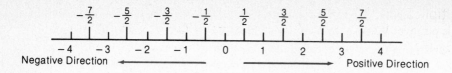

Figure 1.1   The real number line

algebraic sign and consider only the *absolute value* of the number—that is, the value of the number without regard to its algebraic sign. For example, the absolute value of -7, which is denoted $|-7|$, is 7. We know from elementary arithmetic that, as we go along the real number line in the positive direction, 5 is greater than 4, 6 is greater than 5, and so on. Conversely, going in the negative direction, -5 is less than -4, -6 is less than -5, and so on. However, in absolute values, $|-7| = |+7| = 7$.

The real numbers are all positive and negative numbers from $-\infty$ to $+\infty$. The absolute value of any real number is the value of the number without regard to its algebraic sign.

The rules for the arithmetic operations are defined and discussed in the following paragraphs, using the numbers $W = -10$, $X = +15$, $Y = +5$, and $Z = -25$. (Make a note of these values for the variables W, X, Y, and Z. We will use them repeatedly in this chapter.)

Addition
When two numbers have the same sign, add the absolute values of the numbers and retain the sign.

$$W + Z = (-10) + (-25) = -35$$
$$X + Y = (+15) + (+5) = +20$$

When two numbers have opposite signs, subtract the absolute value of the smaller from the absolute value of the larger and retain the sign of the larger.

$$W + X = (-10) + (+15) = +5$$
$$Y + Z = (+5) + (-25) = -20$$

Subtraction
When subtracting numbers, change the sign of the number being subtracted, and then apply the rules for addition.

$$W - Y = (-10) - (+5) = (-10) + (-5) = -15$$
$$W - Z = (-10) - (-25) = (-10) + (+25) = +15$$

Multiplication

When we multiply two numbers that have the same sign, the product is positive. When we multiply two numbers that have different signs, the product is negative.

$$(W)(Z) = (-10)(-25) = +250$$

$$(X)(Y) = (15)(5) = +75$$

$$(W)(X) = (-10)(15) = -150$$

$$(Y)(Z) = (5)(-25) = -125$$

Division

When we divide two numbers that have the same sign, the quotient is positive. When we divide two numbers that have different signs, the quotient is negative.

$$X/Y = +15/+5 = +3$$

$$W/Z = -10/-25 = +0.40$$

$$X/W = +15/-10 = -1.50$$

When performing arithmetic operations with real numbers, consider the signs of the numbers carefully.

Rounding Numbers

In our use of statistical methods, we are often required to divide one number by another. For example, when we divide Y by X (given $X = 15$ and $Y = 5$), we have $5/15$, or $1/3$. If we were to express this fraction as a decimal, the result would be $0.3333+$. From this example, it should be obvious that this decimal can be extended indefinitely. However, we must determine (1) where the decimal should be terminated and (2) which value should be assigned to the last number in the decimal.

In a strict mathematical sense, these two concerns are addressed by several rules for determining the number of significant digits. In the study of statistics, however, the following rule[1] is usually adopted:

After every arithmetic operation that yields a decimal fraction, carry the decimal to three places and round to two places more than there were in the original data.

For example, if the original data were in whole numbers, we would carry the decimal to three places and round to the second decimal place. If the data were recorded in

---

[1] Instructors may have established their own conventions for rounding to significant digits. Any convention that maintains an adequate degree of accuracy is appropriate.

tenths, we would carry the decimal to four places and round to the third decimal place, and so on.

After we have decided on the number of decimal places, we must determine the value of the last digit. The following rule is generally adopted:

If the number beyond the last digit to be reported is greater than 5, increase the last digit to the next higher number. If the number beyond the last digit is less than 5, the last digit remains the same.

Consider the following examples, in which the decimal is to be carried to three places and then rounded to two places.

$$3.826 \text{ rounds to } 3.83$$

$$4.842 \text{ rounds to } 4.84$$

If the number beyond the last digit *is* 5, our rule must be extended as follows:

If the number beyond the last digit to be reported is exactly 5 with no remainder, round the last digit to the nearest even number. If the last digit is already an even number, it remains the same; if the last digit is odd, add 1 to this digit.

Consider the following examples.

$$7.3650001 \text{ rounds to } 7.37$$

$$4.8650000 \text{ rounds to } 4.86$$

$$9.5856901 \text{ rounds to } 9.59$$

$$6.3550000 \text{ rounds to } 6.36$$

The rule for rounding numbers is to carry the decimal to three places and round to two places more than there were in the original data.

Postulates for Real Numbers

A *postulate* is a characteristic of a number system that is accepted without proof. Real numbers have three important characteristics that are often used in statistical formulas and calculations. These characteristics are the *commutative postulate*, the *associative postulate*, and the *distributive postulate*.

Commutative Postulate The commutative postulate tells us that two numbers (X and Y) may be added (or multiplied) in any order to achieve the same result.

$$\text{For addition:} \quad X + Y = Y + X$$

$$15 + 5 = 5 + 15 = 20$$

For multiplication: $(X)(Y) = (Y)(X)$

$$(15)(5) = (5)(15) = 75$$

Associative Postulate   The associative postulate tells us that when three numbers are added (or multiplied), we can add (or multiply) X and Y first and then W or we can add (or multiply) W and X first and then Y. The result will be the same.

For addition: $W + (X + Y) = (W + X) + Y$

$$-10 + (15 + 5) = (-10 + 15) + 5$$

$$-10 + 20 = +5 + 5$$

$$10 = 10$$

For multiplication: $W(X \cdot Y) = (W \cdot X)Y$

$$(-10)(15 \cdot 5) = (-10 \cdot 15)(5)$$

$$(-10)(75) = (-150)(5)$$

$$-750 = -750$$

Distributive Postulate   The distributive postulate tells us that the product of a number (Y) with the sum of two numbers $(W$ and $X)$ is the same as the sum of the products of Y with W and Y with X.

$$Y(W + X) = (Y)(W) + (Y)(X)$$

$$5(-10 + 15) = (5)(-10) + (5)(15)$$

$$5(5) = -50 + 75$$

$$25 = 25$$

In addition to these postulates, there are two special numbers in the real number system, 0 and 1. The arithmetic operations with these two numbers deserve special consideration. When the number 0 is added to or subtracted from any number, the result is that number: $X + 0 = X$ and $X - 0 = X$. When any number is multiplied by 0, the result is 0: $(X)(0) = 0$. Now consider the number 1. When any number is multiplied by 1, the result is that number: $(X)(1) = X$.

Three postulates of the real number system are used in statistical formulas and calculations: the commutative, associative, and distributive postulates.

Special Indicators of Arithmetic Operations

Two special indicators of arithmetic operations commonly appear in statistics. These are the exponent and square root symbols. Squaring a number, which is defined as multiplying a number by itself, is a common calculation in statistics. For example, if we want to find the square of Y, we multiply Y times Y: $(Y)(Y) = (5)(5) = 25$ (given that $Y = 5$). Symbolically, the square is denoted Y^2 or Y to the second power. The number 2 is called the *exponent,* which is the power to which the number is raised. If Y were raised to the fourth power, Y^4, we would have $Y^4 = (Y)(Y)(Y)(Y) = (5)(5)(5)(5) = 625$.

Finding the square root of a number is also a common calculation. The *square root* of a number is a real number such that, when you multiply the square root by itself, the product is the original number. The symbol $\sqrt{}$ over the number indicates that the square root is to be taken. For example, $\sqrt{36} = 6$ because $6^2 = (6)(6) = 36$. The square root of a number can also be indicated by the fractional exponent $1/2$; that is, $\sqrt{36} = (36)^{1/2} = 6$.

Finding the square root of a number involves a very cumbersome procedure. It is unnecessary to review this process, because square roots can be readily obtained by consulting a table of square roots (see Table 11 in the Appendix) or using a hand calculator.

Order of Operations

When we combine two or more operations in a single numerical solution, the order in which the operations are done is important. For example, for the problem

$$5 \times 8 + 16 \div 4$$

do we multiply five times eight (40), add 16 to obtain 56 and divide 56 by 4 for a final solution of 14? No. The operation indicators mean that the 4 is divided only into the 16, then this quotient is added to the product of five times eight. So, the solution is 44.

The rules or order for operations are:

1. Any operations in parentheses (or brackets and braces) are performed first. If we have parentheses within brackets we perform the operations from the inside out.
2. Operations with exponents, such as squares or square roots are performed next, if they were not contained within the parentheses.
3. When these operations are complete, which clear the expression of parentheses, do the multiplications and divisions, then the additions and subtractions.

Consider another example:

$$\sqrt{49} + 3(7 + 8) - (10 + 6) \div 2^2$$

$\sqrt{49} + (3)(15) - 16 \div 2^2 \quad$ operations in parentheses

$7 + (3)(15) - 16 \div 4 \quad$ operations with exponents (and the square roots)

$7 + 45 - 4 \quad$ multiplication and division operation

$48 \quad$ addition and subtraction operation

The Summation Operator

One of the more important operations of statistics, which may be new to the reader, is the *summation operator,* denoted by the Greek capital letter sigma (Σ). The Σ indicates the summing of whatever follows immediately in the expression. For example, suppose we have the following five numbers:

$$X_1 = 3 \qquad X_2 = 7 \qquad X_3 = 4 \qquad X_4 = 2 \qquad X_5 = 8$$

The subscripts 1 through 5 on the X's simply indicate the different numbers. Then, if we want to sum the five numbers, we proceed as follows:

$$\sum_{i=1}^{5} X_i = X_1 + X_2 + X_3 + X_4 + X_5$$

$$= 3 + 7 + 4 + 2 + 8$$

$$= 24$$

The X_i (called cap X, subscript i) is the general symbol for the number. The notation under the Σ indicates the beginning of the summation with the number X_1, and the 5 above the Σ indicates its continuation through the number X_5. In general, the notation

$$\sum_{i=1}^{N} X_i$$

which in running text is written $\Sigma_{i=1}^{N} X_i$ means that summation begins with the first number and concludes with Nth. Often the notations above and below the Σ are omitted; when they are, Σ means summation from the first through the Nth number.

The summation operator is often convenient for simplifying statistical formulas, and certain rules govern the summation operation. Three useful rules are presented here.

Rule 1   Applying the summation operator, Σ, to the products resulting from multiplying a set of numbers by a constant is equal to multiplying the constant by the sum of the numbers. Symbolically,

$$\sum_{i=1}^{N} CX_i = C \sum_{i=1}^{N} X_i$$

Let 2 be the constant, and let the numbers be $X_1 = 1$, $X_2 = 3$, $X_3 = 4$, and $X_4 = 7$. To apply this operation, these numbers could be written 2(1), 2(3), 2(4), and 2(7). Finding the sum of the numbers, we get

$$\sum_{i=1}^{4} (2)X_i = 2(1) + 2(3) + 2(4) + 2(7)$$

$$= 2 \sum_{i=1}^{4} X_i = 2(1 + 3 + 4 + 7)$$

$$= 30$$

Rule 2 Applying the summation operator, Σ, to a series of constant scores is equal to taking the product of N times the constant score. Symbolically,

$$\sum_{i=1}^{N} C = NC$$

Suppose $X_1 = 4$, $X_2 = 4$, and $X_3 = 4$; that is, all three scores are equal to the constant 4. Then

$$\sum_{i=1}^{3} X_i = 4 + 4 + 4$$

$$= \sum_{i=1}^{3} C = 3(4)$$

$$= 12$$

Rule 3 Applying the summation operator, Σ, to the algebraic sum of two (or more) scores of a single individual, and then summing these sums over the N individuals, is the same as summing each of the two (or more) scores separately over the N individuals and then summing the scores. Symbolically,

$$\sum_{i=1}^{N} (X_i + Y_i) = \sum_{i=1}^{N} X_i + \sum_{i=1}^{N} Y_i$$

Suppose we have four individuals with the following X and Y scores:

Individual	1	2	3	4
X	2	5	3	1
Y	7	9	6	5

We find that

$$\sum_{i=1}^{4} (X_i + Y_i) = (X_1 + Y_1) + (X_2 + Y_2) + (X_3 + Y_3) + (X_4 + Y_4)$$

$$= 9 + 14 + 9 + 6$$

$$= 38$$

and

$$\sum_{i=1}^{4} X_i + \sum_{i=1}^{4} Y_i = (X_1 + X_2 + X_3 + X_4) + (Y_1 + Y_2 + Y_3 + Y_4)$$

$$= (2 + 5 + 3 + 1) + (7 + 9 + 6 + 5)$$

$$= 38$$

The summation operator, Σ, is one of the most widely used symbols in statistics. It indicates the summing of the numbers that immediately follow it in the expression.

Variables

Earlier we introduced the term "data" in describing the meaning of statistics, but what are data and where do they come from? *Data* are bits of information that are gathered on some characteristic of, for example, a group of individuals. If the characteristic can take on different values for different individuals, then the characteristic is referred to as a *variable*. For example, a group of college sophomores will be found to differ in sex, height, intelligence, and attitudes and in many other ways. If 25 college sophomores performed on a physical agility test, scores on this test would be the variable.

On the other hand, if a characteristic is the same for every member of the group, it is called a *constant*. For example, if only men are involved in a study, sex of the individual is constant. Similarly, grade level is a constant in a study involving only fifth-grade students.

A variable is a characteristic that can take on different values for different members of the group under study. A constant is a characteristic that assumes the same value for all members of the group.

Continuous and Discrete Variables

Numerous descriptive terms can be used to categorize variables. One such categorization distinguishes between continuous and discrete variables. A *continuous variable* is one that can take on any value on the measurement scale being used. Weight, height, and temperature are examples of continuous variables. It is important to note that, even though a scale measures weight only to the nearest pound or quarter of a pound, any weight *can* occur. A continuous variable is said to have underlying continuity. This means that there are no breaks, at least theoretically, in the possible values that the variable can assume.

Unlike a continuous variable, a *discrete variable* can take on only designated values. For example, the number of patrons in a theater at a given time is a variable that can take on only integer values: 1, 2, 3, and so on. There can be no fractional value of a patron at any time. Another example of a discrete variable is the possible sums obtained from repeated throws of two dice. The only possible values of these sums are the integers 2 (double 1) through 12 (double 6), inclusive.

A continuous variable has underlying continuity and can take on any value on the scale of measurement. A discrete variable can assume only designated values.

Independent and Dependent Variables

In research studies dealing with possible relationships among variables, a distinction is often made between independent and dependent variables. *Independent variables* are those over which the researcher exercises control. In every experiment, the researcher manipulates at least one independent variable in accordance with the purpose of the experiment. For example, suppose the purpose of an experiment is to determine the effects of the level of drug dosage on the performance of rats running a maze. The researcher controls the dosage and observes the variation in performance that results as the dosage is varied. In some studies, the independent variable is simply a classifying variable; it classifies the individuals under study. Sex of the individual is a classifying variable that takes on only two values, male and female.

A *dependent variable* is a variable that is affected by the independent variable. It is sometimes said to be (or is presumed to be) the "result" of manipulation of the independent variable by the researcher. In the drug dosage example, the dependent variable is performance in running the maze. This variable might be measured as the time required to negotiate the maze successfully. As the independent variable is changed, or varied, the researcher observes the changes in the dependent variable to determine how these changes are related to or associated with the changes in the independent variable.

An independent variable is one that is controlled or manipulated by the researcher; a dependent variable is one that is affected by the independent variable.

Measurement Scales

Considering the number of different variables that we might measure in an experiment, it soon becomes apparent that not all measurement is the same. There are different levels of measurement. Measuring people's height is not the same as measuring their opinions about political parties. And classifying people as short, medium, or tall is not as precise as measuring their height to the nearest inch.

Measurement, broadly considered, is the process of assigning numbers to characteristics according to a defined rule. If a student is measured on knowledge of history via a written test, the test score represents the measurement of that characteristic. The score is determined through the rule by which numbers are assigned to responses on the test. What we intend to come up with in this case is a number with quantitative meaning. In this example, the higher the score, the greater the student's knowledge of history.

The level of measurement has important implications for the use of statistics. The measurement scale of the dependent variable is one of the more important factors used in determining the statistical methods that are appropriate to analyze data. Measurement scales are distinguished according to the degree of "precision" the measurement exhibits. To label an individual as tall is not so precise as saying that the individual is 6 feet, 5 inches in height. Some variables are easier than others to measure precisely. Weight and height are examples of variables that can be measured very precisely. On the other hand, it is more difficult to measure precisely such variables as attitude toward a profession and level of anxiety.

The measurement of all characteristics is classified into a hierarchy of measurement scales. This hierarchy includes the nominal scale, the ordinal scale, the interval scale, and the ratio scale.

Nominal Scale

The first and simplest scale in the measurement hierarchy is the *nominal scale*. Nominal measurement is the process by which different objects are classified into categories on the basis of some defined characteristic. For example, consider the classification of tennis rackets by their various manufacturers. Following identification of the manufacturer, the number of tennis rackets produced by each is counted. It is important to note that, in nominal measurement, there is no logical ordering of the categories. That is, one brand of tennis racket is *not* assumed to be better than any other. This property of nominal measurement is of utmost importance when the categories are given some numerical designation. For example, sex of an individual is also a nominal variable, and we often assign a 1 to males and a 2 to females. This numerical designation has no quantitative meaning; we could have assigned a 1 to females and a 2 to males without changing the measurement or the meaning of the variable.

Data obtained from nominal-scale measurement are called nominal data. The properties of nominal data are as follows:

1. Data categories are mutually exclusive (an observation can belong to only one category).
2. Data categories have no logical order.

In summary, a nominal scale simply classifies without ordering.

Ordinal Scale

At the next level of measurement in the hierarchy, the *ordinal scale,* an additional property is present that was lacking in the nominal scale—*a logical ordering of the categories.* The measurement process is essentially the same, but the variable measured is such that differences in the amount of the characteristic possessed are discernible. And, when numbers are assigned to the categories, they are assigned according to the amount of the characteristic possessed. An example of an ordinal scale is the letter grading system: A, B, C, D, and F. If these grades are assigned in some academic course, we know that an individual who received an A demonstrated a higher level of performance than an individual who received a B. However, we cannot say that the difference in performance between an A and a B is the same as the difference between a B and a C.

The properties of ordinal data are as follows:

1. Data categories are mutually exclusive.
2. Data categories have a logical order.
3. Data categories are scaled according to the amount they possess of the characteristic being considered.

In summary, the categories that make up an ordinal scale exhibit the properties of distinctiveness and order.

Interval Scale

The third level in the measurement hierarchy is the *interval scale.* In addition to all the properties of the preceding scales, interval data have an additional property— *differences between the various levels of the categories on any part of the scale reflect equal differences in the characteristic measured.* That is, an equal unit is established in the scale. For this reason, interval-scale measurement is also called equal-unit measurement. The temperature scale on thermometers is a commonly cited example of an equal-unit scale. Equal differences between any two points on the scale are the same regardless of their positions in the scale. For example, the difference between temperatures of 75 and 80 degrees Fahrenheit is the same as the difference between temperatures of 55 and 60 degrees Fahrenheit. It is important to note that the point zero is just another point on the scale. It does not reflect the starting point of the scale, nor does it indicate total absence of the characteristic.

The properties of interval data are as follows:

1. Data categories are mutually exclusive.
2. Data categories have a logical order.
3. Data categories are scaled according to the amount they possess of the characteristic being considered.
4. Equal differences in the characteristic are represented by equal differences on the scale.
5. The point zero is just another point on the scale.

In summary, an interval scale has distinctive and ordered categories with equivalence of interval differences.

Ratio Scale

The highest level of measurement in the hierarchy is the *ratio scale*. Ratio data have all the properties of the preceding scales and one additional property—*a true zero point that reflects an absence of the characteristic*. This property enables us to make statements about proportional amounts of the characteristic possessed by two or more objects. Consider height, weight, and age; these are examples of ratio variables. With this additional property, not only can we say that the difference between 150 and 155 pounds is the same as the difference between 70 and 75 pounds, but we can also say that a person who weighs 150 pounds weighs twice as much as a person who weighs 75 pounds.

The properties of ratio data are as follows:

1. Data categories are mutally exclusive.
2. Data categories are ordered.
3. Data categories are scaled according to the amount they possess of the characteristic being considered.
4. Equal differences in the characteristic are represented by equal differences on the scale.
5. The point zero reflects an absence of the characteristic.

In summary, the ratio scale has distinctive and ordered categories with equivalence of interval differences and a true zero point.

The four levels of measurement can be summarized as follows:
The nominal scale categorizes without order.
The ordinal scale categorizes with order.
The interval scale categorizes with order and establishes an equal unit in the scale.
The ratio scale categorizes with order, establishes an equal unit, and contains a true zero point.

As can be seen, at each level of the measurement hierarchy, one additional property yields additional precision of measurement. However, it is possible to use a lesser degree of precision in measuring a variable, even though the higher degree of precision is available. For example, we could measure precisely the weight of an individual in a physiological research study but then categorize the individual as "very light," "light," "moderate in weight," "heavy," or "very heavy." In this case, we have reduced the measurement scale to a lower-level scale—from ratio to ordinal. This practice is not uncommon, but remember that reducing the precision of the measurement results in a loss of some information about the variable under investigation.

Populations and Samples

In the study of statistics, the terms "population" and "sample" are frequently used. A *population* consists of all members of some defined group. For example, the population of United States citizens consists of all residents of the 50 states at any one time. Another population could be defined as all the registered voters in Scranton, Pennsylvania. Often we tend to assume that populations contain a large number of members. This is not always true. A smaller defined population might be, for example, the students enrolled in an introductory psychology course at a specific state university during the spring semester. The distinguishing characteristic of a population is that it includes *all* members who meet the specific criteria used to define the population.

On the other hand, a *sample* is defined as a subset of a population. A sample is drawn from the population when it is impossible to include all members of the defined population in a particular research study. In statistics, we generally refer to random samples. This term and the criteria used to select the members of the population that are to be included in a random sample will be discussed in a later chapter. At this point it is important to note that the selection of a sample for a research study is not a haphazard process. The sample is selected in such a way that, when data are collected and analyzed using the statistical procedures discussed in this book, valid statements about the nature of the entire population can be made on the basis of the data collected from the sample.

Suppose we want to determine what proportion of the population of registered voters in Scranton is female. This proportion is sometimes called a measure or characteristic of the population and is defined as a *parameter*. By contrast, suppose we draw a random sample of registered voters and determine what proportion of this sample is female. Such a measure or characteristic of the sample is called a *statistic*. To distinguish between descriptive measures of a population and descriptive measures of a sample, we need two different sets of symbols. Greek letters are used to denote parameters (population measures), whereas Latin letters are used to denote statistics (sample measures). For example, μ (mu) is the symbol for the mean of a population and $\bar{X}$ is the symbol for the mean of a sample.

Descriptive and Inferential Statistics

The study of statistics is commonly divided into two broad categories: descriptive statistics and inferential statistics. These categories are exactly what their names imply. *Descriptive statistics* are procedures used for classifying and summarizing data; in essence, the data are described. Various measures are computed from the data, and these measures provide descriptions that help us understand the data.

Inferential statistics are procedures for making generalizations about a population by studying a subset of the population, called a sample. For example, suppose an economist is interested in conducting a survey in a city area about the spending habits of the population of junior high school students. It is not feasible (or necessary) to interview or collect data from every junior high schooler in the city. So a sample (presumably random) is selected, and data are collected from the members of the sample. On the basis of the sample data, the economist will make *inferences* about the population—hence the term "inferential statistics." The measures or statistics of the sample reflect the corresponding measures of the population (within the limits allowed for sampling fluctuation). In this case, it is not the sample that is of primary interest but the population. However, because the entire population cannot be measured, the sample data are used to draw conclusions about the population.

Descriptive statistics are procedures for summarizing and describing data. Inferential statistics are procedures for making inferences about measures of a population (parameters) from measures of a sample (statistics).

The essential difference between descriptive and inferential statistics is one of purpose—that is, the use to be made of the statistics. Describing data is the purpose of descriptive statistics; extrapolating from samples to populations is the purpose of inferential statistics. Both types of statistics involve numerous procedures used in understanding data and making decisions.

Summary

This introductory chapter is intended to acclimate the reader to the study of statistics. It contains some new terminology, perhaps unfamiliar, and some basic ideas about the meaning of statistics. Familiarity with this terminology is necessary for learning about and understanding statistics.

Some basic mathematical operations are reviewed in this chapter, and the summation operator is introduced. Although the mathematics required for using this text is not complex, it is important to understand it. The summation notation, though probably new to the reader, is straightforward.

Why do we bother with statistics, anyway? The answer is quite simple. It enables us to manipulate, understand, and interpret data that might otherwise be far less meaningful. It would do little good to carry data around as numbers on sheets of paper or on computer cards. The data must be organized, summarized, and manipulated. Statistics are the procedures that enable us to do this.

Numerous factors, such as level of measurement, influence the investigator's choice of what statistical procedures to use. Of course, the purpose of the study and the reason for acquiring the data in the first place have great bearing on these procedures.

The remaining chapters of this text deal with specific statistical procedures, their rationales, the conditions under which they apply, and the way they are computed.

Key Concepts

Statistics

Basic arithmetic operations

Real numbers

Signed numbers

Absolute value

Commutative postulate

Associative postulate

Distributive postulate

Exponent

Square root

Summation operator

Data

Variable

Constant

Continuous variable

Discrete variable

Independent variable

Dependent variable

Measurement

Nominal scale

Ordinal scale

Interval scale

Ratio scale

Population

Sample

Descriptive statistics

Inferential statistics

Parameter

Statistic

Exercises

1.1 For each of the following, perform the basic arithmetic operations indicated.

 a. $5 + 8 + 97 =$ 110

 b. $107 - 32 =$ 75

 c. $(7 + 12)(8 + 3) =$ (19)(11) 209

 d. $(15 + 10)/(2 + 3) =$ 5

 e. $(4)(3) + 7 =$ 19

 f. $82 + 391 - 61 =$ 412

 g. $7936/8 =$ 992

 h. $27 - (-10) =$ 37

1.2 Solve the following, using absolute value as indicated.

 a. $10 + |-7| =$ 3

 b. $|-21| - |-32| =$ -53

 c. $|735|/5 =$ 147

 d. $|8(-12)| + |-100| =$ -196

1.3 Solve the following, using the exponents indicated.

 a. $5^3 =$ 125

b. $(16)^2 =$ 256
c. $\sqrt{81} = 9$
d. $(100)^{1/2} = 10$
e. $2^6 = 64$ $x^3 y^3$
___ f. $(XY)^3 =$
g. $(3 + 2)^3 = 125$
h. $3 + 2^3 = 11$

1.4 Use the rules for order of operations to solve the following.
a. $(12)(9) + 32/8 = (108) + 4 = 112$
b. $7^2 - (5 + 3) + (6)(2) = 49 - 8 + 12 = 53$
c. $\sqrt{64}/(3 + 1) = 2$
. d. $(X + Y)^2 - X - Y = XY$
e. $75 - (3 + 12)/5 = 72$

1.5 Find $\sum\limits_{i=1}^{N} X_i$ when:
a. $N = 5$, such that $X_1 = 7, X_2 = 5, X_3 = 21, X_4 = 2, X_5 = 12$ 47
b. $N = 3$, such that $X_1 = 15, X_2 = -3, X_3 = 0$ 12
c. $N = 4$, such that $X_1 = X_2 = X_3 = X_4 = 9$ 36
d. $N = 2$, such that $X_1 = (12 + 5), X_2 = (6 + 5)$ 28

1.6 Multiply each of the following numbers by 2 and then use the first rule of summation to show that the sum of these products is equal to the sum of the original numbers multiplied by 2.
54
8, 3, 11, 4, 1

1.7 Find $\sum\limits_{i=1}^{N} (X_i + Y_i + Z_i)$ when: 145

$X_1 = 3$	$Y_1 = 32$	$Z_1 = 1$
$X_2 = 4$	$Y_2 = 40$	$Z_2 = 0$
$X_3 = 7$	$Y_3 = 27$	$Z_3 = 3$
$X_4 = 6$	$Y_4 = 21$	$Z_4 = 1$
20	120	5

1.8 Classify each of the following as a discrete or a continuous variable.
a. Intelligence C
b. Level of anxiety about taking a test C
c. Number of cars in a parking lot at various hours of the day D
d. Number of pushups performed in 2 minutes D
e. Mathematical ability C

1.9 Identify the level of measurement appropriate for each of the following.
a. Classifying the animals in a zoo by species
b. Scores on an essay exam
c. Measuring anxiety using the Manifest Anxiety test

d. Consumption of gasoline for different model cars *Interval*
e. Scores on a science exam with 100 objective items *ratio*
f. Major in college *Nom*
g. Number of fatal accidents during 1970–1979 *Ratio*
h. General intelligence
i. Marital satisfaction
j. Political party affiliation *Nom*
k. Grading system (A, B, C, D, F) *Interval*
l. Level of sugar in the blood *Ratio*
m. Order of finish in the Boston Marathon *Interval*
n. Size of family
o. Time required to complete a maze *Interval*

1.10 Let $A = +6$, $B = -13$, $C = +2.55$, and $D = -0.5$. Carry out the following arithmetic operations, taking into consideration the rules of significant digits discussed in this chapter.
   a. $A + B$ $6 + -13 = -7$
   b. $A - B$ $6 - -13 = 19$
   c. $C(A + D)$ $2.55(6 + -0.5) = 2.55(5.5 = 14.025$
   d. $(A)(C)(D)$ $(6)(2.55)(-.5) = -7.65$
   e. $|(A)(B)|$ $= -78$
   f. A/D -12
   g. $B/C = -5.098$
   h. $|A(C - D)|$ $6(0.55 \pm .5) = 6(3.05) = 18.3$
   i. $|C/D|$ -5.1
   j. $(A)(C)$ 15.3
   k. $|(A)(D)|$ 3

1.11 Using the values of A, B, and D given in exercise 1.10, illustrate the commutative, associative, and distributive postulates.
   a. $A + (B + D) = (A + B) + D$
   b. $A(B \cdot D) = (A \cdot B)D$
   c. $D(A + B) = (D)(A) + (D)(B)$

1.12 Let $X_1 = 4$, $X_2 = 2$, $X_3 = 6$, $X_4 = 5$, $X_5 = 3$, and $X_6 = 8$. Use the rules of summation to find the following.
   a. ΣX_i 4
   b. ΣX_i^2 6
   c. Show that $(\Sigma X_i)^2 \neq \Sigma X_i^2$.
   d. If $c = 3$, show that $\Sigma c X_i = c \Sigma X_i$.

1.13 Distinguish between a variable and a constant. List three examples of each.

▦ Key Strokes for Selected Exercises

	Value	Key Stroke	Display
1.1c $(7+12)(8+3) =$			
	7	+	7
	12	=	19
		STO	19
	8	+	8
	3	=	11
		×	11
		RCL	19
		=	<u>209</u> answer
1.4a $(12)(9) + 32/8 =$			
	12	×	12
	9	=	108
		STO	108
	32	÷	32
	8	=	4 save
		RCL	108
		+	108
	4	=	<u>112</u> answer

1.12c Show that $(\Sigma X_i)^2 \neq \Sigma X_i^2$.
Step 1: Determine $(\Sigma X_i)^2$.

Value	Key Stroke	Display
4	+	4
2	+	6
6	+	12
5	+	17
3	+	20
8	=	28
	X^2	<u>748</u> $= (\Sigma X_i)^2$

Step 2: Determine ΣX_i^2.

Value	Key Stroke	Display
4	X^2	16
	+	16
2	X^2	4
	+	20
6	X^2	36
	+	56
5	X^2	25
	+	81
3	X^2	9
	+	90
8	X^2	64
	=	<u>154</u> $= \Sigma X_i^2$

2

Frequency Distributions

Suppose a school psychologist has the academic ability test scores of 138 high school students from low socioeconomic home environments. The range of possible scores on the test is from a minimum of 0 to a maximum of 150. At this point, the school psychologist is thinking about how to organize these scores in order to enhance their meaning. A simple alphabetical listing of the scores according to the students' last names does little to give meaning to these scores. Some other means of organizing and describing these scores is required.

The term data was defined in Chapter 1 as bits of information that represent some characteristic exhibited by a group of individuals. For example, the academic ability scores for the 138 high school students are data. Other examples of data are the

49	50	49	54	51	46
58	49	54	41	50	40
50	31	36	28	46	49
55	52	42	39	41	50
57	33	51	42	52	55
49	40	56	44	38	47
42	50	48	45	46	32
40	55	47	46	52	51
54	36	36	41	42	47
35	45	41	55	51	51
59	50	40	53	53	50
55	30	40	40	55	41
47	54	35	50	50	46
45	45	57	41	38	54
46	59	50	33	50	44
53	41	36	40	53	43
36	43	41	47	58	45
44	46	45	47	53	59
39	57	57	47	47	59
41	49	36	57	63	45
46	49	59	43	56	39
56	48	38	54	57	47
57	42	34	53	42	37

Table 2.1   Academic ability scores of 138 high school students

socioeconomic status of the registered voters in a given locality and crop yield with various mixtures of fertilizers. Numbers that represent the measurement of the variable(s) under consideration constitute the data. The data that we will use in this chapter are the academic ability scores of these 138 high school students from low socioeconomic home environments. The score for each student represents that student's academic ability on this standardized ability measure.

Suppose we have the scores listed in alphabetical order of the students' last names (this listing is found in Table 2.1). It is almost impossible to make any sense out of this listing without finding some logical procedure for organizing and summarizing the scores. In this chapter, the frequency distribution will be discussed as a procedure for organizing and summarizing data to represent them meaningfully. Some elementary descriptive measures of data will also be introduced.

Frequency Distributions—
A Method for Organizing Data

A *frequency distribution* is an arrangement of data that shows the number of times given scores or groups of scores occur. Such an arrangement of data offers the re-

63	55	50	47	43	40
59	54	50	47	43	39
59	54	50	47	42	39
59	54	50	47	42	39
59	54	50	46	42	38
59	54	50	46	42	38
58	54	50	46	42	38
58	53	50	46	42	37
57	53	50	46	41	36
57	53	49	46	41	36
57	53	49	46	41	36
57	53	49	46	41	36
57	53	49	45	41	36
57	52	49	45	41	36
57	52	49	45	41	35
56	52	49	45	41	35
56	51	48	45	41	34
56	51	48	45	40	33
55	51	47	45	40	33
55	51	47	44	40	32
55	51	47	44	40	31
55	50	47	44	40	30
55	50	47	43	40	28

Table 2.2   Rank distribution of the academic ability scores

searcher insight into their nature. A frequency distribution does not disclose all the meaning that can be derived from the data, but it is a first step in understanding them.

A frequency distribution is a tabulation of data that indicates the number of times given scores or groups of scores occur.

Consider the 138 academic ability scores shown in Table 2.1. The first step in organizing these scores is to reorder them from highest to lowest. This reordering is called *rank distribution* and is illustrated in Table 2.2. This first step results in identification of the highest and lowest scores and the grouping of all identical scores so that the frequency of each score can be determined. Knowing the highest and lowest scores, we can determine the *range* of the scores as follows:

$$\text{Range} = (\text{highest score} - \text{lowest score}) + 1 \qquad (2.1)$$

The difference is increased by 1 so that both the highest and the lowest scores are included. For the data shown in Table 2.2, the range would be

$$\text{Range} = (63 - 28) + 1 = 35 + 1 = 36$$

Score	f	Score	f
63	1	44	3
59	5	43	3
58	2	42	6
57	7	41	9
56	3	40	7
55	6	39	3
54	6	38	3
53	6	37	1
52	3	36	6
51	5	35	2
50	11	34	1
49	7	33	2
48	2	32	1
47	9	31	1
46	8	30	1
45	7	28	1

Table 2.3   Frequency distribution of academic ability scores

The range of scores in a distribution is the number of units on the scale of measurement for the variable necessary to include both the highest score and the lowest score.

The second step in organizing and summarizing the academic ability scores is to develop the frequency distribution. The tabulation for this distribution is presented in Table 2.3. Essentially, this tabulation consists of keeping the scores ordered and listing the frequency for each score. That is, there was 1 score of 63, 5 scores of 59, and so on. Thus the data have been classified into as many categories as there are different scores in the distribution. However, it is not always convenient to retain as many categories as there are scores. So we reduce the number of categories by combining two or more (usually several) scores into an interval of scores—for example, all scores between 39 and 43, inclusive. These intervals of scores are called *class intervals* and, in this case, the interval width is 5. Suppose we develop a frequency distribution for the ability scores by using class intervals of width 5. This distribution is presented in Table 2.4. It is important to emphasize that, whereas this reclassification may give us a more meaningful insight into the data, we sacrifice some of the specifics of the data when we reclassify them in this way. For example, the frequency distribution in Table 2.4 no longer contains the number of scores of 39, 40, and so on. We know only that there are 28 scores within the interval 39–43.

Class Interval	f
59–63	6
54–58	24
49–53	32
44–48	29
39–43	28
34–38	13
29–33	5
24–28	1

Table 2.4   Frequency distribution of academic ability scores using class intervals

Combining scores into class intervals reduces the number of categories and may facilitate manipulation of data. However, such reclassification results in a loss of certain information about the actual scores in the distribution.

Exact Limits of the Class Interval

In Chapter 1 a distinction was made between discrete variables and continuous variables. Academic ability score is considered a continuous variable, because we can assume that the variable has underlying continuity and that it is theoretically possible to score anywhere along the scale of measurement. However, the scores were recorded as whole numbers (34, 35, 36, and so on). Because we consider this variable to be continuous, we assume that each score actually represents a value that falls within certain limits. For example, if we assume that a score of 34 actually represents a score somewhere between 33.5 and 34.5, the values 33.5 and 34.5 represent the *exact limits* of the score 34.

The concept of exact limits of a score can be extended to frequency distributions by distinguishing between the exact limits of a given class interval and its *score limits*. Consider the class interval 39–43 in Table 2.4. The interval 39–43 represents the score limits, whereas the interval 38.5–43.5 represents the exact limits. The exact limits are thus 0.5 unit below and 0.5 unit above the score limits of the class interval. Table 2.5 illustrates the frequency distribution of the academic ability scores with both the score limits and the exact limits.

Now consider the exact limits of the class interval of variables in a situation wherein measurement is more precise. The times in the 800-meter run are often recorded to the nearest tenth of a second, such as 1:51.8. Under the assumption of continuity, this score actually represents times between 1:51.75 and 1:51.85. Extending this concept of the exact limits of a class interval, we find that, if the score limits of a

Class Interval	Exact Limits	Midpoint	Frequency	
				Cum Fr
				139
59–63	58.5–63.5	61	6	*132*
54–58	53.5–58.5	56	24	*108*
49–53	48.5–53.5	51	32	*96*
44–48	43.5–48.5	46	29	*47*
39–43	38.5–43.5	41	28	*19*
34–38	33.5–38.5	36	13	
29–33	28.5–33.5	31	5	*6*
24–28	23.5–28.5	26	1	*1* *1/138*

20.
9.42
3.6290
.72%

Table 2.5 Frequency distributions of academic ability scores, including exact limits and midpoints

class interval were 1:49.6–1:53.2, the exact limits would be 1:49.55–1:53.25. Thus the exact limits would be 0.05 unit below and 0.05 unit above the score limits.

The use of exact limits in the development of a frequency distribution is based on the assumption that the variable under consideration is continuous and, theoretically, can take on any value on the scale of measurement.

Rules for Developing Class Intervals

What are the steps in developing a frequency distribution? In the initial step, selecting the score limits for the class intervals, two general rules of thumb should be applied. The first rule is that *the width of the class interval should be such that it will take from 8 to 15 class intervals to cover the total range of scores in the distribution*. Recall that the range of ability scores was 36. If we had used an interval width of 3, then 12 intervals would have been required; 36 ÷ 3 = 12. For an interval width of 4, 9 intervals would have been required. Note that, because we used an interval width of 5, we needed 8 intervals; 36 ÷ 5 = 7.2, which is rounded to 8. The purpose of this first rule is to provide a sufficient number of intervals so that the general shape of the distribution can be determined using the graphing procedures that will be discussed in a later section of this chapter.

The second rule is that, *whenever possible, the width of the class interval should be an odd number*. When we use this rule, if the data are integers (which is often the case), the midpoint of the interval will be a whole number rather than a fraction. The *midpoint* of the interval is defined as the point on the scale of measurement that is halfway through the interval. Consider the frequency distribution given in Table 2.5; the midpoint of the first interval is 61. Note that the midpoint is the same regardless of whether we use the exact limits or the score limits.

Assumptions for Class Intervals

As we have mentioned, when the width of the class intervals of a frequency distribution is greater than 1, some specific information about the values of the actual scores of the distribution is lost. We lose the exact number of times that each score appears in the distribution. In addition, some scores that were originally different, such as 55 and 56, are grouped together in the same interval. We must make two assumptions in order to represent the frequency distribution graphically as well as to compute certain statistics for the distribution.

The first assumption is that, when we use a single score to represent the class interval, we use the midpoint of the interval. The second assumption is that, for any class interval, the scores within the interval are uniformly distributed between the exact limits of the interval. To illustrate these assumptions, consider the interval 38.5–43.5. Under the first assumption, we use the midpoint 41 as the single score to represent all 28 scores in the interval. Second, we assume that the 28 scores are uniformly distributed over the interval, as follows:

Interval	Frequency
38.5–39.5	5.6
39.5–40.5	5.6
40.5–41.5	5.6
41.5–42.5	5.6
42.5–43.5	5.6
Total	28.0

The scores within any class interval are assumed to be uniformly distributed throughout the interval, and all are assumed to be adequately represented by the midpoint.

Graphing Data

Practically everyone encounters graphs at some time or another. The word "graph" comes from a Greek word meaning "to be drawn or written." For our purposes, we define a *graph* as a pictorial representation of a set of data. Common examples are the graphs used in newspapers and weekly news magazines to depict changes in the prime interest rate over the past twelve months. Such graphs show the relationship between two variables, the variables in this case being the prime interest rate and time across the twelve-month period. Used in this way, graphs serve to enhance the meaning of data.

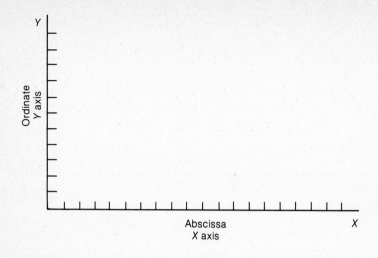

Figure 2.1   General layout of a graph

The Nature of Graphs

The first step in developing a graph is to draw the coordinate axes, which are two straight lines, one horizontal and one vertical. The scales of measurement for the independent and dependent variables (see Chapter 1) are marked off along these two axes. Generally, the scale of measurement of the independent variable is placed on the horizontal axis and the scale of measurement of the dependent variable is placed on the vertical axis. Using traditional mathematical terminology, we call the horizontal axis the *X-axis* or the *abscissa* and the vertical axis the *Y-axis* or the *ordinate*. The general layout of a graph with the coordinate axes is illustrated in Figure 2.1. Consider an example with level of education as the independent variable and annual income as the dependent variable. For this example, level of education has been categorized as follows:

Had fewer than 8 years of formal education = 1
Had more than 8 years of formal education, but no high school diploma = 2
Finished high school with diploma = 3
Completed post-secondary vocational training or associate's degree program = 4
Completed Bachelor's Degree program = 5
Completed Master's Degree program = 6
Completed post-Master's degree (M.D., Ph.D., etc.) = 7

Note that the categorization of this variable is an example of measurement on an ordinal scale. (See Chapter 1 for descriptions of measurement scales.) Thus, even though the numbers 1 through 7 are equally spaced along the *X*-axis in Figure 2.2,

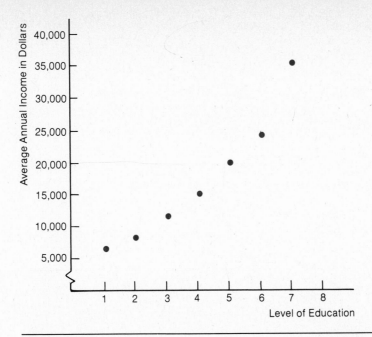

Figure 2.2 Graph of average annual income by level of education

these numbers do not necessarily reflect equal differences between the levels of education.

A range of average yearly income values from $5,000 to $40,000 is marked off on the Y-axis of Figure 2.2. The measurement of this variable is an example of the ratio scale. The average yearly incomes for persons in the respective categories of education levels are found in Table 2.6; the dots on Figure 2.2 illustrate the relationship between level of education and average yearly income. For example, those persons who had less than an eighth grade education had an average yearly income of $7,150, those who had more than eight years of education but no high school diploma had an average income of $8,775, and so on. As can be seen, these data indicate that the more education a person has, the greater the average income.

The graphing of data can be used to illustrate the relationship between the independent variable and the dependent variable.

Bar Graph

When the independent variable is measured on a nominal scale, the graphing procedure for illustrating the relationship between it and the dependent variable is the

Level of Education Category	Average Yearly Income
1	$ 7,150
2	8,775
3	12,125
4	15,650
5	20,275
6	24,850
7	35,525

Table 2.6   Average yearly income for various levels of education

Make of Automobile	Repair Rate (per 1000 sold)
A	4.2
B	6.8
C	3.3
D	0.4
E	1.2

Table 2.7   Transmission repair rates for domestic automobiles

bar graph. Suppose a consumer advocate is interested in the relative transmission repair rate for five medium-priced domestic automobiles. The data from this investigation are found in Table 2.7 and are graphically illustrated in Figure 2.3. The placement of the types of automobiles on the *X*-axis is quite arbitrary. For convenience, we simply list the makes as A through E. The scale of the relative repair rate is placed on the *Y*-axis. It is important to note that, when graphing nominal data using a bar graph, the individual bars are separated indicating the discrete nature of the nominal scale of measurement.

Bar graphs are used to illustrate the relationship between two variables when the scale of measurement of the independent variable is nominal.

Data Curves

In the previous example of the relationship between level of education and average yearly income, we could have connected the data points in the graph to further illustrate the trend in the data, as shown in Figure 2.4. By connecting the data points, we portray the data as a *data curve* rather than just as a set of data points. Such a technique has the advantage of providing a more intuitive feeling for the data. However, a data

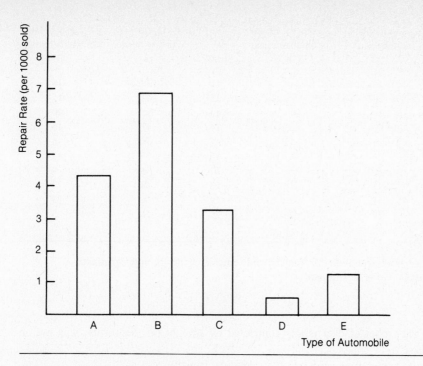

Figure 2.3   Repair rate by type of automobile

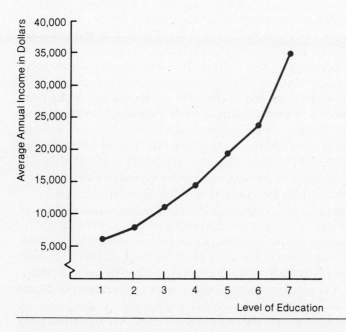

Figure 2.4   Data curve for average annual income (using data of Table 2.6)

Student	Academic Ability Score, X	Semester Hours of Mathematics, Y	Student	Academic Ability Score, X	Semester Hours of Mathematics, Y
1	54	18	11	39	18
2	29	3	12	42	15
3	42	14	13	55	20
4	60	23	14	47	15
5	33	15	15	50	16
6	28	7	16	29	12
7	56	22	17	34	9
8	48	18	18	48	22
9	55	19	19	56	25
10	59	25	20	46	23

Table 2.8 Academic ability scores (X) and semester hours of undergraduate mathematics (Y) for 20 students

curve is only a set of data points connected by straight line segments; it is not an actual curve.

Data curves are formed by connecting the data points of a graph with straight line segments, illustrating the relationship between two variables.

Scattergrams

For the graphs we have discussed so far, the scale of measurement of the independent variable has had only a few discrete data points. Suppose, however, that we are interested in the relationship between academic ability and number of semester hours of mathematics taken as an undergraduate student. The data for 20 students are found in Table 2.8. Note that we have arbitrarily designated academic ability as variable X and number of semester hours of undergraduate mathematics as variable Y. (For these data, neither variable is designated as independent or dependent.)

The data for this example can be graphically displayed in the *scatter diagram* or *scattergram* shown in Figure 2.5. Each dot represents one pair of observations for one student. For example, the dots for students 1, 2, and 3 are identified in the figure. For student 1, the dot corresponds to the score of 54 on the X variable (academic ability) and the score of 18 on the Y variable (number of semester hours of undergraduate mathematics). The dots for students 2 and 3 correspond to scores of 29 and 3 and scores of 42 and 14 on the X and Y variables, respectively.

In terms of the relationship between the two variables, inspection of the data given in Table 2.8 indicates that low academic ability scores tend to be associated with a

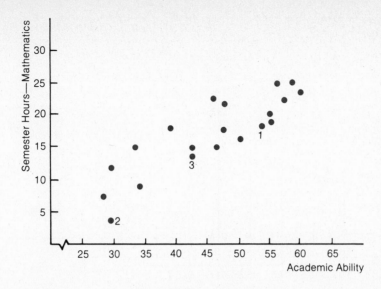

Figure 2.5 Scattergram of academic ability and semester hours in mathematics for 20 students

lesser number of semester hours of undergraduate mathematics. Similarly, high academic ability scores tend to be associated with a greater number of semester hours of undergraduate mathematics. Note that, in Figure 2.5, the data points tend to locate from the lower left to the upper right of the graph. This trend is an example of a positive relationship between two variables. We will discuss the relationship between two variables in greater detail in Chapters 6 and 7.

A scattergram is a graph illustrating the relationship between two continuous variables.

Graphing Frequency Distributions

So far in this chapter, we have discussed organizing and summarizing a set of scores in a frequency distribution and have illustrated the procedures for graphing data. We now combine these two concepts to illustrate how to graph frequency distributions. The major purpose of graphically illustrating a frequency distribution is to enhance our understanding of the data by looking at the shape of the distribution, which reveals how the scores are distributed throughout the range. For example, are the scores evenly distributed throughout, or are they heavily concentrated in certain

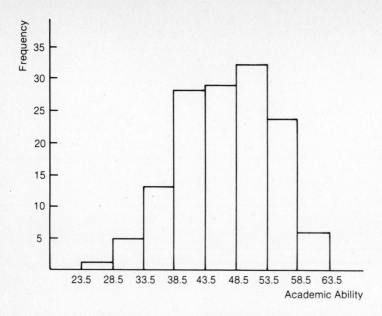

Figure 2.6   Histogram of academic ability scores

parts of the distribution? Although the amount of information that we can obtain by graphing a distribution is limited, combining information about the shape of the distribution with additional statistical measures often yields greater understanding of a set of scores. The purpose of this section is to present the procedures for graphing frequency distributions and to illustrate how one frequency distribution differs from another.

Histograms

A *histogram* is a type of bar graph that depicts the frequencies of individual scores or scores in a class interval by the length of the bars. The scale of measurement on the Y-axis is the frequency of the scores; the scale of measurement on the X-axis is the range of scores of the variable under consideration. The histogram of the academic ability scores of our 138 high school students is shown in Figure 2.6. Note that the plot points on the X-axis are the exact limits of the class intervals and that the bars are not separated. This illustrates that the academic ability scores represent a continuous variable.

Frequency Polygons

When we use a histogram to represent a frequency distribution graphically, we assume that the scores in each class interval are equally distributed throughout the

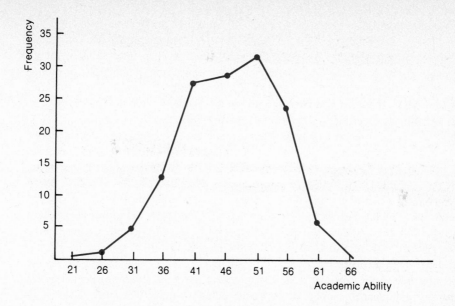

Figure 2.7   Frequency polygon of academic ability scores

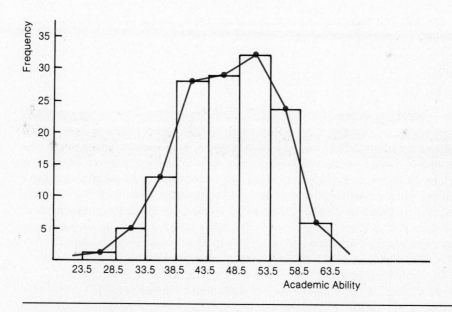

Figure 2.8   Frequency polygon superimposed on histogram of academic ability scores

interval. We could also graphically depict the frequency distribution by using a frequency polygon. For the *frequency polygon,* the assumption is that the scores in each class interval can be represented by the midpoint. For each interval, we plot the frequency on the graph at the midpoint and then connect the midpoints with straight lines. The frequency polygon for the 138 academic ability scores is shown in Figure 2.7. In order to show the relationship between the histogram and the frequency polygon, the latter is superimposed on the former in Figure 2.8.

A simple rule of thumb to follow when using either histograms or frequency polygons in graphing frequency distributions is the three-quarter-high rule. This rule states that the height of the *Y*-axis, which would incorporate the maximum number of frequencies in any interval, should be approximately equal to three-quarters of the length of the *X*-axis, which would include the range of scores in the distribution. The three-quarter-high rule has been used in Figures 2.6, 2.7, and 2.8. Using this rule reduces the possibility that the data will be misinterpreted or misrepresented. Because designation of the scales of measurement along the respective axes is quite arbitrary, it is possible to exaggerate relatively small differences. Using graphs deceptively by exaggerating the scales, analysts with different (and not very objective) intentions can enlist the same data to support two opposite points of view. The three-quarter-high rule helps assure consistency and avoid distortion in histograms and frequency polygons.

Frequency distributions can be graphically displayed by using either a histogram or frequency polygon. In constructing either, the three-quarter-high rule should be employed to ensure appropriate representation of the data.

The Cumulative Frequency Distribution and Its Graph

The *cumulative frequency distribution* is used extensively in describing a distribution of scores; it is developed by adding the frequency of scores in any class interval to the frequencies of all preceding class intervals on the scale of measurement. As indicated in Table 2.9, each class interval in the cumulative frequency distribution contains the frequency for that interval plus the frequencies for all the preceding class intervals. The cumulative percentage is the corresponding information in percentage form; it is computed by dividing the cumulative frequency by the total frequencies of the distribution and then multiplying by 100. Table 2.9 contains both the cumulative frequencies and the cumulative percentages for the distribution of our academic ability scores. The graph of a cumulative percentage distribution is called an *ogive;* the ogive for the distribution given in Table 2.9 is presented in Figure 2.9. As we will see in Chapter 4, such an ogive can be used in determining various percentile points of a distribution of scores.

Class Interval	Exact Limits	Midpoint	Frequency	Percentage	Cumulative Frequency	Cumulative Percentage
59–63	58.5–63.5	61	6	4.35	138	100.00
54–58	53.5–58.5	56	24	17.39	132	95.65
49–53	48.5–53.5	51	32	23.19	108	78.26
44–48	43.5–48.5	46	29	21.01	76	55.07
39–43	38.5–43.5	41	28	20.29	47	34.06
34–38	33.5–38.5	36	13	9.42	19	13.76
29–33	28.5–33.5	31	5	3.62	6	4.34
24–28	23.5–28.5	26	1	0.72	1	0.72

Table 2.9 Frequency distribution of academic ability scores, including cumulative frequencies and cumulative percentages

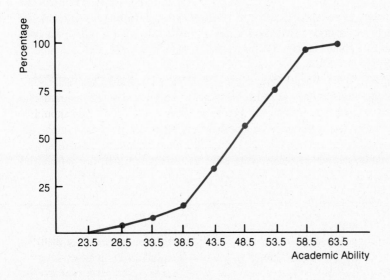

Figure 2.9 Ogive of academic ability scores

The ogive is the graph of a cumulative frequency distribution. It is useful for determining the various percentile points in a distribution of scores.

Various Shapes of Frequency Distributions

Frequency distributions have an unlimited number of possible shapes depending on how the scores are distributed on the scale of measurement. For example, if the scores are evenly distributed throughout, the frequency polygon looks something like distribution (a) in Figure 2.10. This distribution is referred to as a uniform or rectangular distribution. Suppose a frequency distribution has many scores located toward the lower end of the scale of measurement and progressively fewer scores toward the upper end, as illustrated by distribution (b) in Figure 2.10. This distribution is said to be skewed to the right, or *positively skewed*. On the other hand, if there are many scores at the upper end of the distribution and progressively fewer scores at the lower end, as shown in distribution (c), the distribution is said to be skewed to the left, or *negatively skewed*.

We refer to a distribution of scores as being bell-shaped when there are many scores in the middle of the scale of measurement and progressively fewer scores at the extremes. If the graph of such a distribution were folded along its middle line, and two halves of the graph would coincide; this is a *symmetric distribution*. Undoubtedly the most familiar symmetric distribution is the normal distribution, which we will discuss in detail in a later chapter. Distribution (d) in Figure 2.10 is an example of a normal distribution. Distributions (e) and (f) are examples of symmetric distributions that vary in degree of peakedness. (The degree of peakedness is called *kurtosis*.) For distribution (e), the vast majority of the scores tend to be located at the center of the distribution. Such a distribution is called *leptokurtic*. When the scores are more uniformly distributed but many scores still cluster at the center of the distribution, the distribution may look like distribution (f), which is called *platykurtic*. Distributions that have a moderate degree of peakedness, such as distribution (d), are referred to as *mesokurtic*.

The shape of a frequency distribution depends on how the scores are distributed throughout the range of scores. Distributions may be uniform, skewed, or symmetric in shape, and symmetric distributions may exhibit different degrees of kurtosis.

Summary

This chapter introduces procedures for organizing, classifying, and summarizing data in frequency distributions. Frequency distributions illustrate the number of times a given score or group of scores occurs, and they indicate where the scores are grouped on the scale of measurement.

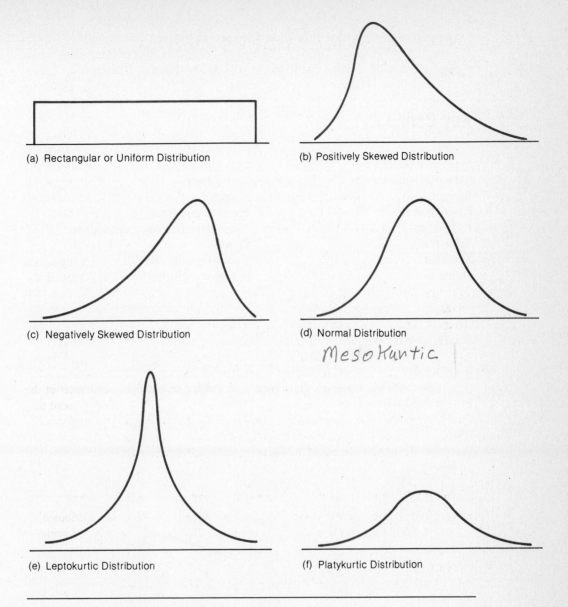

(a) Rectangular or Uniform Distribution

(b) Positively Skewed Distribution

(c) Negatively Skewed Distribution

(d) Normal Distribution

Mesokurtic

(e) Leptokurtic Distribution

(f) Platykurtic Distribution

Figure 2.10   Various shapes of frequency polygons

Graphs are defined as pictorial representations of data sets that illustrate the relationship between two variables. Histograms and frequency polygons are used to depict frequency distributions graphically. The ogive, a graph displaying the cumulative frequency distribution, will be used again in Chapter 4. The final section of the chapter discusses the various shapes that frequency distributions may take.

The emphasis in this chapter is on tabulating and graphing data to provide a more meaningful representation. The next step in interpreting data, that of using quantitative measures to further describe distributions, will be introduced in Chapter 3.

Key Concepts

Frequency distribution
Rank distribution
Range
Class interval
Exact limits
Score limits
Midpoint
Graph

Data curve
Scattergram
Histogram
Frequency polygon
Cumulative frequency distribution
Ogive
Skewed distribution
Symmetric distribution

Exercises

2.1 The following list gives the numbers of Sunday newspapers purchased on the past 52 Sundays.

77	73	75	81
75	72	82	86
71	74	73	73
79	76	102	74
75	82	98	75
88	78	77	75
85	81	83	98
66	68	76	87
80	81	71	74
70	88	74	79
85	77	84	83
72	79	67	68
71	85	72	74

a. Using the methods introduced in this chapter, develop a frequency distribution with 65–69 as the lowest class interval.
b. Construct a histogram and a frequency polygon for this frequency distribution.
c. Using these same data, develop a frequency distribution with 66–68 as the lowest class interval.
d. Superimpose the histogram and the frequency polygon that result from this frequency distribution on those you developed in part (b).

2.2 Consider the following frequency distribution of scores, which reflect the amount of time (number of seconds) required for a group of experimental rats to complete a certain maze.

Class Interval	Frequency		
150–154	2	*194*	
145–149	3	*192*	*97.42*
140–144	4	*189*	
135–139	6	*185*	
130–134	9	*179*	*82.63*
125–129	18	*170*	*78.35*
120–124	22	*152*	*67.01*
115–119	34	*130*	*49.48*
110–114	35	*96*	*31.44*
105–109	28	*61*	*17.01*
100–104	15	*33*	*9.28*
95– 99	10	*18*	
90– 94	6	*8*	*4.12*
85– 89	2	*2*	*1.03*

a. Complete the frequency distribution and the cumulative frequency distribution, using the methods introduced in this chapter.
b. Draw a histogram and a frequency polygon for this distribution.

2.3 Consider the following frequency distributions of final test scores for two groups of college students in an introductory calculus course.

Class Interval	Group I Frequency	Group II Frequency
90–94	4	4
85–89	12	2
80–84	17	1
75–79	22	8
70–74	28	12
65–69	24	16
60–64	18	18
55–59	15	28
50–54	12	33
45–49	9	30
40–44	1	15
35–39	4	10
30–34	3	4

a. On the same axis, construct a frequency polygon for each of the two groups.
b. Describe and compare the two distributions.

2.4 Two general rules have been given for developing frequency distributions when using class intervals other than the intervals of the original scores. These rules concern the number and the size of class intervals. Suppose we had used the following intervals for developing the frequency distribution for the data in Table 2.2.

Class Interval

61–63
58–60
55–57
52–54
49–51
56–48
43–45
40–42
37–39
34–36
31–33
28–30

a. Construct the frequency distribution for these data, using the foregoing class intervals.
b. Construct a frequency polygon and compare it with the frequency polygon shown in Figure 2.7.
c. Construct the cumulative frequency polygon and compare it to Figure 2.10.

3

Describing Distributions

In Chapter 2, we discussed the procedures that the school psychologist could use to organize academic ability scores into a frequency distribution and then to illustrate this distribution graphically with a frequency polygon. These procedures were said to be useful in organizing and summarizing scores to give a more meaningful representation of the data. The frequency polygon graphically showed how these scores were distributed throughout the range of scores on the scale of measurement. The frequency distribution and the corresponding graph can be helpful in describing a distribution of scores, but they do not permit us to make quantitative statements about the distribution or to compare it with other distributions.

Suppose this school psychologist wanted to compare the academic ability scores of the 138 high school students from low socioeconomic home environments with the scores of another group of high school students. This second group of students were from high socioeconomic home environments. Using the methods introduced in Chapter 2, the school psychologist could develop a frequency distribution and draw a frequency polygon for this second group. Using the procedures we have discussed thus far to compare the two groups, the psychologist would be limited to superimposing one frequency polygon on the other. This, of course, would represent only a crude comparison of the two groups.

To enhance the description of a distribution of scores, as well as the comparison of two or more distributions, we must determine certain quantitative characteristics of the distribution(s). Two of these characteristics are the central tendency and the variation of the scores. *Central tendency* is defined as a central value, between the extreme scores in the distribution, around which the scores are distributed. *Variation* is defined as a quantitative measure of the extent to which the scores are dispersed throughout the distribution. The more commonly used measures of central tendency and variation are described in this chapter. These two quantitative measures, as well as the shape of the frequency distribution, are necessary to describe adequately and to understand a distribution of scores.

Three characteristics necessary for adequately describing a distribution are the shape of the distribution, the central tendency, and the variation of the scores.

Measures of Central Tendency

Measures of central tendency are points on the scale of measurement that locate the distribution. In the behavioral sciences, many variables have distributions with heavy concentrations of scores in the middle of the distribution and fewer scores trailing out toward the ends (tails) of the distributions. The measures of central tendency are the quantitative measures that identify where the concentrations of scores are located. These measures are commonly referred to as averages, because they are points located between the extreme values on the scale of measurement of a variable. There are different averages, however, so use of this term has led to some confusion when the specific measure of central tendency is not identified. For example, officials of labor unions and their counterparts in management may refer to ''the average salary of employees'' and cite numerical values that are quite different. Both are citing accurate values, but each is using a different measure of central tendency. In this section we will define three commonly used measures of central tendency—the mode, the median, and the mean—and discuss the procedures for computing them. In addition, we will show the relationship among the three measures for frequency distributions of various shapes.

63	55	50	47	43	40
59	54	50	47	43	39
59	54	50	47	42	39
59	54	50	47	42	39
59	54	50	46	42	38
59	54	50	46	42	38
58	54	50	46	42	38
58	53	50	46	42	37
57	53	50	46	41	36
57	53	49	46	41	36
57	53	49	46	41	36
57	53	49	46	41	36
57	53	49	45	41	36
57	52	49	45	41	36
57	52	49	45	41	35
56	52	49	45	41	35
56	51	48	45	41	34
56	51	48	45	40	33
55	51	47	45	40	33
55	51	47	44	40	32
55	51	47	44	40	31
55	50	47	44	40	30
55	50	47	43	40	28

Table 3.1   Rank distribution of 138 academic ability scores

Measures of central tendency are points on the scale of measurement that locate the distribution.

Mode

The *mode* is the simplest index of central tendency; it is defined as the most frequent score in a distribution of scores and is determined by inspection or counting rather than by computation. Consider the distribution of academic ability scores of the 138 high school students from low socioeconomic home environments. This distribution is reproduced in Table 3.1. The mode of this distribution is the score 50. Note that, when the data are grouped into a frequency distribution of a certain interval width, the mode is the midpoint of the class interval with the greatest frequency. Consider the frequency distribution given in Table 3.2. Here the mode would be 51, the midpoint of the interval 49–53.

It is possible that two or more scores in a distribution may have the same frequency and that this frequency may be the greatest for any scores. When this occurs, the i = 5

N = 138

Class Interval	Exact Limits	Midpoint	f	cf
59–63	58.5–63.5	61	6	138
54–58	53.5–58.5	56	24	132
49–53	48.5–53.5	51	32	108
44–48	43.5–48.5	46	29	76
39–43	38.5–43.5	41	28	47
34–38	33.5–38.5	36	13	19
29–33	28.5–33.5	31	5	6
24–28	23.5–28.5	26	1	1

Table 3.2   Frequency distribution of academic ability scores, including exact limits and midpoints

distribution is said to be multimodal. Specifically, the term ''bimodal'' denotes a distribution that has two modes.

The mode is a rather unsophisticated measure of central tendency. It provides little information beyond identifying the most frequent score(s). In skewed and nonsymmetric distributions, the mode may be positioned near the extreme of the scale of measurement. In addition, the mode does not readily lend itself to mathematical manipulation and thus is of limited practical value as a statistic. We use it most often in conjunction with other descriptive measures of the distribution when we are working with distributions that consist of large numbers of scores.

The mode is the most frequent score in a distribution.

Median

The *median* is defined as the point on the scale of measurement below which 50% of the scores fall. It is the 50th percentile, a concept discussed in the following chapter. The basic concept of the median is relatively simple, but calculating it can be formidable when there is a large number of observations and the data are grouped into a frequency distribution. On the other hand, calculating the median for ungrouped data is not complicated.

The Median of Ungrouped Data
Computing the median of ungrouped data essentially consists of identifying the middle score. If there is an odd number of scores, the median is the middle score in the distribution. If the number of scores is even, the median falls between the two middle scores. Therefore, for ungrouped data, we compute the median by taking the following steps:

Arrange the scores in descending order (highest to lowest).

If the number of scores is odd, the median is score number $(N + 1)/2$. For example, consider the following distribution of scores: 23, 21, 19, 18, 12, 6, 3. There are seven scores ($N = 7$), so the median is score number $(7 + 1)/2 = 4$, or 18. If the number of scores is even, the median is halfway between score number $N/2$ and score number $(N/2) + 1$. For example, consider the following distribution of scores: 46, 44, 40, 29, 28, 27, 23, 18. There are eight scores ($N = 8$), so the median is halfway between 28 and 29, or 28.5.

The Median of Grouped Data   The process of computing the median becomes more complex when there are a large number of scores and when these scores are grouped into class intervals. Consider the frequency distribution of academic ability scores given in Table 3.2. For these data, the median is defined as the point below which 50% of the scores fall—that is, 138/2 or 69 scores. Note that the point 48.5, which is the upper limit of the class interval 43.5–48.5, is the point below which 76 scores fall. According to the definition, the median is a point in the interval 43.5–48.5. With 47 scores below 43.5, we need only 22 more scores to find the median ($47 + 22 = 69$). Thus, under the assumption that the 29 scores in the interval 43.5–48.5 are uniformly distributed in the interval (see Chapter 2), we need to go 22/29 of the distance through the width of the interval to find the median. Because the width of the class interval for this example is 5, the median (abbreviated Mdn) is computed as follows:

$$\text{Mdn} = 43.5 + \left(\frac{22}{29}\right)(5)$$

$$= 43.5 + 3.79$$

$$= 47.29$$

In general, the formula for the median is

$$\text{Mdn} = ll + \left(\frac{N(.50) - cf}{f_i}\right)(i) \tag{3.1}$$

where

ll = lower exact limit of the interval containing the $N(.50)$ score

N = total number of scores

cf = cumulative frequency of scores below the interval containing the $N(.50)$ score

f_i = frequency of scores in the interval containing the $N(.50)$ score

i = width of class interval

To illustrate the use of this formula, let us again use the data given in Table 3.2.

$$\text{Mdn} = 43.5 + \left(\frac{(138)(.50) - 47}{29} \right)(5)$$

$$= 43.5 + \left(\frac{69 - 47}{29} \right)(5)$$

$$= 43.5 + \left(\frac{22}{29} \right)(5)$$

$$= 47.29$$

The median is the point below which 50% of the scores in a distribution lie.

Mean

The *mean* is by far the most frequently used measure of central tendency. It is defined as the arithmetic average of a distribution of scores. It is computed by summing the scores and dividing this sum by the number of scores. Symbolically, the mean[1] is given by

$$\mu = \frac{\Sigma X i}{N} \tag{3.2}$$

where

X_i = any score in the distribution

N = total number of scores in the distribution

Consider the distribution of academic ability scores. The mean of that distribution would be computed as follows:

$$\mu = \frac{63 + 59 + 59 + \cdots + 28}{138}$$

$$= \frac{6456}{138}$$

$$= 46.78$$

The mean is the arithmetic average of the scores in a distribution.

---

[1]At this point we will use μ as the symbol for the mean. When inferential statistics are discussed, we will make a distinction between the mean of a sample ($\overline{X}$) and the mean of a population (μ).

X_i	$x_i = (X_i - \mu)$	$x_i^2 = (X_i - \mu)^2$	$(X_i - 8)^2$
9	3	9	1
12	6	36	16
7	1	1	1
5	−1	1	9
2	−4	16	36
3	−3	9	25
4	−2	4	16
$\Sigma 42$	0	76	104

$N = 7$
$\mu = 6$

Table 3.3 Data illustrating the properties of the mean

Properties of the Mean

Two properties of the mean are important in the development of more complex statistical concepts. The first property is that *the sum of all deviations around the mean is zero*. A *deviation* is defined as the difference between a given score and the mean. Consider the data given in Table 3.3; the deviation score is denoted in Table 3.3 by the lower-case *x*. Symbolically, this property of the mean is indicated as follows:

$$\Sigma(X_i - \mu) = \Sigma x_i = 0$$

The algebraic proof[2] of this property, using the rules of summation, is found in Appendix D.

If we take a deviation score $(X_i - \mu)$ and square it, we have $(X_i - \mu)^2$, a deviation score squared. Then, if we sum these squared deviations for all scores in a distribution, we have $\Sigma(X_i - \mu)^2$, the sum of squares of deviations about the mean, or simply the *sum of squares*. This leads to the second property of the mean, which is that *the sum of squared deviations about the mean is smaller than the sum of squared deviations about any other value*.

Consider the distribution given in Table 3.3 The sum of squared deviations about the mean, $\Sigma(X_i - \mu)^2$, is 76. Select any other value than the mean, say 8. The sum of the squared deviations around 8 is 104, a number greater than 76. This will be true for any value selected. This property shows that the mean is the measure of central tendency in the *least-squares* sense; that is, the sum of the squared deviations around the mean is a minimum. The proof of this property is also found in Appendix D.

Two properties of the mean are as follows:
1. The sum of the deviations around the mean is equal to 0.
2. The sum of the squared deviations is a minimum.

---

[2]Brief proofs are presented in Appendix D for the interested reader.

Weighted Mean—
The Mean of Combined Groups

Suppose that, of the 138 high school students whose scores appear in Table 3.2, 87 were female ($N_F = 87$) and 51 were male ($N_M = 51$). If the mean academic ability score for the females (μ_F) is $4090/87 = 47.01$ and the mean for the males (μ_M) is $2366/51 = 46.39$, then, to determine the mean of the total group, we must compute the *weighted mean*. The procedure involves weighting the individual means for the two groups on the basis of the number of observations in each group and dividing by the number of individuals in the combined group. For this example, the mean of the total group would be computed as follows:

$$\mu = \frac{N_F\mu_F + N_M\mu_M}{N_F + N_M}$$

$$= \frac{(87)(47.01) + (51)(46.39)}{87 + 51}$$

$$= \frac{4089.87 + 2365.89}{138}$$

$$= 46.78$$

Note that the mean is the same as that computed on page 52 (within rounding error).

The computation of the weighted mean can be extended to apply to more than two groups. Suppose we are given the means for the students on the basis of their high school curriculum. The mean for the 63 academic students ($N_a = 63$) is $3118/63 = 49.49$, the mean for the 48 general students ($N_g = 48$) is $2206/48 = 45.96$, and the mean for the 27 vocational students ($N_v = 27$) is $1132/27 = 41.93$. The mean for the total group would be computed as follows:

$$\mu = \frac{(N_a)(\mu_a) + (N_g)(\mu_g) + (N_v)(\mu_v)}{N_a + N_g + N_v}$$

$$= \frac{(63)(49.49) + (48)(45.96) + (27)(41.93)}{63 + 48 + 27}$$

$$= \frac{3117.87 + 2206.08 + 1132.11}{138}$$

$$= 46.78$$

In general, the formula for the weighted mean is given by

$$\mu = \frac{\sum_{j=1}^{k} n_j\mu_j}{N} \tag{3.3}$$

where

n_j = number of scores in each of the jth groups

μ_j = mean of each of the jth groups

N = number of scores in the total group; that is, $\Sigma n_j = N$

Comparison of Mode, Median, and Mean

What is the best measure of central tendency? To some extent, selection of the most appropriate measure depends on the scale of measurement of the variable. If the data are nominal, only the mode is appropriate. If the data are ordinal, the median and mode may be appropriate. Any of the three measures of central tendency can be used when the data are either interval or ratio; in fact, it may be advantageous to report more than one measure.

For example, consider the following distribution of salaries for employees in a small manufacturing company (numbers of employees for each position are given in parentheses).

President (1)	$180,000	
Executive vice president (1)	60,000	
Vice president (2)	40,000	
Comptroller (1)	22,800	mean
Senior salesperson (3)	20,000	
Junior salesperson (4)	14,800	
Foreman (1)	12,000	median
Machinist (12)	8,000	mode

Note that this distribution is skewed to the right and illustrates the influence of an extreme score on the mean.

The use to be made of the measure of central tendency influences the choice of the specific measure. If the use is primarily descriptive, the measure that best describes the data should be used. If there are a few extreme scores, as in the foregoing example, the median is preferred to the mean. If the scale of measurement is interval or ratio, all three measures of central tendency can be used. In combination, the three measures completely describe the location of the distribution on the scale of measurement.

If, on the other hand, the measures are to be used to infer from samples to populations (that is, in inferential statistics), the choice of a specific measure depends on the inferential methods to be used. The mean has a distinct advantage for use in most inferential methods. Because the mean requires an equal-unit or interval scale, it can be manipulated mathematically in ways not appropriate for the median or the mode.

We can compare the positions of the three measures by looking at various frequency distributions. When the distribution is symmetric and unimodal, the mean, the median, and the mode coincide, as shown in Figure 3.1(a). When the distribution is symmetric and bimodal, the mean and the median coincide, but two modes are present, as shown

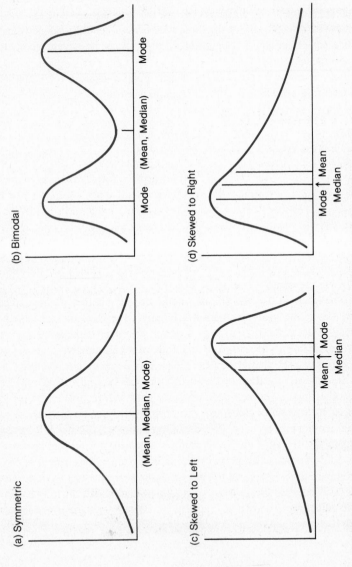

Figure 3.1 Comparisons of the mode, median, and mean in four distributions

in Figure 3.1(b). Asymmetric distributions are illustrated in Figures 3.1(c) and 3.1(d). Curve (c) is skewed to the left (negative), and curve (d) is skewed to the right (positive). Note that in curve (c) the mean is less than the median, which in turn is less than the mode. The opposite order appears in curve (d). Both the mode and the median are generally unaffected by an extreme score in a distribution, whereas the mean *is* influenced by extreme scores. Thus, when the frequency distribution is asymmetric, reporting all three measures of central tendency provides a more descriptive picture of the distribution than reporting only one measure.

The mode can be used with nominal-scale data, the median requires at least ordinal-scale data, and the mean requires at least interval-scale data. Extreme scores in a distribution affect the mean, but they generally do not affect the median or mode.

Measures of Variation

Another characteristic useful in describing a distribution is its dispersion or variation—that is, how the scores are dispersed throughout the distribution. Consider the two distributions shown in Figure 3.2. Distribution (a) represents the distribution of physical fitness scores for 1000 management personnel, and distribution (b) represents the distribution of scores for 1000 construction workers. Note that both distributions are symmetric and have the same mean ($\mu = 60$). However, the range of scores in distribution (a) is much greater than that in distribution (b). Consider the score 72 in both distributions. In distribution (a), there are many persons whose fitness scores are greater than 72, as indicated by the shaded area to the right of the score. In distribution (b), the fitness scores are located more centrally around the mean and there are fewer scores greater than 72. When scores are compactly distributed around the central tendency, as in distribution (b), the distribution is said to be *homogeneous*. On the other hand, when the scores are more widely dispersed throughout the scale of measurement, as in distribution (a), the distribution is said to be *heterogeneous*.

Measures of the dispersion or variation of scores in a distribution are important not only for enhancing the description of a distribution but also, as we will see, for distinguishing between different distributions. By contrast to measures of central tendency, which are points, the measures of variation are intervals or distances on the score scale that indicate how the scores are dispersed throughout the distribution.

Measures of variation are intervals on the scale of measurement that are indicators of the dispersion or variability of the distribution of scores.

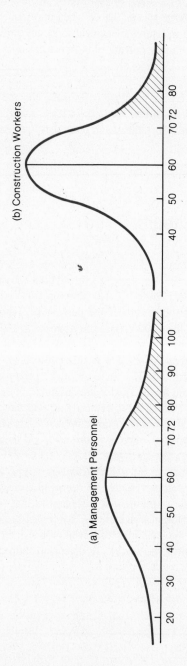

(a) Management Personnel

(b) Construction Workers

Figure 3.2 Distributions with the same mean and different variances

In this section, we will discuss four measures of variation: (1) range, (2) mean deviation, (3) variance, and (4) standard deviation. Whereas the first is appropriate when the data are ordinal, the last three require that the data represent at least interval-scale measurement. Generally, the standard deviation and the variance (the square of the standard deviation) are the most widely used measures of variation for both descriptive and inferential statistics.

Range

The *range* is the simplest measure of variation. In Chapter 2, it was defined as follows:

$$\text{Range} = (\text{highest score} - \text{lowest score}) + 1$$

This formula reveals that the range is the scaled distance between the highest and the lowest scores. The difference is increased by 1 in order to include both the highest and the lowest scores in this range. Consider the following two distributions.

Distribution 1	11	16	18	23	29	31	37
Distribution 2	18	19	21	23	24	26	29

Note that the medians for both distributions are the same, but the ranges are not. The range of distribution 1 ($37 - 11 + 1 = 27$) is greater than the range of distribution 2 ($29 - 18 + 1 = 12$); thus distribution 1 is said to be more variable.

The range is an easy index to compute, but it is generally not adequate by itself for describing the variation of scores in a distribution. Its most serious limitation is that it tends to vary with the size of the group; that is, larger groups tend to result in more extreme values for the range. Therefore, the ranges of two distributions composed of different numbers of scores are not directly comparable. In addition, the range may be greatly affected by adding or deleting a single score when that score is at one end of the distribution.

The range is defined as the number of units on the scale of measurement that are necessary to include the highest and lowest scores.

Mean Deviation

Earlier in this chapter, we compared the locations of the mean, median, and mode for distributions of various shapes. At that time, we indicated that the mean is influenced by extreme scores in a distribution, whereas the median and the mode are not. If we want to describe the variation or dispersion of scores in a distribution, we want to incorporate a sensitivity to all scores in a distribution by including the deviation of each score from the mean. Subtracting each score from the mean gives us a *deviation score*. That is,

$$x_i = X_i - \mu$$

If the deviations were summed, we would have an estimate of how much the scores, collectively, vary from the mean. Dividing this sum by N would yield a measure of the average deviation of all scores from the mean. Recall, however, from the first property of the mean, that the sum of the deviations of all scores from the mean is 0; that is, $\Sigma(X_i - \mu) = \Sigma x_i = 0$. Thus, with this property, the process would result in an average deviation equal to 0 regardless of the distribution. On the other hand, the process we have described makes sense intuitively when we can consider these deviations without regard to whether the score was above or below the mean.

Consider the distribution of scores given in Table 3.3. As can be seen, the mean equals 6. The score of 9 has a corresponding deviation score that is 3 score units $(9 - 6 = 3)$ above the mean. Similarly, the score of 3 is also 3 score units from the mean $(3 - 6 = -3)$, but the negative sign indicates that this score is below the mean. Thus, if we ignore the algebraic sign of the deviation score, we can say that both of these scores vary equally from the mean. Mathematically, what we have done is to define the *absolute value* of the deviation score. Symbolically, the absolute value of the number 3 is represented as $|3|$; and $|3| = |-3| = 3$.

Now that we have overcome the difficulty of dealing with the fact that the sum of the deviation scores is 0, we can define the *mean deviation* (abbreviated M.D.) as the average deviation (in terms of absolute values) of a distribution of scores.

$$\text{M.D.} = \frac{\Sigma|X_i - \mu|}{N} = \frac{\Sigma|x_i|}{N} \tag{3.4}$$

where

μ = mean of the population

N = total number of scores in the population

$|x_i|$ = absolute value of x_i

For the data given in Table 3.3,

$$\text{M.D.} = \frac{20}{7}$$

$$= 2.86$$

It is possible to use the mean deviation to compare the dispersions of several distributions; the distributions with the greater mean deviations have the greater dispersions. However due to the lack of a mathematical relationship between the mean deviation and the specific location of scores in the distribution, the mean deviation has limited utility in describing the dispersion within a single distribution. This is a result of using the absolute value of the deviation score. Also, it is difficult to manipulate absolute values algebraically for more advanced mathematical and statistical analyses. This problem is overcome by using the variance and standard deviation, both of which are discussed in the following sections. It is important to note, however, that both are related conceptually to the mean deviation.

The mean deviation is the average deviation from the mean of the scores in a distribution.

Variance

One way to avoid the difficulty associated with using absolute values is to square the deviation scores. Such a procedure does not restrict the algebraic manipulation usually required in more complex analyses. Thus, to obtain a measure of dispersion without mathematical restrictions, we can first square the deviation scores and then sum them. This eliminates negative signs before summing. What we obtain is the sum of squared deviations around the mean, or the *sum of squares* (SS) for short. Symbolically,

$$SS = \Sigma(X_i - \mu)^2 = \Sigma x_i^2 \tag{3.5}$$

Now, if we divide the sum of squares by N, we will have the average or mean of the squared deviations, which is called the *variance*. Symbolically,

$$\sigma^2 = \frac{SS}{N} = \frac{\Sigma(X_i - \mu)^2}{N} = \frac{\Sigma x_i^2}{N} \tag{3.6}$$

where

μ = mean of the distribution

N = total number of scores in the distribution

Because of the slight difference in computation, it is necessary to distinguish between the variance of a population (σ^2) and the variance of a corresponding sample (s^2). Recall that, in Chapter 1, we made a distinction between a parameter (a descriptive measure of a population) and a statistic (a descriptive measure of a sample). The computational formula for the variance of a population is given in formula 3.6. On the other hand, the formula for the variance of a sample is as follows:

$$s^2 = \frac{SS}{n-1} = \frac{\Sigma(X_i - \bar{X})^2}{n-1} = \frac{(\Sigma x_i)^2}{n-1} \tag{3.7}$$

where

$\bar{X}$ = mean of the sample

n = number of scores in the sample

Notice that we have also denoted the sum of squared deviations around the sample mean as $SS = \Sigma(X_i - \bar{X})^2$, μ being replaced by $\bar{X}$, the symbol for the mean of a sample. Furthermore s^2 is computed by dividing $\Sigma(X - \bar{X})^2$ by $n - 1$ rather than n. This is done to obtain an unbiased estimate of the population variance. An estimate is unbiased if the mean of all possible values of the estimate, for a given sample size, equals the parameter being estimated. In the case of variance, s^2 as computed above is an unbiased estimate of σ^2. The concept of unbiased estimate will be used later in the chapters on inferential statistics.

The variance of a distribution is the mean of the squared deviations about the mean.

In order to illustrate the computation of variance, consider the data given in Table 3.3 We will assume these data to be a distribution of a population. That is, we are using variance only in a descriptive sense. The variance would be computed as follows:

$$\sigma^2 = \frac{(9 - 6)^2 + (12 - 6)^2 + \ldots + (4 - 6)^2}{7}$$

$$= \frac{76}{7}$$

$$= 10.86$$

This formula for computing σ^2 is called the *deviation formula*, because the squares of the deviation scores are used in the calculation. The deviation formula is appropriate when the number of cases is small and when the mean is an integer. However, consider the 138 academic ability scores. Calculating the variance by using the deviation formula would be a tedious task. A more useful formula is the raw-score or observed-score formula, which uses the actual scores rather than the deviation scores. This formula is derived algebraically from the deviation formula; the derivation appears in Appendix D. The *raw-score formula* for the variance of a population is

$$\sigma^2 = \frac{SS}{N} = \frac{\Sigma X_i^2 - \dfrac{(\Sigma X_i)^2}{N}}{N} \tag{3.8}$$

Similarly, the raw-score formula for the variance of a sample is

$$s^2 = \frac{SS}{n - 1} = \frac{\Sigma X_i^2 - \dfrac{(\Sigma X_i)^2}{n}}{n - 1} \tag{3.9}$$

Using the raw-score formula for the variance of a population, we compute the variance for the distribution of the scores given in Table 3.3 as follows:

$$\Sigma X_i^2 = 328$$

$$\Sigma X = 42$$

$$N = 7$$

And therefore

$$\sigma^2 = \frac{328 - \dfrac{(42)^2}{7}}{7}$$

$$= \frac{328 - 252}{7}$$

$$= \frac{76}{7}$$

$$= 10.86$$

Note that this formula results in the same value we obtained when we used the deviation formula. Using formula 3.8 for computing the variance for the academic ability scores of the 138 high school students, we find that

$$\sigma^2 = \frac{309{,}530 - \dfrac{(6456)^2}{138}}{138}$$

$$= \frac{7501.48}{138}$$

$$= 54.36$$

Standard Deviation

As can be seen, the variance is a measure of dispersion in squared units of the measurement scale. If $(X_i - \mu)$ is a deviation measured in feet, $(X_i - \mu)^2$ is a deviation measured in square feet. The square root of the variance, then, is a measure of dispersion in the original unit of measurement. The square root of the variance is called the *standard deviation*. Symbolically,

$$\sigma = \sqrt{\sigma^2} = \sqrt{\frac{\Sigma(X_i - \mu)^2}{N}} \tag{3.10}$$

Correspondingly, for the sample standard deviation,

$$s = \sqrt{s^2} = \sqrt{\frac{\Sigma(X_i - \overline{X})^2}{n - 1}} \tag{3.11}$$

The standard deviation is frequently used in interpreting the percentage of scores in a distribution that are 1 standard deviation above the mean and 1 standard deviation below the mean. If the scores have a "normal" distribution, the range between 1 standard deviation below the mean and 1 standard deviation above the mean contains 68.26% of all the scores. Most distributions represent a less than perfectly normal distribution, but we can say that, in general, approximately two-thirds of the scores in most distributions are in this range.

Because the standard deviation is in the original unit of measurement, there are certain advantages to its use. For example, consider modern intelligence tests, which commonly have a mean equal to 100, a standard deviation equal to 15, and a variance equal to 225. If a student is found to have an IQ equal to 115, we can say that the student has an IQ that is 1 standard deviation above the mean. The variance (equal to 225) would not be a very useful or descriptive piece of information in this case, because the variance is not in the same unit of measurement. Both the variance and the standard deviation will be used often in later chapters.

For computing the standard deviation, we generally use the raw-score formula for computing the variance and add the step of taking the positive square root. For the distribution of 138 academic ability scores of our 138 high school students, we have

$$\sigma = \sqrt{\frac{309{,}530 - \dfrac{(6456)^2}{138}}{138}}$$

$$= \sqrt{\frac{7501.48}{138}}$$

$$= \sqrt{54.36}$$

$$= 7.37$$

The standard deviation is the positive square root of the variance. It is given in the same units as the original measurement of the variable.

Summary

This chapter discusses the procedures for computing quantitative measures for a distribution of scores—that is, measures of central tendency and measures of variation. The measures of central tendency (mode, median, and mean) define quantitatively a central value between the extreme values in the distribution. The measures of variation are quantitative values defining the extent to which the scores are dispersed throughout the distribution. We need to know these measures and the shape of the distribution (see Chapter 2) to describe adequately and understand a distribution of scores.

Each measure of central tendency discussed in this chapter is used in describing distributions, but the mean is the most widely used due to its mathematical properties. Likewise, the variance and standard deviation are the most widely used measures of variation. In the next chapter, we will use these measures in describing individual scores in a distribution.

Key Concepts

Central tendency

Variation

Mode

Multimodal

Median

Mean

Deviation score

Sum of squares

Weighted mean

Range

Mean deviation

Variance

Deviation formula

Raw-score formula

Standard deviation

Exercises

3.1 The following are the scores of 25 students who participated in a psychology experiment. The scores represent the number of trials required to complete a given task. Assume the group to be a population.

12	10	12	11	6
15	14	17	9	12
13	8	7	15	14
15	18	19	14	10
14	14	16	8	9

a. Determine the mean, median, and mode.

b. Determine the range, variance, and standard deviation.

3.2 Consider the following scores: 11, 13, 7, 10, 15, 3, 12, 11, 4, 14.

a. Determine the mean.

b. Using these data, show that $\Sigma(X_i - \overline{X}) = 0$.

3.3 Suppose we add 4 to each of the scores given in Exercise 3.2. What effect will this have on the mean and standard deviation of these scores? Assume that these scores constitute a population of scores.

3.4 The following data are the final examination scores of 40 students in a basic statistics class. These scores were randomly selected from the records of all students who have taken the course over the past 10 years and have taken the standardized final examination.

58	86	70	80	82
88	60	80	72	75
89	61	72	76	80
63	73	82	81	89
75	65	82	86	90
75	63	65	84	82
76	68	82	91	94
68	74	79	84	96

a. Determine the mean and median.

b. Determine the variance and standard deviation.

3.5 Provide graphical illustrations for the frequency distributions from which the following measures of central tendency were derived. Describe the distributions in terms of symmetry, indicating the direction of skewness of asymmetric distributions.

Mean = 46	Median = 43	Mode = 40
Mean = 43	Median = 43	Mode = 43
Mean = 40	Median = 43	Mode = 46

3.6 A marital satisfaction inventory was given to a population of married persons with and without children. The following data were obtained.

	N	μ
Male, no children	48	84.3
Female, no children	63	76.8
Male, with children	56	58.8
Female, with children	67	62.6

a. Find the mean for the total group.

b. Find the mean for the males and that for the females.

c. Find the mean for the married persons with children and the mean for those without children.

3.7 Consider the data given in Exercise 3.4. What would be the effect of removing the five highest and the five lowest scores? Describe this new distribution in terms of the range, mean, and standard deviation.

3.8 Consider the data given in Table 2.3 and Table 2.9.

a. Compute the mean for the data given in Table 2.3.

b. Assuming that the midpoint of a class interval can be used to represent all scores in the interval, compute the mean for the data given in Table 2.9.

c. Compare these two means.

3.9 For the following set of scores, show that the sum of squares of deviations about the mean ($\mu = 7.4$) is smaller than the sum of squares of deviations about the value $X = 7$.

$$5 \quad 8 \quad 11 \quad 3 \quad 10 \quad 9 \quad 6 \quad 7 \quad 9 \quad 6$$

3.10 The mean of a set of 8 scores is 37. The first 7 scores are 40, 29, 33, 43, 39, 35, and 40. What is the eighth score?

▦ Key Strokes for Selected Exercises

3.1a. Use the key strokes described in Appendix A to determine ΣX and ΣX^2

$$\Sigma X = 312$$
$$\Sigma X^2 = 4182$$

	Value	Keystroke	Display
Step 1: Determine $\overline{X} = \dfrac{\Sigma X}{N}$	312 (ΣX)	$\div$	312
	25 (N)	$=$	$\underline{12.48} = \overline{X}$

Step 2: Determine variance
and standard deviation.

$$\sigma^2 = \frac{\Sigma X^2 - \dfrac{(\Sigma X)^2}{N}}{N}$$

$$\sigma = \sqrt{\sigma^2}$$

	312 (ΣX)	X^2	97,344
		$\div$	97,344
	25 (N)	$=$	3893.76
		STO	3893.76
	4182 (ΣX^2)	$-$	4182
		RCL	3893.76
		$\div$	288.24
	25(N)	$=$	$\underline{11.5296} = \sigma^2$
		$\sqrt{}$	$\underline{3.3955} = \sigma$

3.6	Step 1: Determine N	48	$+$	48
		63	$+$	111
		56	$+$	167
		67	$=$	$\underline{234} = N$
			STO	234

Step 2: Determine the mean
for the total group.

48	$\times$	48	
84.3	$+$	4046.4	
63	$\times$	63	
76.8	$+$	8884.8	
56	$\times$	56	
58.8	$+$	12177.6	
67	$\times$	67	
62.6	$=$	16371.8	
	$\div$	16371.8	
	RCL	234	
	$=$	$\underline{69.965} = \overline{X}$	

4

Describing Individual Scores

Thus far, we have been concerned with organizing and describing a distribution of scores in terms of (1) shape, (2) location (measures of central tendency), and (3) dispersion (measures of variability). In this chapter, we will describe procedures for interpreting individual scores within the distribution. For example, an expression commonly used by teachers and professors, and often dreaded by their students, is "Grading for my tests will be based on the curve." Though this statement is interpreted and applied differently by different teachers, grading on the curve generally means assigning a grade to a student on the basis of his or her test score in relation to the scores of the other students in the class. Say the student has a test score of 65. This score may have little meaning without an adequate frame of reference—its relative position in the total distribution of scores. In this chapter, two statistical concepts that provide such a frame of reference will be presented; these are the concepts of percentiles and standard scores.

Percentiles

A *percentile* or percentile point is defined as the point on the scale of measurement of the distribution at or below which a given percentage of observed scores is found. For example, the 65th percentile (P_{65}) of a given distribution of scores is the point at or below which 65% of the scores fall. Percentiles are commonly used for interpreting performance on achievement tests and academic ability tests. Suppose a high school counselor is reviewing the results of the Scholastic Aptitude Test (SAT) with a student's parents. The counselor could simply tell the parents the student's raw score, but it would be more meaningful to inform them of the position of the student's score in relation to all other students' scores in selected groups that have taken the SAT. Such groups might be (1) all students who took the SAT on a given administration date, (2) all students in a selected school, (3) all students applying for admission to a selected college, and (4) all students entering engineering curricula. If the student's raw score was at the 20th percentile (P_{20}), indicating that only 20% of all students taking the SAT had scored lower, that student might be advised to enroll in a technical school rather than a 4-year college or university. On the other hand, the student with a raw score at the 85th percentile (P_{85}) would probably be advised to enroll in a 4-year institution.

Consider the academic ability scores introduced earlier (Table 2.1) for the 138 high school students, and suppose we want to determine the 65th percentile—the point at or below which 65% of all the scores fall. The frequency distribution of the academic ability scores, including cumulative frequencies and cumulative percentages, is found in Table 4.1. Note that 48.5 is the upper exact limit of the class interval 44–48 and that it is the point at or below which 76 (55.07%) of the scores fall. Similarly, 53.5 is the point at or below which 108 (78.26%) of the scores fall. Therefore, because the 65th percentile (P_{65}) is defined as the point at or below which $0.65 \times 138 = 89.7$ of the scores fall, P_{65} is a point in the interval 48.5–53.5. Using the data given in Table 4.1, we need 13.7 of the 32 frequencies in the interval 48.5–53.5 to find P_{65}. Under the assumption that the 32 frequencies are uniformly distributed in this interval (see Chapter 2), we know that 13.7/32 of the distance through the width of this interval should give the point at or below which 89.7 of the scores are located. The width of the interval is 5, so P_{65} for this distribution is computed as follows:

$$P_{65} = 48.5 + (13.7/32) \times 5$$

$$= 48.5 + 2.14$$

$$= 50.64$$

Class Interval	Exact Limits	Midpoint	f	%	cf	c%
59–63	58.5–63.5	61	6	4.35	138	100.00
54–58	53.5–58.5	56	24	17.39	132	95.65
49–53	48.5–53.5	51	32	23.19	108	78.26
44–48	43.5–48.5	46	29	21.01	76	55.07
39–43	38.5–43.5	41	28	20.29	47	34.06
34–38	33.5–38.5	36	13	9.42	19	13.76
29–33	28.5–33.5	31	5	3.62	6	4.34
24–28	23.5–28.5	26	1	0.72	1	0.72

Table 4.1   Frequency distribution of academic ability scores

This procedure can be generalized for computing any desired percentile and is summarized by the following formula:

$$Xth \text{ percentile} = P_X = ll + \left(\frac{Np - cf}{f_i}\right)(i) \qquad (4.1)$$

where

ll = lower exact limit of the interval containing the percentile point

N = total number of scores

p = proportion corresponding to the desired percentile

cf = cumulative frequency of scores below the interval containing the percentile point

f_i = frequency of scores in the interval containing the percentile point

i = width of class interval

Applying formula 4.1, we compute the 25th percentile (P_{25}) for the data given in Table 4.1 as follows:

$$P_{25} = 38.5 + \frac{[138(.25)] - 19}{28}(5)$$

$$= 38.5 + \frac{34.5 - 19}{28}(5)$$

$$= 38.5 + 2.77$$

$$= 41.27$$

Thus, the point at or below which 25% of the scores are found is 41.27.

We can also find percentiles by using the ogive—that is, the graph of the cumulative percentage distribution (see Chapter 2). Figure 4.1 is the ogive for the data given in Table 4.1. To find P_{25}, draw a horizontal line from the point 25 on the percentage side to the curve. Then, where this line intersects the curve, draw a vertical line perpendicular to the score scale. This point will be P_{25}. The procedure is illustrated in Figure 4.1.

A percentile is the point in a distribution at or below which a given percentage of scores is located.

Percentile Rank

In the previous section, a percentile was defined as the point on the scale of measurement for a set of scores at or below which a given percentage of scores is located. This concept offers insight into the nature of the distribution, and it also provides the frame of reference we need to interpret an individual score in the distribution. For example, we found that the 65th percentile (P_{65}) for the distribution of academic ability scores was 50.64; however, we have not yet determined the relative position of a specific academic ability score, say 55, in the distribution. One procedure is to determine the *percentile rank* of the score 55 (PR_{55}), which is a point on the percentage scale that corresponds to the percentage of scores in the distribution at or below the score 55.

To find the percentile rank of the score 55 (PR_{55}), consider the distribution shown in Table 4.1. The score is in the class interval with exact limits from 53.5 to 58.5, and 108 scores are below the exact lower limit of 53.5. Under the assumption that the 24 scores in this interval are uniformly distributed, the first step is to determine the number of scores in the interval that are at or below the score 55. The score 55 is $(55 - 53.5)/5$ or 0.3 of the distance through the interval, so there would be 0.3×24 or 7.2 scores at or below 55. By adding the 7.2 scores to the 108 scores below 53.5, we can determine PR_{55} as follows:

$$PR_{55} = [(108 + 7.2)/138] \times 100 = 83.48$$

Thus we say that the percentile rank of the score 55 is 83.48.

This procedure can be generalized for computing the percentile rank of any number in a distribution of scores and is summarized by the following formula:

$$PR_X = \left[\frac{cf + \left(\dfrac{X - ll}{i}\right)(f_i)}{N} \right] \times 100 \qquad (4.2)$$

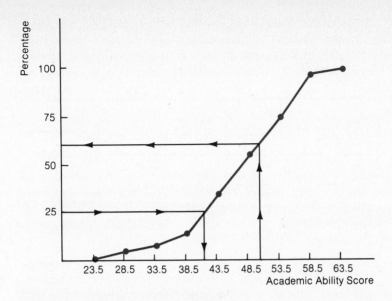

Figure 4.1 Ogive of academic ability scores

where

X = score for which the percentile rank is to be determined

cf = cumulative frequency of scores below the interval containing the score X

ll = lower exact limit of the interval containing X

i = width of class interval

f_i = frequency of scores in the interval containing X

N = total number of scores

Applying formula 4.2 to find PR_{50} for the data given in Table 4.1, we obtain

$$PR_{50} = \left[\frac{76 + \left(\dfrac{50 - 48.5}{5} \right) (32)}{138} \right] \times 100$$

$$= \frac{76 + 9.60}{138} \times 100$$

$$= 62.03$$

Thus the score 50 has a percentile rank of 62.03.

We can also find the percentile rank by using the ogive. For example, to find PR_{50}, draw a vertical line from the point 50 on the score scale up to the curve. Then, where this line intersects the curve, draw a horizontal line over to the percentage scale. This point on the percentage scale is PR_{50}. The procedure is illustrated in Figure 4.1.

The percentile rank of a score in a distribution is a transformed score, one transformed from the original measurement scale to the percentile scale.

The difference between a percentile point and a percentile rank is illustrated in the following example. Suppose we have a distribution of scores in which a score of 83 on the measurement scale of the variable corresponds to the 72nd percentile. We say that 83 is the 72nd percentile point. On the other hand, the percentile rank of a score of 83 is 72.

Use of Percentiles and Percentile Ranks

Percentiles and percentile ranks are widely used in reporting the results of standardized tests, because percentiles are relatively easy to interpret in terms of indicating an individual's position within a group. However, percentiles do have one drawback. Because many variables in the behavioral sciences tend to be normally distributed, differences between percentile points are not uniform throughout the scale of measurement. That is, a difference between two percentile points in the middle of the scale, say P_{50} and P_{51}, represents a smaller difference on the scale of measurement than a difference between two percentile points near the ends of the scale, say P_{91} and P_{92}. In essence, percentiles are an example of an ordinal scale of measurement. This point is illustrated in Figure 4.2. Note that the raw score 48 corresponds to P_{50}, whereas the scores 45 and 51 correspond to P_{40} and P_{60}, respectively. Thus a difference of 6 raw-score units in the middle of this distribution is equivalent to a difference of 20 percentile points. Due to this piling up of percentile points in the middle of the distribution, little distinction should be made between an individual who scores at P_{49} and one who scores at P_{52}.

As illustrated, in the center of the distribution, using percentiles tends to exaggerate small differences. In the tails of the distribution, on the other hand, using percentile scores tends to underestimate actual differences. Percentiles are unequal units of measurements and hence should not be arithmetically manipulated. Thus there is no justification for summing, combining, averaging, or manipulating them as we might manipulate scores that have interval-scale or equal-unit-scale measurement. Percentiles should be used only for describing points in a distribution. If statistical manipulations are to be performed, it is necessary to convert the percentiles into some other kind of scores, such as normal curve equivalent scores (which will be discussed in the next chapter), and then perform the manipulations. Anyone using percentiles to report standardized test scores should be aware of the ordinal characteristics of percentiles and the corresponding limitations that these characteristics place on the manipulation of percentiles.

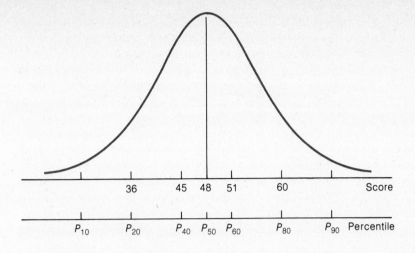

Figure 4.2 **Positions of percentiles on the scale of measurement for a normal distribution**

The ordinal nature of percentiles places limitations on the statistical operations appropriate for them.

Standard Scores

Suppose a high school counselor has the results of the various subtests of a standardized test on which to base advice about specific areas of academic strengths and weaknesses. Converting a raw score to a percentile rank would enable us to establish the relative positions of the subtest scores in their respective distributions. Whereas this procedure does provide a frame of reference for reporting the nature of individual subtest scores, the ordinal scale of measurement of percentile ranks prohibits our summing the percentile ranks of the various subtest scores to obtain a composite score reflecting the overall academic performance of the student. Some tests may contain such a composite score, but it is *not* based on the average of the percentiles of the subtests.

Consider another example, in which a student receives a score of 85 on an algebra test and a score of 60 on a chemistry test. Assuming that both tests are comprehensive in nature, do these scores reveal whether this student performed better on the algebra test or the chemistry test? We cannot answer this question without additional information about the distribution of scores for both tests and the relative positions of the two test scores in their respective distributions. We could convert both test scores to percentile ranks. Or we could use a different frame of reference that would enable us to arithmetically manipulate the two scores in order to obtain a composite score. This procedure requires us to convert the raw scores to *standard scores* which, in effect, standardize each of the distributions so that we can locate a particular student's scores on the two tests in the respective distributions and compare and/or combine them.

The use of standard scores provides an equal-interval-scale measurement through the use of the standard deviation as the unit of measurement. The mean and the standard deviation of a distribution of scores were defined and discussed in Chapter 3; we know that the mean is a point on the measurement scale and that the standard deviation is an interval. Computing standard scores involves using the mean and the standard deviation to describe the relative position of a single score with respect to the entire distribution. The standard score that corresponds to a given raw score indicates how many standard deviations the raw score is above or below the mean. The lower-case z is used as the symbol for a standard score, and a *z score* is computed as follows:

$$\text{Standard score} = \frac{\text{raw score} - \text{mean}}{\text{standard deviation}} \quad (4.3)$$

or

$$z = \frac{X - \mu}{\sigma}$$

Note that we are using the population parameters μ and σ. If we were using the sample statistics $\overline{X}$ and s, the standard score would be computed as follows:

$$z = \frac{X - \overline{X}}{s} \quad (4.4)$$

For example, in a distribution with a mean of 40 and a standard deviation of 8, the z scores that correspond to raw scores of 36, 42, and 50 would be -0.50, 0.25, and 1.25, respectively.

$$z_{36} = \frac{36 - 40}{8} = -0.50$$

$$z_{42} = \frac{42 - 40}{8} = 0.25$$

$$z_{50} = \frac{50 - 40}{8} = 1.25$$

A negative standard score indicates that the raw score is below the mean, whereas a positive standard score indicates that the raw score is above the mean. A raw score equal to the mean has a standard score of 0.

In the foregoing examples, the z score of -0.50 for the raw score of 36 indicates that this raw score is half a standard deviation below the mean. The raw scores 42 and 50 are 0.25 and 1.25 standard deviations above the mean, respectively.

Subject	Raw Score X	Standard Score z
A	10	1.29
B	9	.97
C	3	− .97
D	10	1.29
E	9	.97
F	2	− 1.29
G	2	− 1.29
H	10	1.29
I	5	− .32
J	5	− .32
K	1	− 1.61
L	6	.00
M	8	.64
N	6	.00
O	6	.00
P	1	− 1.61
Q	3	− .97
R	6	.00
S	10	1.29
T	8	.64
$N = 20$		
Mean (μ)	6.0	0
Standard Deviation (σ)	3.10	1.0

Table 4.2 Distribution of raw scores and z scores

The standard score, or z score, indicates the number of standard deviations a corresponding raw score is above or below the mean.

Properties of Standard Scores

A standard score can be calculated for each raw score in a distribution. Consider the data given in Table 4.2. Each of the 20 raw scores has been converted to a z score via formula 4.3. What happens to the distribution of scores when each raw score is converted to a standard score? The first property of standard scores is that *the shape of the distribution of standard scores is identical to that of the original distribution of raw scores*. The second property is that *the mean of the distribution of z scores will always equal 0, regardless of the value of the mean in the raw score distribution.* This property is illustrated using the data given in Table 4.2; the algebraic proof is found in Appendix D.

The third property of standard scores is that *the variance of the distribution of z scores equals 1.0.* Because the standard deviation is the square root of the variance, the standard deviation of the distribution of z scores also equals 1.0. This property is illustrated using the data given in Table 4.2; the algebraic proof of this property is found in Appendix D. Thus, by calculating the z score for each raw score in a distribution, we *transform* the original distribution of scores into a distribution that has an identical shape but a mean equal to zero (0) and a standard deviation equal to one (1.0).

The properties of a distribution of z scores are as follows:
1. The shape of the original distribution is retained.
2. The mean is zero (0).
3. The variance and standard deviation are one (1.0).

Transformed Standard Scores

Attractive as z scores may be for purposes of comparison, they have some undesirable characteristics that make them difficult to manage and to report. For example, one must decide how many decimal places to retain. The usual practice of reporting two decimal places implies a precision that the raw scores usually do not possess. And for negative scores, the omission of a minus sign can drastically change the meaning of a z score. For these reasons, z scores are commonly transformed into a different distribution of scores with a new scale of measurement, often arbitrarily chosen, with a predetermined mean and standard deviation.

Consider an example. Suppose the publisher of an aptitude test wants to report test scores in some standardized format so that they can be interpreted in a meaningful way by high school students and their parents. Because negative scores tend to have a negative connotation, the reporting of a negative z score may be poorly received. And because the mean of the distribution of z scores is 0, an average score on the test might be misinterpreted as a *zero-correct* performance. Thus some method of expressing these scores is desirable. One approach is to transform the z scores into a distribution with a mean of 50 and a standard deviation of 10. With this transformation, there are no negative scores, and, with a standard deviation of 10, there is a spread of scores sufficient to lessen concern regarding the number of decimal places. The transformed distribution of scores with a mean of 50 and a standard deviation of 10 is called a *T distribution*, and the scores are called *T scores*. The process of converting a distribution of raw scores to a distribution of *transformed standard scores* such as the *T* distribution is frequently used in the behavioral sciences for reporting test scores. This process involves two steps. The first step is to transform the original distribution of raw scores into a distribution of z scores. These z scores are then transformed into T scores by multiplying each z score by 10 (the standard deviation) and then adding 50 (the mean). In other words,

$$T = (10)(z) + 50 \qquad (4.5)$$

Consider the data given in Table 4.2. If we were to transform the raw score for individual A into a T score, we would first compute the z score:

$$z = \frac{10.00 - 6.00}{3.10} = 1.29$$

Transforming this z score to a T score, we find that individual A's T score would be

$$T = 10(1.29) + 50$$

$$= 62.9, \text{ or } 63$$

Similarly, the remaining T scores in Table 4.2 are rounded to the nearest whole number.

We interpret the T score for individual A exactly as we would interpret the z score, but now all decimal points and minus signs have been eliminated. The purpose of such transformations is to facilitate the interpretation of scores while retaining a standardized format.

A T distribution is a "standard" distribution with a mean of 50 and a standard deviation of 10.

In addition to the T distribution, other standardized distributions are used in the behavioral sciences for reporting test results. Most IQ test scores, for example, are reported on the basis of a mean of 100 and a standard deviation of 15. The College Entrance Examination Board (CEEB) reports a mean of 500 and a standard deviation of 100. It must be emphasized that these standardized distributions do not magically emerge from the distribution of raw scores. They must first be converted to z scores and then transformed into a distribution with the desired mean and standard deviation. The general formula for this latter transformation is

$$X' = (\sigma')(z) + \mu' \qquad (4.6)$$

where

X' = new or transformed score for a particular individual

σ' = desired standard deviation of the distribution

z = standard score of a particular individual

μ' = desired mean of the distribution

To transform a distribution of scores to a distribution with a desired mean and standard deviation, simply multiply the z scores by the standard deviation and add the mean: $X' = (\sigma')(z) + \mu'$.

Weighted Averages for Several Tests

Test scores are often combined into a total score that represents some earlier established set of weightings. For example, a grade in a course might be determined from performance on three tests—two short tests and one long examination. The instructor could choose to weight the three test scores either equally or differently. Combining percentiles into such an average would be inappropriate because of the ordinal nature of percentiles. Using raw scores would not be advisable, because the means and especially the standard deviations of these tests would probably be unequal. Averaging the raw scores would give more weight or influence to the test that had the greatest standard deviation. Instead, if averaging is done, it should be done in terms of standard scores, because the distributions would then all have the same mean (0) and the same standard deviation (1.0). Applying any previously selected set of weights is then not a problem. The desired *weighted average score* for an individual would be found with the following formula:

$$\text{Weighted score}_j = \frac{\Sigma W_i z_{ij}}{\Sigma W_i} \tag{4.7}$$

where

W_i = weight of each test

z_{ij} = standard score for person j on test i

Suppose an instructor gives three tests and wishes to weight them 1:1:2. That is, the final test is to be weighted twice as heavily as either of the first two tests. If a particular student's z scores on the three tests are $+0.25$, -0.50, and -0.20, the weighted score for this student is

$$\frac{1(0.25) + 1(-0.50) + 2(-0.20)}{1 + 1 + 2} = -0.16$$

If these composite scores are to be discussed with students, parents, or the public, it might be well to transform this distribution of weighted scores into one that eliminates the negative signs and the decimal places, such as a T distribution.

When developing a composite score from two or more individual scores, we transform the individual scores into standard scores. Then we apply the weights to generate the weighted score.

Summary

This chapter focuses on interpreting an individual score, primarily the location of that score in a distribution. Percentiles and standard scores are discussed. A percentile is a point on the scale of measurement indicating the percentage of scores that fall at or below the point. If, in a distribution, the score 78 is at the 57th percentile, 57% of the scores fall at or below 78. We can also find percentile rank, which addresses the question, "Given a score on the scale of measurement, what percentage of scores fall at or below this score?" Percentiles locate scores in distributions in terms of percentages of scores, a concept that has considerable appeal for interpreting test scores.

Standard scores refer to the transformation of raw scores into a distribution with a specified mean and a specified standard deviation. The "basic" standard score is the z score, the distribution of which has a mean of 0 and a standard deviation of 1.0. The z score is widely used in statistical work but, due to decimal points and negative scores, it is somewhat inconvenient for general interpretation. Hence, transformation into another standard score is often performed. A commonly used score is the T score, the distribution of which has a mean of 50 and a standard deviation of 10. Regardless of the transformed score used, the idea is the same: we transform each score into a score whose distribution has a specified mean and standard deviation.

Percentiles are scores on an ordinal scale, a fact which is often overlooked. This means that they are not manipulable, as scores on an interval scale are. Standard scores, however, retain the interval scale (assuming that such a scale existed in the raw scores), so we can subject them to such statistical manipulations as computing averages. Two or more scores in standard-score form can also be weighted to generate a composite or total score.

Key Concepts

Percentile
Percentile rank
Standard score
z scores

Transformed standard score
T score
Weighted average score

Exercises

4.1 Consider the frequency distribution given in Exercise 2.2 of Chapter 2.
    a. Determine P_{15}, P_{45}, and P_{80}.
    b. Determine PR_{141}, PR_{115}, and PR_{96}.

4.2 Find the percentiles and percentile ranks listed in Exercise 4.1 by using the ogive for this frequency distribution.

4.3 Consider the following set of scores (assume that they represent a sample).

23	18	28
20	15	24
25	12	26
18	14	20

   a. Determine the mean and the standard deviation.
   b. Convert each raw score to a standard score.
   c. Show that this set of standard scores has a mean equal to 0 and a standard deviation equal to 1.0.

4.4 For the data given in Exercise 4.3, multiply each of the standard scores by 10 and add 50. Show that the mean of this transformed distribution of scores has a mean of 50 and standard deviation equal to 10.

4.5 As one of his decision-making criteria, a college admissions director uses a composite score based on SAT verbal performance, SAT mathematics performance, and high school grade point average. The three variables are weighted 2:1:3 in developing the composite score. Assume that high school grade point averages are normally distributed with a mean of 2.50 and a standard deviation of 0.50, and that SAT scores are normally distributed with a mean of 500 and a standard deviation of 100. Find the composite z score for an applicant with an SAT verbal score of 450, an SAT mathematics score of 575, and a high school grade point average of 2.35. Do likewise for an applicant with an SAT verbal score of 650, an SAT mathematics score of 625, and a high school grade point average of 3.18.

4.6 John took three standardized tests with the following results:

	John's Score	μ	σ
English	70	60	10
Mathematics	80	75	5
General science	90	95	10

   a. Compute a T score for each test.
   b. What appears to be John's strongest subject area?

4.7 Add 4 units to each score in Exercise 4.3, and similarly convert the raw scores to standard scores. Compare these standard scores with those obtained from the original set of scores.

4.8 A distribution has a mean of 18 and a standard deviation of 4.2. What is the raw score that corresponds to a T score of 72?

▦ Key Strokes for Selected Exercises

		Value	Keystroke	Display
4.1a	Determine P_{15}.			
		194	×	194
		.15	=	29.1
			−	29.1
		18	=	11.1
			÷	11.1
		15	=	0.74
			×	0.74
		5	=	3.7
			+	3.7
		99.5	=	$\underline{103.2} = P_{15}$
4.1b	Determine PR_{141}.			
		141	−	141
		139.5	=	1.5
			÷	1.5
		5	=	0.3
			×	0.3
		4	=	1.2
			+	1.2
		185	=	186.2
			÷	186.2
		194	=	.9598
			×	.9598
		100	=	$\underline{95.98} = PR_{141}$

5

The Normal Curve

In previous chapters, we have referred to distributions of scores that are, or are assumed to be, "normal." What do we mean when we say that the distribution of scores of a certain natural phenomenon, such as general intelligence, is normally distributed. The image that immediately comes to mind is a "bell-shaped" curve representing the distribution of scores. In the physical and behavioral sciences, many variables are normally distributed, so studying the normal distribution and its properties is important. In addition, the assumption of a normal distribution will be important in the later chapters of this book when we discuss inferential statistics.

The Nature of the Normal Curve

To begin with, we must distinguish between what we generally refer to as the normal curve and data that we say are normally distributed. The *normal curve* is a mathematical equation that has been generated to depict the nature of the frequency distribution of a set of scores. Although it is not determined by any specific event in nature and does not reflect a specific law of nature, the normal curve is helpful in describing a set of data that are appropriately distributed.

The general equation for the normal curve was developed by the French mathematician Abraham DeMoivre and is based on his observations of the outcomes of games of chance. Consider a two-outcome situation wherein each outcome has an equal probability of occurring, such as the flipping of a fair coin. (The concept of probability will be discussed more thoroughly in Chapter 8.) Probability, in terms of relative frequencies, can be thought of as the number of possible successes divided by the total number of possible outcomes. So, if getting "tails" on a single flip of a coin is deemed a success, then the probability of success would be 1/2—one possibility of getting "tails" divided by the two possible outcomes. Probabilities can range from zero (0), or no chance of success, to one (1), or certainty of success; and the sum of the probabilities for all possible outcomes is 1.0.

Consider all possible outcomes of flipping 5 fair coins simultaneously, when success is defined as "getting tails." The probabilities are illustrated in Figure 5.1. Note that the distribution is symmetric. Though it is not apparent in the figure, the probability of getting no tails is 0.031, the probability of getting one tail is 0.156, and the probability of getting two tails is 0.313.[1] Now consider all possible outcomes of flipping 10 coins simultaneously; the probabilities of success (obtaining tails) are illustrated in Figure 5.2. Again, the distribution is symmetric.

DeMoivre sought a mathematical equation that would approximate the frequency polygons shown in Figures 5.1 and 5.2 as well as all possible outcomes of flipping any number of coins. The equation developed by DeMoivre is

$$u = \frac{1}{\sqrt{2\pi}\sigma} e^{\frac{-(X - \mu)^2}{2\sigma^2}}$$

(5.1)

where

u = height of the curve (ordinate) for any given value of X in the distribution of scores

---

[1]The probabilities were computed from what is called the *binomial distribution*. The formula for these computations is

$$\binom{n}{n_1} p^{n_1}(1 - p)^{n - n_1}$$

We are not concerned at this point with the calculations of these probabilities but rather with the symmetry of the distribution of probabilities for all possible outcomes in this example. The binomial distribution will be discussed in Chapter 8.

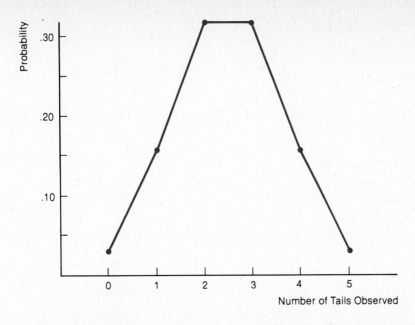

Figure 5.1   Probability of getting tails when flipping five coins simultaneously

π = mathematical value of the ratio of the circumference of a circle to its diameter (3.1416)

e = base of the system of natural logarithms, approximately equal to 2.7183

μ = mean of the distribution of scores

σ = standard deviation of the distribution of scores

With this equation, it is possible to determine the probabilities either directly or from a table based on the equation.

Formula 5.1 represents the *normal distribution* for a specific mean (μ) and standard deviation (σ). The formula is somewhat complicated and is included here only to show that there is a general formula that defines the normal curve. More important, we know from experience that many variables in the behavioral sciences are normally distributed (or approximately so). Although the shapes of their distributions are generally similar, they are not exactly the same because their means and/or standard deviations differ. Consider the three pairs of normal distributions shown in Figure 5.3.

The distributions of pair (a) have the same standard deviations but different means, whereas, for pair (b), the means are the same but the standard deviations differ. The

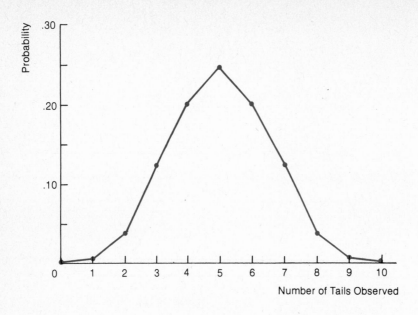

Figure 5.2 Probability of getting tails flipping ten coins simultaneously

distributions of pair (c) have neither mean nor standard deviation in common. Thus, rather than having one curve for what is generally referred to as a normal distribution, we have a family of curves. Note, however, that all members of this family of curves have similar properties. The more obvious of these properties are as follows:

1. A normal curve is unimodal, symmetrical, and bell-shaped with the maximum height at the mean.
2. A normal curve is continuous. There is a value of u (the height) for every value of X where X is assumed to be a continuous variable rather than a discrete variable. (The scale of measurement for X is the horizontal axis.)
3. A normal curve is asymptotic to the X axis when graphed. This is basically a theoretical property. It means that the curve never touches the X axis no matter how far a particular score is from the mean of the distribution. (The farther out the curve goes from the mean, the closer it gets to the X axis.) We will see that it is highly unlikely that there will be a score more than 3 standard deviations away from the mean. However, because it is theoretically possible, it is allowed for by the asymptotic characteristic of the curve.

The normal curve is not a single curve but a family of curves, each of which is determined by its mean and standard deviation.

(a)

(b)

(c)

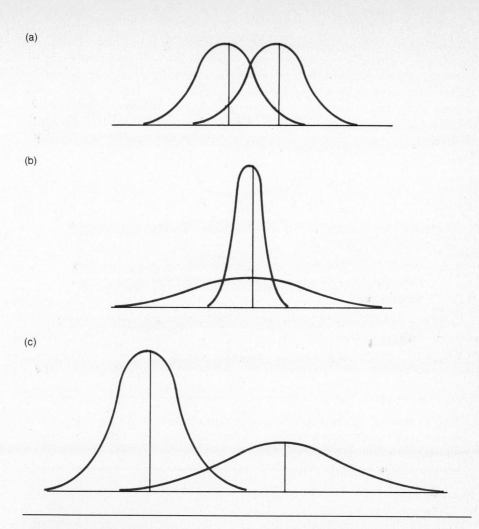

Figure 5.3   Normal distributions with the same and different mean and/or the same and different standard deviation

The Standard Normal Curve

The concept of the family of normal curves is important in that it enables us to work with any number of variables in the behavioral sciences that are normally distributed, regardless of their means and standard deviations. However, in order to work effectively with this variety of normal distributions, we cannot have a table for every possible combination of means and standard deviations. What we need is a *standardized* normal distribution that can be used for any normally distributed variable. Such a normal distribution is available and is based on the concept of standard scores. The *standard normal distribution*, sometimes referred to as the *unit normal distribution*, is the distribution of normally distributed standard scores. For example, consider a distribution of scores from a psychological inventory that is normally distributed. If we were to transform each of the scores into standard scores (z scores), the distribution of these scores would then be the standard normal distribution. The distribution of the original scores can be represented by formula 5.1. But, because the mean of the transformed or standard scores (z scores) equals 0 and the standard deviation equals 1.0, the distribution of standard scores can be represented by the following formula:

$$u = \frac{1}{\sqrt{2\pi}} e^{-z^2/2} \qquad (5.2)$$

where

u = height of the curve (ordinate) for any given z score in the standard normal curve

z = the given standard score, which represents the number of standard deviations that the original score is from the mean of the original distribution

π = mathematical value of the ratio of the circumference of a circle to its diameter (3.1416)

e = base of the system of natural logarithms, approximately equal to 2.7183

The standard normal distribution, or normal distribution, is the distribution of normally distributed standard scores with a mean equal to zero (0) and a standard deviation equal to one (1.0).

Using the Standard Normal Curve

When a variable is normally distributed with a given mean and standard deviation, we can use the *standard normal curve* (unit normal curve) in describing the distribution. For example, the standard normal curve can be used to determine the proportion of scores that are between two points in the distribution, the Xth percentile of the distribution of scores, or the percentile rank of a given score in the distribution.[2] Suppose we want to know the proportion of scores that are below the mean. Mathematically, the process requires integral calculus. Rather than carrying out this complex procedure for each example, statisticians have developed the table of values found in Table 1 in the Appendix. This table, entitled ''Ordinates and Areas of the Normal Curve,'' contains the values of the height of the standard normal curve (the ordinate)

[2]The standard normal curve is also used in determining the probability of certain natural events. This use of the curve will be considered in Chapters 8, 9, and 10.

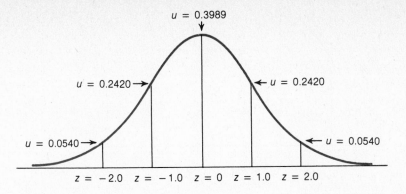

Figure 5.4 Ordinates of the normal curve for $z = 0$, $z \pm 1$, and $z \pm 2$

for a given z score and the proportion of the area under the curve that falls between the mean and a given z score. (The symbol for the z score in the table is under the column heading $\frac{x}{\sigma}$.)

Figure 5.4 illustrates the use of the table for determining the ordinate or the height of the standard normal curve for several z scores. For example, the ordinate for the mean ($z = 0$) is 0.3989. For $z = +1.0$, $u = 0.2420$; for $z = +2.0$, $u = 0.0540$. Because the curve is symmetric, the value for $z = -1.0$ would be the same as that for $z = +1.0$, and the value for $z = -2.0$ equals that for $z = +2.0$. Note also the asymptotic property of this curve; it approaches the horizontal axis but never reaches it.

Figure 5.5 illustrates the use of the table in determining the proportion of the area under the curve that lies between the mean and a given standard score. Suppose we want to determine the proportion of the area between $z = 0$ and $z = +1.0$. From Table 1, we find the proportion to be 0.3413. Thus approximately 34% of the total area falls between the mean and 1 standard deviation above the mean. Similarly, we find 0.4772, or 47.7%, of the area between $z = 0$ and $z = +2.0$. The proportion of the area between $z = 0$ and $z = +3.0$ is 0.4987, or approximately 49.9%. Usually, when discussing the standard normal curve, we let the area under the curve be 1.0 (hence the name "unit normal") and simply refer to area rather than proportion of area.

Again, because the curve is symmetric, the area between $z = 0$ and $z = +1.0$ is the same as the area between $z = 0$ and $z = -1.0$; that is, it is 0.3413. Therefore, the area between $z = -1.0$ and $z = +1.0$ is $0.3413 + 0.3413 = 0.6826$, or approximately 68% of the area. Similarly, the area between $z = +2.0$ and $z = -2.0$ is $0.4772 + 0.4772 = 0.9544$, or approximately 95.4% of the area, whereas the area between $z = +3.0$ and $z = -3.0$ is $0.4987 + 0.4987 = 0.9974$, or approximately 99.7% of the area. Note that the area under the curve for z values

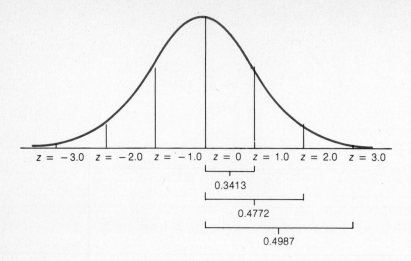

0.3413

0.4772

0.4987

Figure 5.5   Areas under the normal curve for $z = 0$, $z \pm 1$, $z \pm 2$, and $z \pm 3$

greater than $+3.0$ or less than -3.0 is quite small (0.0026). Thus, for all practical purposes, the normal curve is considered to extend from $z = +3.0$ to $z = -3.0$, or 3 standard deviations on either side of the mean.

Area in the standard normal curve is given from the mean to the z score. The curve is symmetric, so only areas for positive z scores are given.

Approximately 99.7% of the area in a normal distribution is contained within 3 standard deviations of the mean (between $z = -3.0$ and $z = +3.0$).

Determining Proportions   Suppose we have a distribution of standardized test scores on a psychological inventory (assumed to be normally distributed) for 15,000 students in a large metropolitan area. We want to determine the proportion of the scores that are between 0.5 standard deviation below the mean and 1.5 standard deviations above the mean. In standard-score terminology using the normal curve, we want to determine the proportion of the area between $z = -0.5$ and $z = +1.5$. This example is illustrated in Figure 5.6. Referring to Table 1, we find that the area under the standard normal curve between $z = -0.5$ and $z = 0$ is 0.1915. (Recall that the distribution is symmetric and that the area between $z = -0.5$ and $z = 0$ is the same as the area between $z = 0$ and $z = +0.5$.) The area between $z = 0$ and $z = +1.5$ is 0.4332. The z values under consideration lie on opposite sides of the mean, so we must *add* these two values to obtain the desired proportion. Thus $0.1915 + 0.4332$

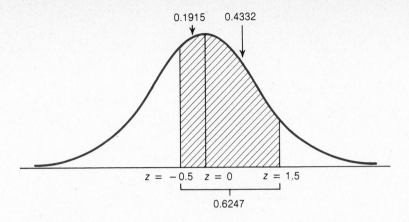

Figure 5.6 Areas under the normal curve between $z = -0.5$ and $z = +1.5$

$= 0.6247$, or approximately 62.5%, of all test scores are between 0.5 standard deviation below the mean and 1.5 standard deviations above the mean. Applying this value to the specific example, we can say that $0.6247 \times 15,000 = 9370.50$, or approximately 9370, students had test scores in this range.

In determining proportions using the normal curve, it is helpful to sketch the normal curve and shade the area representing the desired proportion. For the foregoing example, the area fell on both sides of the mean (see Figure 5.6), so we added the two area segments. However, if the area were between two z scores on the same side of the mean, we would have to *subtract* to find the desired proportion. For example, suppose we want to determine the proportion of students who had test scores between 1 and 2 standard deviations above the mean—that is, between $z = +1.0$ and $z = +2.0$. This example is illustrated in Figure 5.7. Referring to Table 1, we find that the area under the standard normal curve between $z = 0$ and $z = +1.0$ equals 0.3413 and that the area between $z = 0$ and $z = +2.0$ equals 0.4772. Because these z scores lie on the same side of the mean, we subtract the area for $z = +1.0$ from that for $z = +2.0$. Subtracting 0.3413 from 0.4772, we find that 0.1359, or approximately 13.6%, of all the test scores are between 1 and 2 standard deviations above the mean. In other words, $0.1359 \times 15,000 = 2038.50$, or approximately 2038, students had test scores in this range.

To find areas between z scores in the standard normal curve, we add the areas if the z scores are on opposite sides of the mean. If the z scores are on the same side of the mean, we subtract the area of the z score that is closest to the mean from the area of the z score that is farthest from the mean.

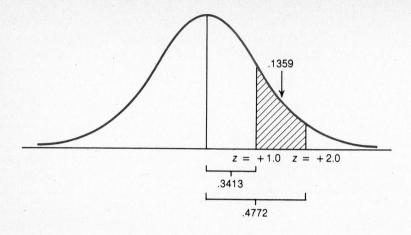

Figure 5.7   Areas under the normal curve between $z = +1.0$ and $z = +2.0$

Determining Percentiles and Percentile Ranks

We can also use Table 1 to find the Xth percentile of a given distribution of scores as well as the percentile rank of a particular score in the distribution. Recall from Chapter 4 that the Xth percentile of a given distribution is the point on the score scale at or below which X percent of the scores are located and that the percentile rank of a given score is the point on the percentage scale that corresponds to the percentage of scores that are at or below the given score. Suppose we want to determine the 84th percentile of the distribution of standardized test scores for the 15,000 students. Referring to Table 1, we know that 50% of all the scores are below the mean ($z = 0$) and that 34.13% of the scores are between $z = 0$ and $z = +1.0$. Therefore, 84.13% of all scores are below $z = +1.0$. We want the z score below which 84% of the scores fall, so we must interpolate the numbers in the table.

The process of interpolation is outlined as follows:

Area	Difference	z
0.3389		+0.99
	0.0011	
0.3400		approx. +0.995
	0.0013	
0.3413		+1.00

Referring to the normal curve table, we find that 83.89% of the area under the normal curve is below a z score of 0.99. Thus the z score we want is between $z = +0.99$ and $z = +1.00$. Note that the difference in area between $z = +0.99$ and $z = +1.00$ is $0.3413 - 0.3389 = 0.0024$. Because we need only 0.0011 of this 0.0024 to make the area equal to 0.3400, the desired z score is approximately $+0.995$. Thus the z score of $+0.995$ is the 84th percentile in the standard normal curve. Percentiles below the 50th percentile have negative z scores. The z score of 0 is the 50th percentile.

In determining percentiles, we find the z score that corresponds to the desired percentage of area that lies to the left of the z score.

Suppose we want to find the 84th percentile on the original scale of measurement for our distribution of 15,000 psychological test scores. Assuming that the mean of this distribution is 85 and that the standard deviation is 20, we use the general formula for transforming a distribution of standard scores into a distribution of scores with a given mean and standard deviation (formula 4.6). For this example, we find the 84th percentile[3] to be

$$X' = (\sigma')(z) + \mu'$$

$$= (20)(0.995) + 85$$

$$= 104.90, \text{ or } 105$$

Similarly, we can find the 28th percentile. The area in the table is given between the mean and the z score, so we must always be careful to note on which side of the mean the z score is located. Again, applying interpolation to the values appearing in the table, we find that 22% of the area under the standard normal curve is between $z = -0.583$ and $z = 0$. Therefore, 28% of the area is to the left of $z = -0.583$. Using the general transformation formula, we find the 28th percentile to be

$$X' = (\sigma')(z) + \mu'$$

$$= (20)(-0.583) + 85$$

$$= 73.34, \text{ or } 74$$

If we know a percentile in z-score form, we can perform a transformation and determine the percentile in raw-score form, provided we know the mean and the standard deviation of the raw-score distribution.

The foregoing process is reversed when we use the standard normal curve to find the percentile rank of a given score in a distribution. For example, suppose we want to find the percentile rank of the score 102 on the standardized psychological test we have been discussing—the one with a mean of 85 and a standard deviation of 20. The first step is to determine the standard score or z score for this raw score and then refer

[3]When using the standard normal curve to determine percentiles, we round fractional values up to the next whole number, if that is the level of accuracy in the scale of measurement.

to Table 1 to determine the proportion of the area that is below this z score. For this example,

$$z = \frac{102 - 85}{20} = +0.85$$

Referring to Table 1, we find that 0.3023 of the area is between $z = 0$ and $z = +0.85$ and that 0.5000 of the area is below $z = 0$. Therefore, 0.8023 of the area is below $z = +0.85$, which means that the percentile rank of the score 102 is 80.23, or 80.

In a similar manner, we can find the percentile rank of a raw score below the mean. We know that such a score would have a percentile rank less than 50. Consider the raw score 80. The z score is found to be

$$z = \frac{80 - 85}{20} = -0.25$$

The table indicates that 0.0987 of the area is between $z = -0.25$ and the mean. Therefore, $0.5000 - 0.0987 = 0.4013$ of the area under the standard normal curve is to the left of $z = -0.25$, and we can say that the percentile rank of the raw score 80 is 40.13, or 40.

To determine the percentile rank of a raw score, we transform the score into a z score and determine the proportion of area below that z score.

Normalized Standard Scores

In Chapter 4, we indicated that the ordinal nature of percentiles and percentile ranks limits their use in subsequent mathematical manipulations and statistical analyses, such as summing across subtest scores. While using z scores overcomes this limitation when distributions are normal or near normal in shape, *normalized standard scores* can be generated that incorporate the advantages of both percentile ranks and standard scores. Normalized z scores have not only the characteristic properties of percentile ranks and percentiles but also the additional advantage of an equal-interval scale.

An example of normalized standard scores is the the *normal curve equivalent scores*, or *NCE scores*, that are being used to interpret standardized achievement test results. The NCE score scale ranges from 1 to 99 with a mean of 50 and a standard deviation of 21.[4] A score of 1 corresponds to the first percentile, a score of 50 to the 50th percentile, and a score of 99 to the 99th percentile. However, for other scores, the NCE and the percentile do not necessarily coincide. Figure 5.8 shows a normal curve and the relative positions of NCE scores, percentiles, and z scores. Note the inequality of spacing on the percentile scale, which indicates that (as we know) percentiles are

[4]The standard deviation is actually 21.38. However, for our purposes, we will use $s = 21$.

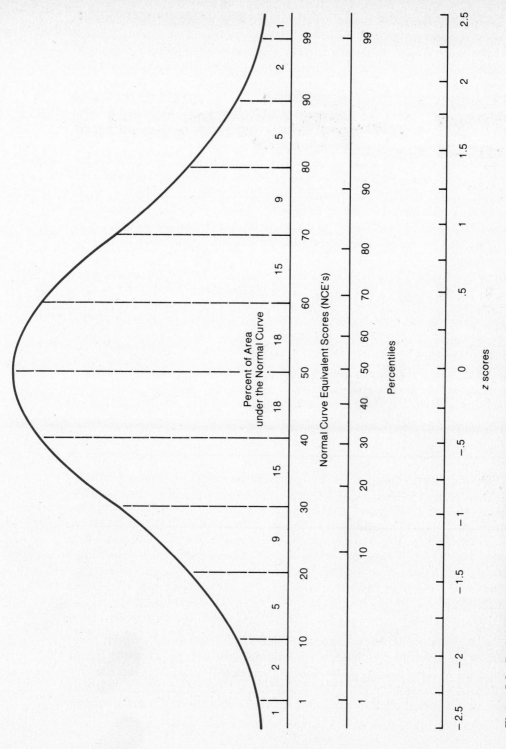

Figure 5.8 Positioning of NCE's, percentiles, and z scores for the normal curve

not an equal-interval scale. On the other hand, the NCE score scale and the z score scale *are* equal-interval scales. Because NCE scores have both the characteristics of percentile ranks and an equal-interval scale, they can be used to make meaningful comparisons between different test batteries and/or between different subtests within the same test battery.

Computing an NCE score involves three steps. First, we convert the raw score to a percentile rank. Second, we convert the percentile rank to a standard score, using Table 1. Third, we convert the standard score to the NCE score via the general transformation formula used earlier in this chapter, formula 4.6.

To illustrate the computation of an NCE score, we will use a score from the distribution of academic ability scores given in Table 4.1. Suppose we want to determine the NCE score that corresponds to a raw score of 50. In Chapter 4 we found the percentile rank for a raw score of 50 to be 62.03 ($PR_{50} = 62.03$), so the first step is completed. Next we use Table 1 to determine the z score that corresponds to this percentile rank. Interpolating the areas under the standard normal curve of Table 1, we find that

$$\text{For } z = +0.30, \text{ area} = 0.1179$$
$$\text{For } z = +0.31, \text{ area} = 0.1217 \quad \text{Difference} = 0.0038$$

To get to 0.1203, we need 0.0024 of the 0.0038 difference, which is 24/38 or 0.63 of the difference. Therefore, the corresponding z score is $+0.306$. For the final step, we apply formula 4.6:

$$NCE_{50} = (21)(0.306) + 50$$

$$= 6.43 + 50$$

$$= 56.43, \text{ or } 56$$

The normal curve equivalent (NCE) score is a standard score with a mean of 50 and a standard deviation of 21. Scores range from 1 to 99, and an equal unit is retained in the scale.

Summary

The normal curve is a very important concept in the behavioral sciences; many variables are normally distributed or approximately so. Because a specific normal distribution or curve depends on its mean and standard deviation, it is necessary for us to have a "standard" curve. This is the standard normal curve or unit normal curve. It is a normal curve in standard-score form with a mean of 0 and a standard deviation of 1.0. Given the mean and standard deviation for any normal distribution, we can transform it into the standard normal curve.

The standard normal curve is presented in Table 1 of the Appendix, with ordinates and areas given in terms of z scores. The table contains values for positive z scores only, but the curve is symmetrical, so corresponding negative z scores have the same values. Area is given from the mean ($z = 0$) to the given z score. Examples of finding areas between z scores are given in the chapter. It is a good idea, at least while learning to use the table, to sketch the areas you are seeking so that you will add or subtract areas properly. Percentiles and percentile ranks with respect to the standard normal curve are also discussed in the chapter.

The concept of normalized standard scores is introduced, specifically through normal curve equivalent (NCE) scores. These were devised in an attempt to provide relative positioning in the normal distribution yet to retain an equal unit in the measurement scale. NCE's essentially consist of a standard score, normally distributed, with a mean of 50 and a standard deviation of approximately 21. NCE scores can take on values from 1 to 99, inclusive.

It is extremely important for the reader to know how to use the normal curve table. Not only is it necessary for transforming a normal distribution into the standard normal curve, but the technique of using a statistical table will also be applied to other distributions later in the book.

Key Concepts

Normal curve
Unit normal distribution
Standard normal curve

Normalized standard scores
Normal curve equivalent (NCE) score

Exercises

5.1 Consider a set of 500 scores that is normally distributed with a mean equal to 50 and standard deviation equal to 8.
   a. What are the z scores corresponding to the raw scores 56, 38, and 63?
   b. How many scores lie between the values 48 and 54? Between 56 and 63? Between 37 and 48?
   c. How many scores exceed the values of 58, 44, and 64?
   d. How many scores are less than the values 54, 39, and 60?
   e. Find P_{35}, P_{80}, PR_{55}, and PR_{36}.

5.2 Use the data given in Exercise 5.1 to answer the following questions.
   a. What are the T scores corresponding to z scores of 2.14, -1.73, 0.00, and -0.42?

b. What are the z scores corresponding to T scores of 64, 38, 53, and 45?

c. What would the raw scores have been for the T scores given in Exercise 4.2(b)?

5.3 The norms for a standardized mathematics test were as follows:

National norms	$\mu = 75$	$\sigma = 12$
Large-city norms	$\mu = 68$	$\sigma = 15$

John had a score of 80 and Mary had a score of 65. What are their percentile ranks in terms of the national norms? In terms of the large-city norms?

5.4 A statistics instructor tells the class that grading will be based on the normal curve. He plans to give 10% A's, 20% B's, 40% C's, 20% D's, and 10% F's. If the final examination scores have mean of 75 and a standard deviation of 9.6, what are the ranges of scores for each grade?

5.5 For the data given in Table 2.9, convert the following scores to NCE scores: 60, 43, and 33.

5.6 Within an ancient culture, the average male life span was 37.6 years, with a standard deviation of 4.8 years. The average female life span was 41.2 years, with a standard deviation of 7.7 years. Use the properties of the normal distribution to find:

a. What percentage of men died before age 30?

b. What percentage of women lived to an age of at least 50?

c. A male death at age 35 could be equated to a female death at what age?

▦ Key Strokes for Selected Exercises

		Value	Keystroke	Display
5.1a	Determine z_{56}.	56	−	56
		50	=	6
			÷	6
		8	=	$.75 = z_{56}$
	Determine z_{38}.	38	−	38
		50	=	−12
			÷	−12
		8	=	$−1.50 = z_{38}$
5.2a	Determine the T score for $z = 2.14$.	2.14	×	2.14
		10	=	21.4
			+	21.4
		50	=	$71.4 = T$

5.2b   Determine the z score for $T = 64$.

64	−	64
50	=	14
	÷	14
10	=	1.4 $= z$

5.2c   Determine the raw score for
$T = 64$.

1.4	×	1.4
8	=	11.2
	+	11.2
50	=	61.2 $= X$

6

Correlation

A sociologist is interested in the social adjustment and the perception of family life quality among children aged 10–12. Inventories are available to measure each of these variables, and data from both inventories have been collected on 20 children. These data are found in Table 6.1. The distributions of scores for this group of children on both of the inventories could be described separately via the methods introduced in Chapters 2, 3, and 4. For example, the frequency distributions could be graphed, using either a frequency polygon or histogram, and measures of central tendency (mode, median, and mean) and measures of dispersion (variance and standard deviation) could be computed.

Child	X	Y			
A	8	6	$\Sigma X = 305$	$\Sigma Y = 278$
B	11	12		
C	19	12	$\Sigma X^2 = 5455$	$\Sigma Y^2 = 4530$
D	22	9		
E	11	16	$\Sigma XY = 4631$	
F	7	14		
G	24	15		
H	15	16		
I	12	20	$\overline{X} = 15.25$	$\overline{Y} = 13.9$
J	6	5		
K	20	16	$s_x = 6.50$	$s_y = 5.92$
L	28	18		
M	22	20		
N	14	11		
O	6	9		
P	18	20		
Q	10	4		
R	14	8		
S	23	24		
T	15	23		

Table 6.1   Data for calculating the Pearson product-moment correlation coefficient for scores on social adjustment (X) and scores and perception of family life quality (Y)

But suppose the sociologist is interested in the relationship between social adjustment and perception of family life quality for children in this age range. Suppose further that this relationship is assumed to be such that students who have high scores on the social adjustment inventory tend to have high scores on the family life quality inventory. Nevertheless, the sociologist cannot simply determine whether such a relationship exists by "eyeballing" the data given in Table 6.1. What is needed is a number, computed from the data, that will indicate whether such a relationship does exist. More specifically, what the sociologist needs is a numerical index that will provide information about the *nature* of the relationship, not simply whether the relationship exists. Such an index is called a correlation coefficient.

Many research studies in the behavioral sciences deal with the relationship between two or more variables. This is not surprising, considering the diversity of these studies and the number of variables involved. In these investigations, the nature of the relationship is explored. Does a relationship exist? If so, are high scores on one variable associated with high scores on the other variable, and vice versa? Or are high scores on one variable associated with low scores on the other variable, and vice versa? In this chapter, we will develop the concept of correlation and illustrate the computational procedures used to determine the index of the relationship between two variables: the correlation coefficient.

The Concept of Correlation

On the surface, the concept of correlation seems rather simple—that is, determining the nature of the relationship between two variables. In the study of statistical procedures, however, we need a precise definition of correlation so that its use will be consistent. *Correlation*, then, deals with the extent to which two variables are related, and the *correlation coefficient* is the index of this relationship.

The correlation coefficient is an index of the relationship between two variables.

Scattergrams

The first step in determining the nature of the relationship between two variables is to graph the data, using the methods described in Chapter 2. Recall that such a graph is called a scatter diagram or scattergram. A *scattergram* is a plot of pairs of scores for each individual on the two variables. Consider the data given in Table 6.1; the scattergram for these data is found in Figure 6.1. For this example, we have arbitrarily called variable X the scores on the social adjustment inventory, and we have called variable Y the scores on the inventory of perception of family life quality. The point that represents child D ($X = 22$; $Y = 9$) has been denoted to illustrate how these points have been plotted.

Note that the points plotted in Figure 6.1 tend to run from the lower left to the upper right. This pattern represents a *positive correlation* between the two variables. In other words, a positive relationship is represented when high scores on variable X are associated with high scores on variable Y. On the other hand, points running from the upper left to the lower right would represent a *negative correlation*. These two general patterns are illustrated in Figure 6.2(a) and (b). A perfect relationship between two variables exists when all the points in the scattergram lie on a straight line. A perfect positive correlation and a perfect negative correlation are illustrated in Figure 6.2(c) and (d), respectively. A scattergram in which the points are in a nearly circular pattern illustrates zero or near-zero correlation; see Figure 6.2(e).

The computed value of the perfect positive correlation coefficient is $+1.00$, and the computed value of the perfect negative correlation coefficient is -1.00. When no relationship exists between the two variables, the correlation coefficient is 0.00. The computed value of the correlation coefficient is a function of the *slope* of the general pattern of points in the scattergram and the *width of the ellipse* that encloses the points. If the slope is negative, the sign of the correlation coefficient is negative. If the width

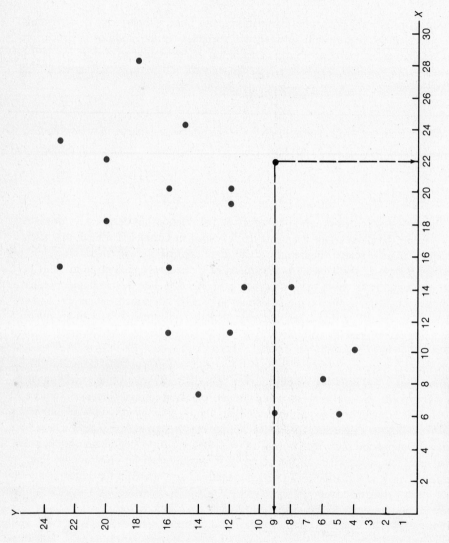

Figure 6.1 Scattergram of data in Table 6.1

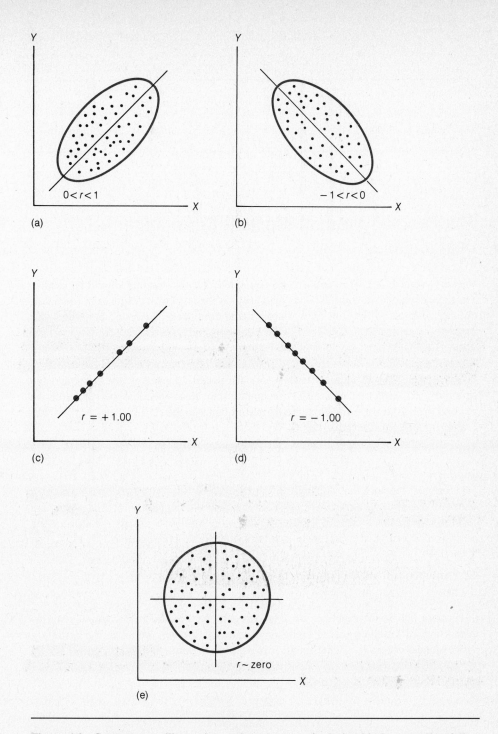

Figure 6.2 Scattergrams illustrating various degrees of relationship between *X* and *Y*

of the ellipse is narrow, the degree of relationship is larger and the correlation coefficient is larger.

The computed value of the correlation coefficient can range from -1.0 to $+1.0$, inclusive. The sign of the coefficient indicates the direction of the relationship. The absolute value of the coefficient indicates the magnitude of the relationship.

The Pearson Product-Moment Correlation Coefficient

Many correlation coefficients can be computed, but the one that should be used in a specific research situation depends primarily on the level of measurement (nominal, ordinal, interval, or ratio) of the variables being correlated. However, the most commonly used correlation coefficient in the behavioral sciences is the *Pearson product-moment correlation coefficient,* or the Pearson r. The level of measurement required for using the Pearson r is that both variables being correlated must be measured on either the interval or the ratio scale. Both the social adjustment inventory and the inventory of perception of family life quality can be assumed to be measured on interval scales, so we can use the Pearson r in our example.

The formula for the Pearson r is

$$r = \frac{\Sigma z_X z_Y}{N} \tag{6.1}$$

The formula indicates that all X and Y scores must be converted into standard scores or z scores and the product of each pair of z scores computed. These products are then summed and the sum is divided by N. More simply, r equals the sum of the cross-products of two variables, in standard score form, divided by N.

Computational Formula for the Correlation Coefficient

Formula 6.1 is not generally used to compute the Pearson r, because it requires converting all observed scores to standard scores. However, by using the definition of the standard score (see Chapter 5), it is possible to derive a computational formula for the Pearson product-moment correlation coefficient that involves only the observed X and Y scores. This formula is

$$r = \frac{N\Sigma XY - \Sigma X \Sigma Y}{\sqrt{[N\Sigma X^2 - (\Sigma X)^2][N\Sigma Y^2 - (\Sigma Y)^2]}} \tag{6.2}$$

Formula

For the data in our example,

$$r = \frac{(20)(4631) - (305)(278)}{\sqrt{[(20)(5455) - (305)^2][(20)(4530) - (278)^2]}}$$

$$= +.54$$

Interpreting the Correlation Coefficient

After we have computed the value of the correlation coefficient, what does it mean? In our example, we found $r = +.54$. Is this high or low? We can interpret a correlation coefficient only when we know more about the properties of the correlation coefficient. These properties are as follows:

1. The range of r is from -1.0 to $+1.0$.
2. The sign of the correlation coefficient indicates the direction of the relationship.
3. The absolute value of the correlation coefficient indicates the strength of the relationship. The greater the absolute value, the stronger the relationship.

Even when we keep these properties in mind, deciding what magnitude of r constitutes a "meaningful" relationship is somewhat arbitrary. It depends specifically on the variables being correlated. However, Table 6.2 contains a range of values that can be used as a guide or rule of thumb for interpreting the magnitude of correlation coefficients. Table 6.2 suggests that a correlation coefficient of $+.85$ would be interpreted as indicating a high positive relationship, whereas a correlation coefficient of $+.55$ would indicate a somewhat lower relationship. It is important to note that the range of values of the correlation coefficient is an ordinal scale, *not* an interval or ratio scale. It is incorrect to say that the difference between $r = +.30$ and $r = +.50$ is the same as the difference between $r = +.50$ and $r = +.70$ or, further, that $r = +.70$ is twice as large as $r = +.35$.

The range of values for the Pearson product-moment correlation coefficient is an ordinal scale.

The Assumption of Linearity

At this point, it is important to emphasize that the Pearson product-moment correlation coefficient is an index of the *linear* relationship between the two variables. That is, the points in the scattergram tend to locate along a straight line. (We will have more to say about this in the next chapter.) This does not mean that the points must fall exactly on a straight line, but that the points on the scattergram display a random scatter about a straight line. The trend of the data is then linear.

The scattergrams presented and discussed so far have exhibited linear trends, whether in a positive or a negative pattern. Consider now the scattergrams shown in

.90 to 1.00 ($-$.90 to $-$1.00)	Very high positive (negative) correlation
.70 to .90 ($-$.70 to $-$.90)	High positive (negative) correlation
.50 to .70 ($-$.50 to $-$.70)	Moderate positive (negative) correlation
.30 to .50 ($-$.30 to $-$.50)	Low positive (negative) correlation
.00 to .30 (.00 to $-$.30)	Little if any correlation

Table 6.2   Rules of thumb of interpreting correlation coefficients

Figure 6.3. The first scattergram, Figure 6.3(a), illustrates a linear trend in the relationship between variables X and Y, whereas Figures 6.3(b) and (c) exhibit curvilinear trends in the relationship between the two variables. Figure 6.3(b) depicts a curvilinear relationship in which increasingly greater values of X are associated with increasingly greater values of Y up to some intermediate point (k). Beyond point (k), as the values of X increase, they are associated with decreasing values of Y. An example of this phenomenon is the relationship between anxiety and performance. As the anxiety level of an individual increases, so does the individual's performance—but only up to a point. At that point, an increase in anxiety is no longer associated with a further increase in performance; in fact, greater anxiety is thereafter associated with a decrease in performance.

The Pearson r is the index of the *linear relationship* between two variables and is *not* appropriate to describe the curvilinear relationships illustrated in Figures 6.3(b) and (c). If the Pearson r is used with these data, the result is an underestimate of the actual relationship between the two variables. This underestimate is substantial when the trend is markedly curvilinear.

The Pearson r is the index of the linear relationship between two variables. If applied to variables that are curvilinearly related, the Pearson r will underestimate this relationship.

Useful in interpreting the correlation coefficient is the *coefficient of determination*, which is defined as the square of the correlation coefficient (r^2). The coefficient of determination indicates the proportion of the *shared variance* between the two variables being correlated. In other words, r^2 is the proportion of variance of variable X that can be attributed to the variance in Y, and vice versa. Consider our example. The variance of the social adjustment scores (X) was $(6.50)^2 = 42.25$, and the variance of the family life quality scores was $(5.92)^2 = 35.05$. Because the correlation was .54, the coefficient of determination is $(.54)^2 = 0.29$. Thus 29% of the variance in X is shared variance with Y. That is, $(0.29)(42.25) = 12.32$ of the variance of the social adjustment scores is shared variance with the family life quality scores.

It must be emphasized that r^2 is only the proportion of shared variance. However, if a cause-and-effect relationship can be established such that we can say that changes in Y are causing changes in X, we can conclude that 29% of the variance in social

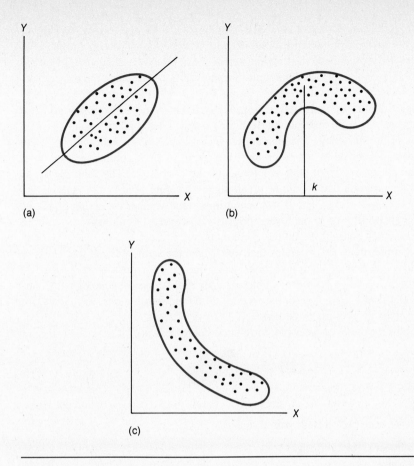

Figure 6.3 Scattergrams illustrating (a) linear and (b, c) curvilinear relationships

adjustment is caused by the variation in family life quality. Cause-and-effect relationships will be discussed in greater detail in Chapter 7.

The Spearman Rho Correlation Coefficient

The Pearson product-moment correlation coefficient is the appropriate coefficient when the level of measurement for the two variables being correlated is either interval or ratio. However, when the level of measurement of either or both variables is ordinal or nominal, the Pearson r is not appropriate. For example, if the level of measurement for both variables is ordinal, the appropriate correlation coefficient is the Spearman rho (ρ_s).

In order to illustrate the use of the *Spearman rho correlation coefficient*, let us consider the following example. Suppose the county mental health director was interested in the relationship between funding priorities for mental health programs as perceived by the professional staff administering these programs and as perceived by the people in the community. A community survey was conducted and the respondents were asked to rank 10 mental health programs from 1 to 10 in order of importance. The professional staff were also asked to provide rankings for the 10 programs. The results of this survey were then summarized, and a final set of rankings were assigned on the basis of the summary. The data appear in Table 6.3. The final rankings of the 10 programs as perceived by the community are found in the X column, and the final rankings for the professional staff are found in the Y column. The rank of 1 was assigned to the program that should be given highest funding priority, 2 to the second highest priority, and so on. Note that is is *not* important whether we assign a rank of 1 to the highest priority or to the lowest priority. However, it *is* important to use the same procedure for both the X and the Y columns.

In this example, there are programs that were given the same ranking; that is, there are ties among the ranks. When *tied ranks* occur, the rank assigned is the average of the ranks occupied. In column X of Table 6.3, Alcohol Abuse, Psychological Counseling, and Child Abuse are tied for the second, third, and fourth ranks. Each therefore receives the average of these three ranks, $(2 + 3 + 4)/3 = 3$.

The formula for the Spearman rho correlation coefficient is

$$\rho_S = 1 - \frac{6\Sigma d^2}{N(N^2 - 1)} \tag{6.3}$$

where

$d =$ difference between the paired ranks

$N =$ number of pairs of ranks

For the data given in Table 6.3,

$$\rho_S = 1 - \frac{6(47.50)}{10(100 - 1)}$$

$$= 1 - 0.29$$

$$= +.71$$

This coefficient indicates that there is a high positive correlation between the perception of the community and that of the professional staff relative to the funding priorities of the county mental health programs. In other words, the rankings given to the funding priorities were similar for both groups.

The Spearman rho is the correlation coefficient that should be used when the level of measurement for both variables being correlated is ordinal.

Mental Health Program	Community, X	Professional Staff, Y	d	d²
Alcohol Abuse	3	6	−3	9
Drug Abuse	1	4	−3	9
Rehabilitative Counseling	8.5	6	2.5	6.25
Spouse Abuse	10	10	0	0
Psychological Counseling	3	1	2	4
Crisis Center	8.5	8.5	0	0
Family Counseling	5.5	2.5	3	9
Outpatient Clinics	7	6	1	1
Rape Counseling	5.5	8.5	−3	9
Child Abuse	3	2.5	.5	.25
				$47.50 = \Sigma d^2$

Table 6.3   Data for the Spearman rho example

Other Correlation Coefficients

As we have seen, the Pearson product-moment correlation coefficient is appropriate for correlating scores on two variables both of which are measured on the interval scale or the ratio scale. The Spearman rho correlation coefficient is appropriate when the level of measurement of both variables is ordinal. However, neither of these procedures is appropriate when the level of measurement for either or both variables is nominal. For example, there are many times when a variable reflects a simple dichotomy, such as voting yes or no on a specific issue. Correlation coefficients have been developed for such variables. In addition, there are correlation coefficients designed for situations in which the two variables exhibit different levels of measurement. Table 6.4 provides a list of the more common correlation coefficients and the level of measurement required for applying each one. Detailed descriptions of these coefficients and others are found in more advanced statistics books.

Factors That Affect Correlations

The size of the correlation coefficient, whether it is the Pearson product-moment or the Spearman rho, is affected by several factors. Of course, the major determinant of the size of r is the degree to which the two variables actually are correlated. IQ test scores of fraternal twins are more highly correlated than IQ scores of cousins. Achievement in mathematics and achievement in science are more highly correlated than psychomotor performance and appreciation of the opera. However, some factors artificially constrain or limit the size of the correlation coefficient that is calculated.

Correlation Coefficient	Measurement of Variables (minimum level required)
1. Pearson product-moment	1. Both variables on interval or ratio scales.
2. Spearman rho	2. Both variables on ordinal scales.
3. Point biserial	3. One variable on an interval scale; the other a genuine dichotomy on a nominal or an ordinal scale.
4. Biserial	4. One variable on an interval scale; the other an artificial dichotomy on an ordinal scale. The dichotomy is artificial because there is an underlying continuous distribution.
5. Coefficient of contingency	5. Both variables on nominal scales.
6. Phi (ϕ)	6. Both variables on dichotomous, nominal scales.

Table 6.4   Correlation coefficients and minimum required level of measurement of the variables

Taken in part from William Wiersma, *Research Methods In Education, An Introduction, 3rd ed.* (Itasca, Ill.: F. E. Peacock Publishers, 1980), p. 255.

Two such factors are the homogeneity of the group measured and the reliability of the measures.

By *homogeneity* of the group, we mean the extent to which the members of the group tend to be the same on the variables being correlated. On either or both variables, as the homogeneity of the group increases, the variance decreases. As the group becomes increasingly homogeneous, the correlation tends to be limited. For example, if we correlated height and weight for players in the National Basketball Association, the correlation would tend to be small. Height in the NBA constitutes a *restriction of range,* which leads to a correlation coefficient that is nearer to zero than would be found if the full range of heights were represented.

The effect of the restriction range is shown in Figure 6.4. If we developed a scattergram for height and weight of adult males in general, the points would be contained by an ellipse, which suggests a moderate correlation between height and weight. But when only NBA players are considered, their portion of the total ellipse has a different shape, which is more suggestive of a weak or near-zero correlation for them. The moderate correlation for the whole population may not hold for a unique subgroup in that population.

The homogeneity of the group affects the correlation in such a way that increased homogeneity tends to limit the size of the correlation coefficient.

The reliability of the measures used to measure the variables being correlated can also affect the size of the correlation coefficient. *Reliability* of measurement is the consistency of measurement. An unreliable measure contains varying amounts of measurement error, extraneous influences on the scores. Unreliable scores are inconsistent from one occasion to the next. This kind of inconsistency interferes with our

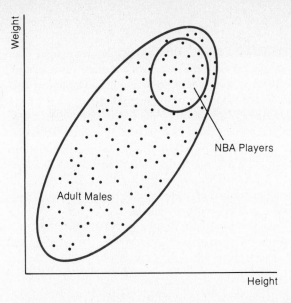

Figure 6.4   Scattergram of adult males and NBA players, illustrating the effect of restriction of range on the correlation of height and weight

obtaining a true reading on the variable of interest. Variables such as self-concept, creativity, and anxiety are measured less reliably than such variables as arithmetic achievement and age.

The effect of using unreliable measures is to cause us to underestimate the size of the correlation coefficient relative to what we would have obtained if we had used better measures. The lack of an anticipated, sizable correlation can sometimes be traced to the use of inadequate measurement procedures.

Unreliable measurement can have the effect of limiting the size of the correlation coefficient.

Uses of the Correlation Coefficient

Researchers use correlations frequently and for a variety of purposes.

1. Researchers may merely wish to describe the *association* of two variables such as IQ and socioeconomic status, reading speed and reading comprehension, or education

115

and income. These correlations support or refute the theories or hypotheses of the researchers.

2. Correlations are used to assess the *reliability* of measurement techniques. The scores of tests given to the same people on different occasions are often correlated to assess how consistent the set of scores is from one time to another. The higher the correlation, the higher the consistency or reliability of these measures.

3. The *validity* of many measures is established through the use of correlation coefficients. For example, the Scholastic Aptitude Test (SAT) is used to predict the grade point average of first-year students in a number of selected colleges. The correlation coefficient for SAT score and grade point average is the statistic of primary interest in these analyses.

4. The concept of correlation is used in prediction, which is the process of estimating scores on one variable from known scores on another variable.

Summary

The correlation coefficient is introduced in this chapter and defined as the index of the relationship or association between two variables. The correlation coefficient is a number that indicates the strength and direction of such a relationship. The absolute value of the coefficient indicates the strength of the relationship; the greater the absolute value, the stronger the relationship. The sign of the coefficient indicates the direction of the relationship; plus indicates a positive relationship and minus indicates a negative relationship. The correlation coefficient can take on values of $+1.0$ to -1.0, inclusive. Zero correlation indicates no relationship between the two variables.

A scattergram is a pictorial representation of the data for a correlation coefficient. The scattergram consists of points, each of which is determined by a pair of values from among the variables being correlated. When the points in a scattergram fall in an ellipse, a relationship is indicated. The narrower the ellipse, the greater the correlation between the variables.

The computation of two correlation coefficients is illustrated. The Pearson product-moment coefficient is used when there are two variables, both measured on interval or ratio scales. The Spearman rho coefficient is used when both variables are measured on an ordinal scale. Computational formulas for these coefficients are provided.

Factors that affect the size of the correlation coefficient are discussed. Homogeneity of the group measured and unreliable measurement tend to reduce the size of the correlation coefficient.

Key Concepts

Correlation
Correlation coefficient

Coefficient of determination
Shared variance

Scattergram
Positive correlation
Negative correlation
Pearson product-moment correlation
Linear relationship

Spearman rho coefficient
Tied ranks
Homogeneity
Reliability
Validity

Exercises

6.1 Decide whether the following pairs of variables are positively or negatively correlated.
   a. Scores on a scholastic aptitude exam and number of items missed
   b. Amount of alcohol consumed and reaction time
   c. The batting averages and runs-batted-in (RBI) totals of baseball players

6.2 A school psychologist and a classroom teacher rank eight children on the variables "aggressiveness during play" and "relative popularity among classmates," respectively. Compute the Spearman rho correlation coefficient.

Child	Aggressiveness	Popularity
A	1	7
B	2	3
C	3	8
D	4	6
E	5	2
F	6	1
G	7	5
H	8	4

6.3 A social studies teacher administers 20-question exams on international and national events. The following are the respective scores of 10 of the students. Draw a scattergram and then compute the Pearson r.

Child	X (International)	Y (National)
A	15	20
B	12	15
C	10	12
D	14	18
E	10	10
F	8	13
G	6	12
H	15	10
I	16	18
J	13	15

6.4 A teacher gave a 100-item test to her class and then correlated the number of correct answers with the IQ scores found in the students' records. The correlation was 0.41. What would the correlation have been if she had used the number of incorrect answers instead?

6.5 Adolescents in a summer training program were measured on self-esteem via a Self-Rating and a Real–Ideal Congruence. Use the Pearson r to find the correlation between the two measures.

Self-Rating		Real–Ideal Congruence			
18	1.5	4	22	−2.5	6.25
13	6	10	14	−4	16
9	10	7.5 18		2.5	6.25
18	1.5	9	16	7.5	56.25
17	3.5	2	25	1.5	2.25
12	7.5	6	19	1.5	2.25
12	7.5	5	20	2.5	6.25
16	5	3	24	2	4
17	3.5	1	26	2.5	6.25
11	9	7.5 18		1.5	2.25

108

6.6 If $r_{XY} = -1.0$ and $r_{XZ} = -1.0$, what is r_{YZ}? (*Hint*: Draw the scattergram.)

6.7 Convert the scores from Exercise 6.5 to ranks and find the Spearman ρ coefficient.

6.8 Apply the Pearson r formula to the ranks in Exercise 6.2 and compare the result to that which was found when the Spearman ρ formula was used.

6.9 Find the Pearson r for the following pairs of scores:

X	Y
19	12
26	12
13	12
22	12
11	12
18	12

▦ Key Strokes for Selected Exercises

	Value	Key Stroke	Display
6.2 Step 1: Determine Σd^2	6	$+/-$	-6
		X^2	36
		$+$	36
	1	$+/-$	-1
		X^2	1
		$+$	37
	5	$+/-$	-5
		X^2	25
		$+$	62
	2	$+/-$	-2
		X^2	4
		$+$	66
	3	X^2	9
		$+$	75
	5	X^2	25
		$+$	100
	2	X^2	4
		$+$	104
	4	X^2	16
		$=$	$\underline{120} = \Sigma d^2$
Step 2: Determine $6\Sigma d^2$.		$\times$	120
	6	$=$	$\underline{720} = 6\Sigma d^2$
		STO	720
Step 3: Determine rho.	8 (N)	X^2	64
		$=$	64
		$-$	64
	1	$=$	63
		$\times$	63
	8 (N)	$=$	504
		1/X	.00198
		$\times$	.00198
		RCL	720
		$=$	1.429
		$+/-$	-1.429
		$+$	-1.429
	1	$=$	$\underline{-.429} = \rho_s$

6.3 Use the key strokes described in the Appendix to determine the following:

$$\Sigma X = 119 \qquad \Sigma Y = 143 \qquad \Sigma XY = 1761$$
$$\Sigma X^2 = 1515 \qquad \Sigma Y^2 = 2155$$

Step 1: Determine denominator.

Value	Key Stroke	Display
10 (N)	×	10
1515 (ΣX^2)	−	15150
119 (ΣX)	X^2	14161
	=	989
	STO	989
10 (N)	×	10
2155 (ΣY^2)	−	21550
143 (ΣY)	X^2	20449
	=	1101
	×	1101
	RCL	989
	=	1088889
	$\sqrt{}$	1043.4984
	STO	1043.4984

Step 2: Determine numerator.

10	×	10
1761 (ΣXY)	−	17610
119 (ΣY)	×	119
143 (ΣY)	=	593

Step 3: Determine r.

	÷	593
	RCL	1043.4984
	=	.568 $= r$

7

Regression and Prediction

Much effort in our world is expended in trying to predict things. Short-range and long-range weather conditions are predicted daily on TV. These predictions are based on a relatively complex set of atmospheric variables. "Old timers" and not-so-old timers predict the weather from variability in such factors as the color of the sky. Investment counselors are always trying to predict what the stock market will do. Like predicting the weather, predicting stock market behavior results in varying degrees of success.

Is making predictions in the behavioral sciences like predicting the weather or the stock market? Consider the problem faced by the admissions director of a college in selecting the members of the next incoming class. Suppose that more prospective students apply than can be admitted. If there were some type of test or measure that could be administered to the applicants such that performance on this test were a good indicator (predictor) of later success in college, then the problem would be largely solved. Of course, admission to college is usually based on several factors, but there are tests used, at least in part, for admission purposes. The tests most commonly used for these purposes are the Scholastic Aptitude Test (SAT) and the American College Testing (ACT) Program. Scores from such tests are used, along with other information, in making admission decisions, because these scores do indicate the likelihood of success in college.

Correlation and Prediction

Prediction, in its simplest sense, is the process of estimating scores on one variable, the *criterion variable*, on the basis of knowledge of scores on another variable, the *predictor variable*. The concept of correlation is used in the prediction process. For example, suppose a sociologist wants to predict a person's score on an aggressiveness measure on the basis of this person's score on a measure of alienation. Assume that a moderate-to-high positive relationship exists between these two variables. That is, high scores on the measure of alienation are associated with high scores on the aggressiveness measure. Hence it would be possible for the sociologist to conclude that, if the person's score on the measure of alienation was high, chances would be good that this person's score on the aggressiveness measure would also be high.

In Chapter 6, the concept of linear correlation was graphically illustrated using the scattergram. In the scattergram shown in Figure 6.1, the points tended to locate along a straight line. If we could draw this straight line, it would represent an average picture of how change in one variable is associated with change in another variable. This line is called the *linear regression line*. If we are using the variable X to predict the variable Y, the line is called the regression line of Y on X.

Prediction is the process of estimating scores on one variable from knowledge of scores on another variable.

Consider an example. As part of a research study in cognitive psychology, researchers are investigating the relationship between creativity (variable Y) and logical reasoning (variable X) and want to determine a mathematical equation for predicting creativity scores (Y) from logical reasoning scores (X). Ten middle-school students were administered both a creativity measure and a logical reasoning test. The resulting data are found in Table 7.1, and the scattergram is shown in Figure 7.1. Note that, for these data, there is a positive relationship between the scores on the two measures. Using formula 6.2, we find that the correlation between these two sets of scores is $+.56$.

Logical Reasoning, X	Creativity, Y		
10	8	$\Sigma X = 56$	$\Sigma Y = 58$
4	3	$\Sigma X^2 = 376$	$\Sigma Y^2 = 364$
5	7		$\Sigma XY = 348$
6	7		
2	4		
5	5		
9	6		
8	6	$\overline{X} = 5.6$	$\overline{Y} = 5.8$
3	4	$s_X = 2.63$	$s_Y = 1.75$
4	8		$r = +.56$

Table 7.1   Data for developing the regression equation for predicting creativity scores from logical reasoning scores

The Regression Line

The process of prediction involves two steps. The first step is determining the regression line, which is a mathematical equation. The second step is using this mathematical equation in the actual prediction of scores. Because we are limiting our discussion to *linear* regression, the mathematical equation is the equation of a straight line.

The mathematical equation of a straight line expresses a functional relationship between two variables. In predicting Y scores from X scores, we say that Y is a function of X and use the *"slope–intercept form"* of the equation for a straight line. The general form of this equation is

$$\hat{Y} = bX + a \tag{7.1}$$

where

$\hat{Y}$ = predicted score

b = slope of the line

a = Y intercept

Figure 7.2 shows the graph of a straight line that will be used to illustrate the definition of the slope–intercept form. This line extends indefinitely in both directions but, for the sake of simplicity, we will consider only that portion of the line between $X = 0$ and $X = 5$. The respective plot points are

$$X = 0, \quad Y = 2 \qquad X = 3, \quad Y = 3.5$$
$$X = 1, \quad Y = 2.5 \qquad X = 4, \quad Y = 4$$
$$X = 2, \quad Y = 3 \qquad X = 5, \quad Y = 4.5$$

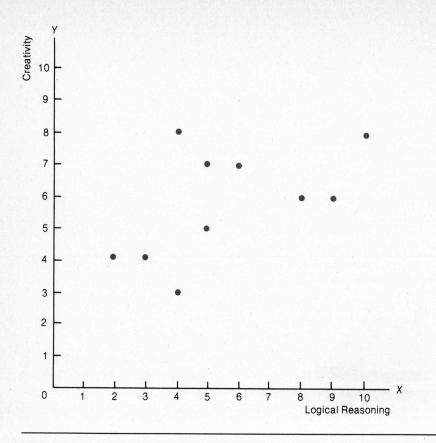

Figure 7.1 Scattergram of logical reasoning and creativity scores

The *slope* of a line is defined as the amount of change in Y that corresponds to a change of 1 unit in X. Note in Figure 7.2 that, corresponding to a 1-unit increase in X from 1 to 2, there is an increase of 0.5 units in Y from 2.5 to 3. This relationship is depicted by dashed lines in Figure 7.2. Hence the slope of this line is $+0.5$. In general, the slope of a line can be positive or negative and can be less than or greater than 1. A slope equal to 0 would indicate that the line is parallel to the X-axis—that is, it is horizontal.

The *intercept* of the line is defined as the value of Y when X equals 0. Note that, in Figure 7.2, the intercept is the value of Y where the line crosses, or intercepts, the Y axis. For this example, the intercept is 2. Now that we know the values of the slope and the Y intercept we can define the mathematical equation of the straight line that appears in Figure 7.2. Using the general formula for a straight line (formula 7.1), we substitute the values for b and a. Thus the equation for this example is

$$\hat{Y} = 0.5X + 2.0$$

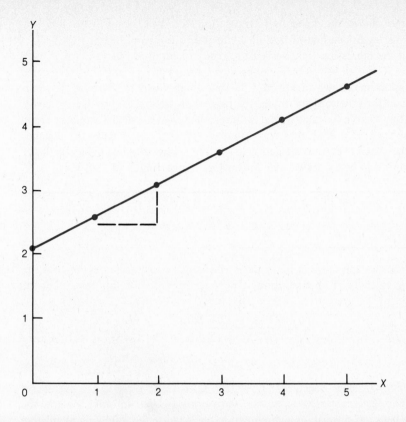

Figure 7.2   Graph of the straight line represented by the equation $\hat{Y} = 0.5X + 2.0$

The general equation for a straight line used in prediction is $\hat{Y} = bX + a$, where b is the slope of the line and a is the Y intercept.

Determining the Regression Line

Now that we know the general form of the equation of a straight line, how do we fit a line to the scattergram of points and then use this line in the prediction process? Consider the data given in Table 7.1 and Figure 7.1. We want to determine the line that describes a trend in the data such that a change in X will be reflected in an "average" change in Y. The specific line in question is "fitted" to the data by what is called the *method of least squares*. In our example of predicting creativity scores (Y) from logical reasoning scores (X), the method of least squares fits the line in such a way that the sum of the squared distances from the data points to the line is a minimum. This regression line of best fit is illustrated in Figure 7.3.

Recall that the regression line is used to predict or estimate the value of Y for a given value of X. For all values of X, the predicted values of Y (denoted $\hat{Y}$) are located on the regression line shown in Figure 7.3. Thus another way to illustrate the least-squares method (or criterion) for fitting the regression line is to consider the errors (e) in predicting Y from X—that is, the difference between the actual value of Y and the predicted value ($\hat{Y}$). Note that this difference ($Y - \hat{Y}$) is actually the distance (parallel to the Y axis) between the data point and regression line. Thus the least-squares criterion can be expressed symbolically as minimizing $\Sigma (Y - \hat{Y})^2$.

We fit the regression line to the data points in a scattergram using the least-squares criterion, which requires $\Sigma (Y - \hat{Y})^2$ to be a minimum.

The formulas for the slope (b) and the Y intercept (a) of the regression line are derived using calculus. This derivation is beyond the scope of this book, but the formulas are

$$b = \frac{N\Sigma XY - (\Sigma X)(\Sigma Y)}{N\Sigma X^2 - (\Sigma X)^2} \qquad (7.2)$$

$$a = \overline{Y} - b\overline{X} \qquad (7.3)$$

The value of b, which is the slope of the regression line, is called the *regression coefficient*. The intercept, a, is called the *regression constant*. In order to show the relationships between correlation and regression, we can use the following alternative formula for the regression coefficient:

$$b = (r)\frac{s_Y}{s_X} \qquad (7.4)$$

where

r = correlation between variables X and Y

s_Y = standard deviation of the Y scores

s_X = standard deviation of the X scores

For the regression equation $\hat{Y} = bX + a$, b is called the regression coefficient and a is called the regression constant.

Now let us apply these formulas to the data given in Table 7.1. For formula 7.2,

$$b = \frac{10(348) - (56)(58)}{10(376) - (56)^2}$$

$$= 0.37$$

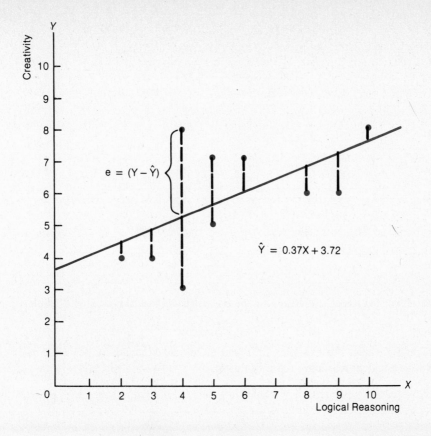

Figure 7.3   Regression line for predicting creativity scores from logical reasoning scores

Alternatively, for formula 7.4,

$$b = (0.56)\frac{1.75}{2.63}$$

$$= 0.37$$

And for formula 7.3,

$$a = 5.8 - (0.37)(5.6)$$

$$= 3.72$$

Thus, on the basis of the data given in Table 7.1, the regression equation for predicting creativity scores from logical reasoning scores is

$$\hat{Y} = 0.37X + 3.72$$

Once the regression equation has been determined, predicting Y scores is straightforward; we simply substitute the X value into the equation and solve it for the

X	Y	$\hat{Y}$	$e = (Y - \hat{Y})$
10	8	7.42	.58
4	3	5.20	− 2.20
5	7	5.57	1.43
6	7	5.94	1.06
2	4	4.46	− .46
5	5	5.57	− .57
9	6	7.05	− 1.05
8	6	6.68	− .68
3	4	4.83	− .83
4	8	5.20	2.80

	X	Y	$\hat{Y}$	$e = (Y - \hat{Y})$
Sum	56	58	57.92	0.08
Mean	5.6	5.8	5.792	0.008
Variance	6.93	3.07		2.108
Standard deviation	2.63	1.75		1.452

Table 7.2 Predicted scores and errors for the creativity and logical reasoning example

corresponding $\hat{Y}$ value. For example, let's find the predicted creativity score ($\hat{Y}$) associated with a logical reasoning score (X) of 5.

$$\hat{Y} = 0.37(5) + 3.72$$

$$= 5.57$$

The predicted score values for each X for our example are found in Table 7.2.

To find predicted scores, substitute the value of the known variable into the regression equation and solve the equation.

Errors in Prediction

If there were a perfect relationship between the two variables, all the data points in the scattergram would lie in a straight line and it would be easy to find the equation of that line. However, when the relationship is less than perfect, the regression line is fitted to the data via the method of least squares. Recall that this process involves determining the line such that the squared distances from the data points to the line are at a minimum. These distances are the differences between the actual Y scores and the predicted scores, $\hat{Y}$. They are defined as the *errors in prediction*. Symbolically,

$$e = (Y - \hat{Y}) \tag{7.5}$$

Because the points in the scattergram are both above and below the regression line, the errors are both positive and negative (see Table 7.2).

Let us now take a closer look at the errors in prediction. First, because we have computed a predicted score $\hat{Y}$ for each X and have then calculated the difference between the actual and the predicted scores, we have a distribution of error scores. Recall from Chapter 3 that, in order to describe a distribution of scores adequately, we need to know (1) the shape of the distribution, (2) the central tendency, and (3) the variation of the scores. As we will discuss later in this chapter, the distribution of the errors in prediction is assumed to be normal. Relative to the central tendency, the mean of the errors equals 0. Note, however, that the mean of the errors in Table 7.2 is 0.008. This difference is due to rounding error.

The variance of the errors in prediction is defined in the same way as the variance of any distribution of scores:

$$s_e^2 = \frac{\Sigma(e - \bar{e})^2}{n - 1} \tag{7.6}$$

The mean of the errors equals 0, so the formula is simplified to

$$s_e^2 = \frac{\Sigma(e)^2}{n - 1} \tag{7.7}$$

The Standard Error of Estimate

The standard deviation of the error scores is defined as the *standard error of estimate*. For the data given in Table 7.2, the standard error of estimate is computed as follows:

$$s_e = \sqrt{\frac{\Sigma(e)^2}{n - 1}} \tag{7.8}$$

$$= \sqrt{\frac{18.9752}{9}}$$

$$= 1.45$$

This formula for the standard error of estimate requires that we compute each error score first. The following formula offers a more direct method:

$$s_e = s_Y \sqrt{1 - r^2} \tag{7.9}$$

where

s_Y = standard deviation of the Y scores

r = correlation between the X and Y scores

For the data in our example,

$$s_e = 1.75 \sqrt{1 - (0.56)^2}$$

$$= 1.45$$

In formula 7.9, we can see the relationship between correlation and prediction. Note that, when the correlation between X and Y is high, the standard error is smaller. In fact, when the correlation equals $+1.0$ or -1.0, the standard error of estimate is 0. Thus a high correlation between X and Y reduces the standard error of estimate and therefore enhances the accuracy of the prediction.

The standard error of estimate is the standard deviation of the errors in prediction. The larger the correlation between the X and Y scores, the smaller the standard error of estimate and the greater the accuracy of prediction.

Another way to look at the standard error of estimate is to consider the standard deviation of the Y scores around the predicted Y for a given value of X. Consider the X score labeled X_1 in Figure 7.4. The predicted score $\hat{Y}$ for X_1 would be on the regression line with a distribution of actual scores above and below the line for this value of X. This is actually a distribution of error scores for this value of X. Note that, for the X values labeled X_2 and X_3, we also have distributions of error scores around the respective predicted Y scores. These distributions are called *conditional distributions* because they are the distributions of error scores for a given value of X. Each of these conditional distributions is assumed to be normal, and the standard deviation of each of these distributions is assumed to be equal. This latter property is referred to as *homoscedasticity*. If we assume that these conditional distributions exhibit normality and homoscedasticity, we can use the properties of the normal distribution to make probability statements about predicted scores.

There are two assumptions underlying the use of the conditional distributions in prediction; those are homoscedasticity and normality.

Predicted Scores and Probability Statements

In our example of predicting creativity scores from logical reasoning scores, the following questions might be asked:

If $X = 6$, what is the probability that Y will be greater than 7?
If $X = 5$, what is the probability that Y will be between 3 and 6?

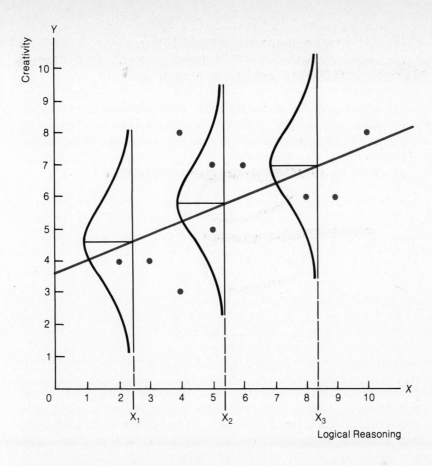

Figure 7.4   Conditional distributions of Y for X_1, X_2, and X_3

The rationale for asking such questions must be couched in the realities of the situation and the decisions that need to be made. For example, a researcher might want to know the probability that a student with a logical reasoning score (X) equal to 6 will have a creativity score (Y) greater than 7. If this probability is quite low, the student might not be recommended for a special program. To answer this question, we need to use the predicted score along with the conditional distribution, given the logical reasoning score of 6. Using the regression equation determined earlier, we find that

$$\hat{Y} = 0.37X + 3.72$$
$$= 0.37(6) + 3.72$$
$$= 5.94$$

Thus the predicted score for a student who had a score of 6 on the logical reasoning measure would be 5.94. Recall that, for the X value of 6, there is a conditional

Predicted Scores and Probability Statements

distribution of Y scores around the predicted score, $\hat{Y} = 5.94$. This distribution is illustrated in Figure 7.5. These scores are assumed to be normally distributed around $\hat{Y} = 5.94$, and, under the assumption of homoscedasticity, the standard deviation is defined as the standard error and was found to be equal to 1.45.

The probability that the actual Y score will exceed 7, given that X equals 6, corresponds to the proportion of the area of the conditional distribution that exceeds $Y = 7$. (See the shaded area in Figure 7.5.) We determine the area under the curve beyond the score of 7 by converting this score to a standard score, using the procedures outlined in Chapter 4, and then consulting the normal curve table (Table 1 in the Appendix). Recall that the general formula for the standard score is

$$z = \frac{X - \mu}{\sigma}$$

For this example using the conditional distribution, the corresponding formula is

$$z = \frac{Y - \hat{Y}}{s_e} \qquad (7.10)$$

For our example,

$$z = \frac{7.0 - 5.94}{1.45}$$

$$= +0.73$$

Thus the probability that Y will exceed 7 if $X = 6$ corresponds to the area under the normal curve beyond a z score of 0.73. From Table 1 in the Appendix, we find that this area equals 0.2327. Therefore, the probability is 0.2327 that Y will exceed 7 if $X = 6$. If the probability required for being admitted to the special program is 0.40 or greater, a student with a score of 6 on the logical reasoning test would not be recommended for admission.

The conditional distribution is used to determine the probability associated with the occurrence of Y, given a specific value of X.

The Second Regression Line

Throughout this chapter, we have discussed the regression equation for predicting Y from X. There is, in fact, a second regression equation—that for predicting X from Y. The general form for this equation is

$$\hat{X} = bY + a \qquad (7.11)$$

It must be emphasized that the regression coefficient (b) and the regression constant

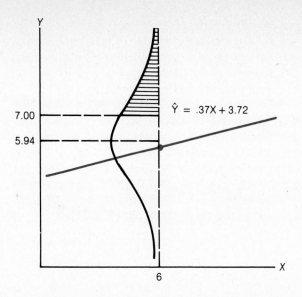

Figure 7.5   Probabilities in the conditional distribution

(*a*) in formula 7.11 are *not* the same as those for predicting *Y* from *X*; the computational formulas are different. The standard convention that has been adopted by most researchers in the behavioral sciences is to use *X* as the predictor variable and *Y* as the criterion (predicted) variable. Therefore, we will not present the second set of formulas.

Correlation and Causation

The mere fact that we use our knowledge about the relationship between two variables, *X* and *Y*, to predict values of *Y* from known values of *X* does not imply that changes in *X* cause changes in *Y*, or vice versa. For example, consider the old saying that "an apple a day keeps the doctor away." Undoubtedly, a moderate correlation could be found between the number of apples consumed during one year and the number of visits to a physician's office. This does not imply that the person had many visits to a physician's office *because* she or he ate an inadequate number of apples. This person could have been severely injured in an automobile accident (even while eating an apple).

There are some variables that simply cannot be caused by other variables. There may be a positive correlation between performance on a physical task and chronological age. But we cannot argue that chronological age is affected by performance on the physical task. Chronological age is affected by only one factor, the passage of time since the individual's birth.

The order of occurrence of variables may indicate whether or not a cause-and-effect relationship is possible. Suppose that X is a variable consisting of treatments such as various levels of drug dosage and that variable Y is performance of a rat in running a maze. X cannot affect Y until it is administered. If the rat runs the maze before receiving the drug dose, the dose cannot have affected its performance.

There are combinations of variables that are highly correlated, and in such cases one variable is an accurate predictor of the other. But again, accurate prediction does not necessarily imply that the predictor variable *causes* the scores on the criterion variable. How, then, does a researcher infer cause and effect? Basically, it is done through understanding the variables under study and the context in which they are operating. The statistics used in correlation and prediction indicate whether a relationship exists and how effective prediction on the basis of that relationship will be. The researcher must still interpret the statistics in the context of the situation.

Summary

This chapter explains how the concept of correlation can be used to allow us to make statistical predictions. Concepts of regression, specifically linear regression, are also introduced. In prediction, we estimate scores on one variable on the basis of our knowledge of scores on another variable. In order to do this, we use the equation of a straight line called the regression line. We fit the line to the scattergram of points by using the least-squares criterion, which requires that $\Sigma (Y - \hat{Y})^2$ be as small as possible.

The equation of the regression line becomes the prediction equation. It links observed scores on one variable with predicted scores on the second variable. Values can be substituted into the equation for the predictor variable to generate predicted scores. The predicted scores all lie on the regression line. These predicted scores are generally *not* equal to the observed scores, so we have error scores of $Y - \hat{Y}$, the difference between observed and predicted scores. These error scores have a mean of 0 and a standard deviation called the standard error of estimate.

Accuracy of prediction is not a matter of attempting to hit certain scores "right on the head" but is concerned with the standard error of estimate. Reducing the standard error of estimate increases the accuracy of prediction. The standard error of estimate is reduced as the correlation between the two variables increases. So the greater the correlation, the greater the accuracy of prediction.

Using the standard error of estimate along with the assumption of homoscedasticity allows us to elaborate on our predictions. Although $\hat{Y}$ is the best single predicted value, we can make probability statements about possible values of Y for a given X. We do so by considering the distribution of possible Y values for the given X value, assuming that this distribution is normally distributed and has a standard deviation equal to the standard error of estimate.

Prediction is a widely used technique in behavioral sciences research, as is the use of correlation coefficients. Correlation is directly related to prediction, and accurate or effective prediction is enhanced by a strong correlation between the predictor and criterion variables. Finally, we must keep in mind the fact that correlation does not necessarily imply a cause-and-effect relationship between the variables.

Key Concepts

Prediction	Regression coefficient
Predictor variable	Regression constant
Criterion variable	Errors in prediction
Linear regression line	Standard error of estimate
Slope	Conditional distribution
Intercept	Homoscedasticity
Method of least squares	

Exercises

7.1 Using the general equation of the regression line ($\hat{Y} = a + bX$), solve for $\hat{Y}$.
   a. $X = 15$, $a = 1.3$, $b = 2$
   b. $X = 1.9$, $a = .4$, $b = .65$
   c. $X = 52$, $a = 21$, $b = -5.6$

7.2 Find the regression line for predicting Y from X for the following data.

X	Y
6	30
12	22
10	18
14	25
11	28
18	32
8	16
17	27
15	22
17	25

7.3 If $s_X = 10$ and $s_Y = 22$, what is the maximum value of b in the equation $\hat{Y} = a + bX$?

7.4 A sociologist is interested in predicting levels of depression from "the number of social events attended per week" for a sample of high school students. Find the prediction equation for the following pairs of scores.

Social Events per Week	Depression
0	15
2	3
2	12
1	11
3	5
1	8
2	15
0	13
3	2
3	4
4	2
1	8
1	10
1	12
2	8

7.5 What is the standard error of estimate for the data given in Exercise 7.4?

7.6 Find the regression equation $\hat{Y} = a + bX$, given the following information.

$$\bar{X} = 110 \qquad\qquad \bar{Y} = 240$$
$$s_X = 16 \qquad\qquad s_Y = 20$$
$$r_{XY} = .5$$

7.7 Using the data given in Exercise 7.6, find the probability that Y is greater than 250 if X equals 100.

7.8 Using the data given in Exercise 7.2, find the probability that Y is less than 25 if $X = 15$.

7.9 Describe the regression line $\hat{Y} = a + bX$ if $r_{XY} = 0$.

7.10 Find the correlation between X and Y if $2s_e = s_Y$.

▦ Key Strokes for Selected Exercises

7.2 Use the Key Strokes described in the Appendix to determine ΣX, ΣY, ΣX^2 and ΣXY.

$$\Sigma X = 128 \qquad\qquad \Sigma Y = 245$$
$$\Sigma X^2 = 1788 \qquad\qquad \Sigma XY = 3200$$
$$N = 10$$

Step 1: Determine b. (Numerator)

	Value	Key Stroke	Display
	10 (N)	$\times$	10
	3200 (ΣXY)	$-$	3200
	128 (ΣX)	$\times$	128
	245 (ΣY)	$=$	640
		STO	640

(Denominator)

	Value	Key Stroke	Display
	10 (N)	$\times$	10
	1788 (ΣX^2)	$-$	17,880
	128 (ΣX)	X^2	16,384
		$=$	1496
		1/X	.00067
		X	.00067
		RCL	640
		$=$	.428 $= b$

Step 2: Determine a.

	Value	Key Stroke	Display
	245 (ΣY)	$-$	245
	.428 (b)	$\times$	.428
	128 (ΣX)	$=$	190.22
		$\div$	190.22
	10	$=$	19.022 $= a$

7.6 Step 1: Determine b.

	Value	Key Stroke	Display
	.5 (r)	$\times$	.5
	20 (s_Y)	$\div$	10
	16 (s_X)	$=$	.625 $= b$

Step 2: Determine a.

	Value	Key Stroke	Display
	240 ($\overline{Y}$)	$-$	240
	.625 (b)	$\times$	.625
	110 ($\overline{X}$)	$=$	171.25 $= a$

8

Probability

The concept of probability arises often in all kinds of situations. The captains of two football teams go to the center of the field before the game for the flip of a coin, which determines who gets the choice of receiving the kick or defending a goal. We say each captain has a 50–50 chance of winning the toss, which means that their probabilities of success are the same. The weather forecaster reports the probability of rain for tomorrow. This probability is based on a record of how frequently rain has fallen in the past under the existing conditions. In the previous chapters, we have been concerned with using statistical procedures to classify and summarize data, to describe the nature of distributions and individual scores, and to determine the relationship between variables. These procedures are generally referred to as *descriptive statistics*, because they are used primarily for describing the data collected. However, scientific inquiry over the past several centuries has required the development of other statistical procedures, techniques that research scientists can use to formulate and test theories about the effects of an experimental treatment on some criterion variable and to generalize these theories to a given population. For example, the health scientist may be interested in determining the effect on cardiovascular health of eating foods containing animal fat. Using only the descriptive statistical procedures we have discussed up to this point would not be adequate for making precise, scientific generalizations from the results of this research investigation. It would be possible for the researcher to say that, based on the data collected, the blood of those subjects who were on the experimental diet containing foods with animal fat had higher levels of certain fatty acids than the blood of those who were on a more conventional diet. But it would not be possible to determine whether this higher level of fatty acids is greater than could be attributed to chance variation alone.

The remaining chapters of this book discuss the statistical procedures used in determining whether the effect of an independent variable on the dependent variable can be attributed to chance variation. In this chapter, we will be concerned with statistical procedures that enable us to investigate the variability of the data from a research study, to generalize the results to a given population, and to draw valid conclusions. These procedures are usually referred to as *inferential statistics*, because the data of the research study are derived from a specific sample and the results are then *inferred* to a larger, more general population. Vital to these procedures are the concepts of sampling and probability; basic sampling procedures and basic probability theory are discussed in this chapter. Subsequent chapters will deal with the role of sampling and probability theory in hypothesis testing and estimation.

Probability—A Definition

We encounter the concept of probability daily in a variety of ways. The weather report informs us of the chances of rain for tomorrow; the professional economist reveals the expected rate of inflation for next year; the agricultural expert reports that the corn and wheat crops will exceed expectations; and the doctor explains to a patient that taking a certain prescribed medication will relieve his or her current symptoms. In these examples and others of a similar nature, the exact probability of the occurrence is not reported, but the use of probability theory is the basis for each of the statements.

Probability, in a mathematical and statistical sense, is defined as the ratio of the number of favorable outcomes to the total number of possible outcomes (both favorable and unfavorable) for an event. For example, if the favorable outcome of the toss of a coin is a head, then the probability of obtaining a head is 1/2. That is, there is only one favorable outcome out of two possible outcomes. Consider another example. Suppose the favorable outcome is a "3" on the roll of a single die. There are six possible outcomes, so the probability of this favorable outcome is 1/6. In general, if we say that event A is the favorable outcome, then the probability of A is defined as follows:

$$P(A) = \frac{\text{number of outcomes favoring A}}{\text{total number of events (favoring A + not favoring A)}} \qquad (8.1)$$

The range of possible values for probabilities is from zero (0.00) to one (1.00), inclusive. A value of 0.00 indicates that there is no chance that the favorable event will occur. A value of 1.00 indicates that the event is absolutely certain to occur. For example, the probability that any of us would be able to run nonstop from Chicago to Boston is 0.00, whereas the probability of an income tax being levied next year, is 1.00.

Probabilities are expressed in decimal form; the larger the decimal, the greater the probability of the favorable event occurring. In our example of the toss of a coin, we said that $P(A) = 1/2$ or 0.50. Note that the probability of getting a tail is also 0.50. In other words, $P(\overline{A})$—which is read "the probability of *not* A"—is 0.50 for the coin example. The sum of $P(A)$ and $P(\overline{A})$ always equals 1.00. In our other example, wherein $P(A)$ was the probability of getting "3" on a roll of the die, $P(A) = 0.167$ and $P(\overline{A}) = 0.833$.

The probability of event A, $P(A)$, is the ratio of the number of outcomes including event A to the total number of possible outcomes, those that include event A and those that do not.

Laws of Probability

Probability theory has a specified set of laws, or rules that enable us to compute the probability of an event or combination of events. In essence, the *laws of probability* describe the behavior of events given certain conditions. To illustrate the laws of probability, consider the whole numbers from 1 to 10. Each of these numbers represents a single event in the total set of possible events, and, in Figure 8.1, all are illustrated in what is called a Euler diagram. Each event represented in the diagram is referred to as a *sample point*, and the total set of possible events is referred to as the *sample space*. For our example, selecting a specified number is defined as a *single event*, and we will assume that there is an equal probability of any of the events occurring. That is, the probability that each of the numbers will be selected equals 1/10 or 0.10.

Addition Rule for Mutually Exclusive Events

By contrast to a single event, we can also define a *compound event*, one that involves two or more single events. For example, we could define the compound

143

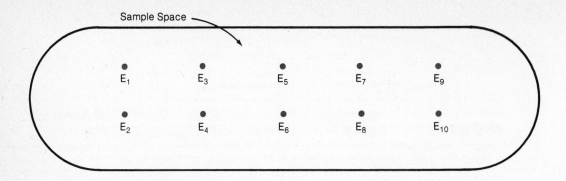

Figure 8.1 **Euler diagram defining the sample space for the numbers 1 through 10**

event A as selecting from the sample space a number greater than 8—that is, selecting the number 9 or the number 10. Note that these two single events are *mutually exclusive,* which means that selecting 9 from the sample space is independent of selecting 10. For our purposes, consider another compound event, event B, defined as selecting an odd number from the sample space. Again, the single events that constitute this second compound event are mutually exclusive. These two compound events are illustrated in Figure 8.2 by circles drawn around the sample points that represent the events.

The probability of a compound event is defined in a similar way to that of a single event. For example, consider the probability of selecting the number 9 *or* the number 10. To determine this probability, we need to use the addition rule for mutually exclusive events:

The probability of the union of events X and Y, where X and Y are mutually exclusive events, equals

$$P(X \text{ or } Y) = P(X) + P(Y) \tag{8.2}$$

Applying this rule to our example, we have

$$P(A) = P(9 \text{ or } 10) = P(9) + P(10)$$
$$= 0.10 \text{ to } 0.10$$
$$= 0.20$$

In like manner, the probability of compound event B is

$$P(B) = P(1 \text{ or } 3 \text{ or } 5 \text{ or } 7 \text{ or } 9)$$
$$= P(1) + P(3) + P(5) + P(7) + P(9)$$
$$= 0.10 + 0.10 + 0.10 + 0.10 + 0.10$$
$$= 0.50$$

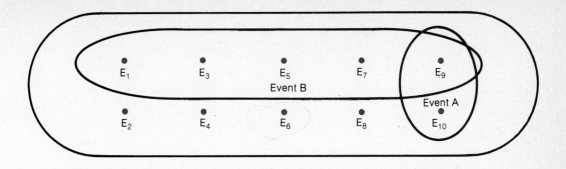

Figure 8.2   Euler diagram defining compound Events A and B for the sample space

The addition rule for mutually exclusive events is given by

$$P(X \ or \ Y) = P(X) + P(Y)$$

Addition Rule for Non–Mutually Exclusive Events

Now consider a more complex example. Suppose we want to know the probability of event A or event B as defined in Figure 8.2. Note that these two events are not mutually exclusive; the number 9 is included in both events. With non–mutually exclusive events, the addition rule is as follows:

The probability of the union of the events X and Y equals

$$P(X \ or \ Y) = P(X) + P(Y) - P(X \ and \ Y) \qquad (8.3)$$

In this more general rule, $P(X \ and \ Y)$ refers to the probability of the occurrence of the sample event(s) in *both* compound events. In our example, $P(A \ and \ B)$ is the probability of selecting the number 9—that is, $P(9)$. Applying formula 8.3 to this example, we find that

$$P(A \ or \ B) = P(A) + P(B) - P(A \ and \ B)$$

$$= 0.20 + 0.50 - 0.10$$

$$= 0.60$$

Actually, this rule is not restricted to non–mutually exclusive events. It also applies to mutually exclusive events, because in that case $P(X \ and \ Y)$ is 0.

The general addition rule that is *not* restricted to mutually exclusive events is given by

$$P(X \text{ } or \text{ } Y) = P(X) + P(Y) - P(X \text{ } and \text{ } Y)$$

Multiplication Rule for Statistically Independent Events

In the previous sections, we were concerned with determining the probability of event A or event B on the basis of a single observation. In the study of probability, however, and especially in its application to statistics, we are confronted with determining the probability of the joint or successive occurrence of two or more events when more observations than one are made. For example, we might want to determine the probability of selecting the number 2 from the sample space and then, after replacing that number, selecting the number 4. The statistical term for this procedure is *sampling with replacement*. In this example, because the first number was returned to the sample space before the second number was selected, the two events are said to be statistically independent. In other words, two events X and Y are *statistically independent* if the probability of one event occurring is not affected by the occurrence of the other. In such cases, we use the multiplication rule for statistically independent events:

Given events X and Y, which are statistically independent, the probability of their joint occurrence is

$$P(X \text{ } and \text{ } Y) = P(X) \cdot P(Y) \tag{8.4}$$

Using the sample space defined in Figure 8.1, let us find the probability of selecting the number 2 from the sample space and then, after replacement, selecting the number 4.

$$P(2 \text{ } and \text{ } 4) = P(2) \cdot P(4)$$

$$= (0.10)(0.10)$$

$$= 0.01$$

To illustrate this rule with a more complex example, consider our compound events A and B. $P(B \text{ } and \text{ } A)$ is defined as the probability of selecting an odd number and then, after replacement, selecting the number 9 or the number 10. Using the above multiplication rule, we find that

$$P(B \text{ } and \text{ } A) = P(B) \cdot P(A)$$

$$= (0.50)(0.20)$$

$$= 0.10$$

The multiplication rule for statistically independent events is given by

$$P(X \text{ and } Y) = P(X) \cdot P(Y)$$

Multiplication Rule for Nonindependent Events

In the previous section, we considered the multiplication rule for events that were statistically independent due to the fact that sampling with replacement was used. The use of *sampling without replacement* changes the multiplication rule. For example, consider again $P(2 \text{ and } 4)$. As before, $P(2) = 0.10$. However, if we do not replace the first number before selecting the second number, the probability of selecting 4 on the second draw is *conditional* on the results of the first draw. In other words, the results of the first draw influence the results of the second draw. These events are said to be *nonindependent*. The multiplication rule for nonindependent events is as follows:

> Given events X and Y, which are nonindependent, the probability of both X and Y occurring jointly is the product of the probability of obtaining one of these events times the conditional probability of obtaining the other event given that the first event has occurred. Symbolically,

$$P(X \text{ and } Y) = P(X) \cdot P(Y|X) \quad \text{or} \quad P(Y) \cdot P(X|Y) \tag{8.5}$$

(The symbol for conditional probability is $P(Y|X)$ and is read "the probability of Y given that X has occurred.") In our example, the probability of selecting the number 2 on the first draw is 0.10. Assuming that we did select the number 2, there are now only 9 numbers left in the sample, and the probability of selecting the number 4 on the second draw is 1/9 or 0.11. Therefore, the probability of selecting the numbers 2 and 4 is

$$P(2 \text{ and } 4) = P(2) \cdot P(4|2)$$
$$= (0.10)(0.11)$$
$$= 0.011$$

Equivalently,

$$P(2 \text{ and } 4) = P(4) \cdot P(2|4)$$
$$= (0.10)(0.11)$$
$$= 0.011$$

The multiplication rule for statistically nonindependent events is given by

$$P(A \text{ and } B) = P(A) \cdot P(B|A)$$
$$= P(B) \cdot P(A|B)$$

Probability and the Underlying Distribution of All Possible Outcomes

The concept of probability plays a vital role in statistics. Probability is best understood when it is considered along with the concept of the *underlying distribution of all possible outcomes*. For example, consider the underlying distribution of all possible sums of the roll of two dice, one green and one red. This distribution is illustrated in Figure 8.3. As illustrated, there are 11 different sums ranging from 2 (double ones) to 12 (double sixes), inclusive. Note that there are three different ways in which the sum can equal 4. The green die could read "3" and the red die "1," the green die could read "1" and the red die "3," or both dice could read "2." There are 36 different possible outcomes of the roll of two dice, so the probability of obtaining a sum equal to 4 on a single roll of the dice is 3/36 or 0.083. Similarly, using the underlying distribution illustrated in Figure 8.3, we find that the probability of obtaining a sum equal to 7 is 6/36 or 0.167.

Once the underlying distribution of all possible outcomes is known, it is a straightforward procedure to determine the probability of any combination of outcomes. For example, consider the probability of rolling the pair of dice for a sum either totaling 3 or less or totaling 10 or greater. In other words, consider the probability of rolling a sum of 2, 3, 10, 11, or 12. Using Figure 8.3, we can determine the probability of each of these independent events:

$$P(\text{sum} = 2) = 1/36 = 0.028$$
$$P(\text{sum} = 3) = 2/36 = 0.056$$
$$P(\text{sum} = 10) = 3/36 = 0.083$$
$$P(\text{sum} = 11) = 2/36 = 0.056$$
$$P(\text{sum} = 12) = 1/36 = \underline{0.028}$$
$$0.250$$

The sum of these probabilities is the probability of rolling a 3 or less or a 10 or greater; this probability is 9/36 or 0.250. We could have counted the squares associated with 3 or less, or with 10 or greater, and we would have found 9 of the 36 meeting this criterion. Note that the sum of the probabilities of all possible outcomes equals 36/36 or 1.0.

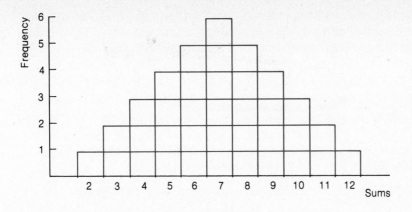

Figure 8.3 Underlying distribution of possible sums of the single roll of two dice

In behavioral science research, we seldom encounter variables that have the underlying distribution of the sum of a single roll of two dice. It is introduced merely to illustrate the meaning of an underlying distribution. In actual research practice, the underlying distributions commonly used do not need to be plotted or generated, because they are known distributions that have been developed through mathematical statistics. The binomial distribution and the normal distribution are examples of commonly used underlying distributions.

The Binomial Distribution as an Underlying Distribution

The binomial distribution was introduced in Chapter 4 in the preliminary discussion of the normal distribution and its applications. This distribution is one of the more important distributions used in statistics. It is used largely with nonparametric procedures (see Chapter 15). Although the various applications of the binomial distribution in inferential statistics are not discussed in the remaining chapters of this book, it is presented here to illustrate yet another underlying distribution.

Consider the toss of two fair coins. The four possible outcomes are illustrated in the following table.

Outcome	First Coin	Second Coin	Probability
1	H	H	1/4 or 0.25
2	H	T	1/4 or 0.25
3	T	H	1/4 or 0.25
4	T	T	1/4 or 0.25

There are actually four possible outcomes, but two of them, in effect, are the same. That is, there are two different ways of obtaining a head and a tail. These results could also have been obtained mathematically by squaring a *binomial*. Symbolically,

$$(X + Y)^2 = (X + Y)(X + Y)$$
$$= X^2 + 2XY + Y^2$$

In our example, if X equals the head and Y equals the tail, then $(H + T)^2 = HH + 2HT + TT$. Now, if we let X equal the probability of obtaining a head and Y the probability of obtaining a tail, we can find the probability of each of these outcomes. Inserting those probabilities in the expanded binomial, we obtain

$$(0.50 + 0.50)^2 = (0.50)(0.50) + 2(0.50)(0.50) + (0.50)(0.50)$$
$$= 0.25 + 0.50 + 0.25$$

These are the respective probabilities of two heads, a head and a tail, and two tails.

Now consider the simultaneous toss of three coins. The possible outcomes are displayed in the following table.

Outcome	First Coin	Second Coin	Third Coin	Probability
1	H	H	H	$1/8 = 0.125$
2	H	H	T	$1/8 = 0.125$
3	H	T	H	$1/8 = 0.125$
4	H	T	T	$1/8 = 0.125$
5	T	H	H	$1/8 = 0.125$
6	T	H	T	$1/8 = 0.125$
7	T	T	H	$1/8 = 0.125$
8	T	T	T	$1/8 = 0.125$

There are eight possible outcomes, but outcomes 2, 3, and 5 are the same (two heads and one tail) and outcomes 4, 6, and 7 are the same (two tails and one head). Thus there are actually only four distinct possible outcomes with probabilities as follows:

H	T	Probability
3	0	$1/8 = 0.125$
2	1	$3/8 = 0.375$
1	2	$3/8 = 0.375$
0	3	$1/8 = 0.125$

We could also have obtained these results by taking the binomial to the third power—that is, $(X + Y)^3 = X^3 + 3X^2Y + 3XY^2 + Y^3$. If we let X equal the probability of obtaining a head and Y the probability of obtaining a tail, we can find the probability of each outcome.

$$(0.50 + 0.50)^3 = (0.50)^3 + 3(0.50)^2 (0.50) + 3(0.50)(0.50)^2 + (0.50)^3$$
$$= 0.125 + 0.375 + 0.375 + 0.125$$

Using these probabilities, we can answer certain questions about the probability of the occurrence of a certain event or outcome. For example, the probability of obtaining *at least 1 head* in the toss of three coins equals the probability of obtaining 3 heads (0.125) plus the probability of obtaining 2 heads (0.375) plus the probability of obtaining 1 head (0.375). The sum of these probabilities equals 0.875.

The *binomial distribution* is defined by taking the binomial $(X + Y)$ and raising it to the *n*th power: $(X + Y)^n$. The binomial distribution can also be used to determine probabilities of certain outcomes in research studies. For example, we can use it in quality control studies to determine the number of defective products that might be expected, or in determining the number of students expected to drop out of high school in a given year. Though our examples have been limited to tosses of a coin, the use of the binomial distribution is readily adapted to other situations. Suppose it is known that 10% of a particular brand of television tubes will burn out before their guarantee has expired. If we select 3 of these tubes, what is the probability that none of them will burn out, that 1 will burn out, that 2 will burn out, and that all 3 will burn out? For this example, we will let X equal the probability that the tube will not burn out and Y the probability that it will burn out. Therefore, the binomial expansion is

$$(0.90 + 0.10)^3 = (0.90)^3 + 3(0.90)^2(0.10) + 3(0.90)(0.10)^2 + (0.10)^3$$

$$= 0.729 + 0.243 + 0.027 + 0.001$$

Listing these probabilities by the respective outcomes, we have

Outcome	Probability
0 burn out	0.729
1 burn out	0.243
2 burn out	0.027
3 burn out	0.001

As might be expected, expanding the binomial to higher powers is quite arduous. Thus we must have a more efficient way to obtain the probabilities when the number of observations increases. We do this by using the binomial theorem, which involves the concept of the *factorial*, a mathematical operator. The symbol 6! (which is read "6 factorial") means the product of all integers from 6 down to 1. Thus 6! = (6)(5)(4)(3)(2)(1) = 270. In general,

$$n! = (n)(n - 1)(n - 2) \cdots (2)(1)$$

that is, the product of all integers n through 1, inclusive. Using the concept of the factorial, we define the *binomial theorem* as follows:

$$(p + q)^n = p^n + \frac{n!}{(n - 1)!(1)!} p^{n-1}q + \frac{n!}{(n - 2)!(2)!} p^{n-2}q^2 \qquad (8.6)$$

$$+ \cdots + \frac{n!}{(2)!(n - 2)!} p^2q^{n-2} + \frac{n!}{(1)!(n - 1)!} pq^{n-1} + q^n$$

The Binomial Distribution as an Underlying Distribution

Now consider the term $\dfrac{n!}{(n-2)!(2)!}$, called a binomial coefficient, from the binomial theorem for $n = 5$.

$$\frac{5!}{(5-2)!(2)!} = \frac{5!}{3!2!} = \frac{(5)(4)(3)(2)(1)}{(3)(2)(1)(2)(1)}$$

This term can be simplified as follows:

$$\frac{(5)(4)(3)(2)(1)}{(3)(2)(1)(2)(1)} = \frac{(5)(4)}{(2)(1)} = \frac{20}{2} = 10$$

It is unnecessary, however, to compute the binomial coefficients each time that the binomial theorem is used; these coefficients for n equalling 1 to 20 are found in Table 8 of the Appendix. To illustrate the use of the binomial theorem, suppose 7 students are enrolled in a course for which we know from past experience that the probability of passing is 0.75. If we let p equal the probability of passing the course and q the probability of failing it, we can use the binomial theorem to determine the probability that all 7 students will pass, that 6 students will pass, and so on, down to the probability that no students will pass. We apply the binomial theorem as follows:

$$(p + q)^7 = (0.75 + 0.25)^7 = (0.75)^7$$

$$+ \frac{7!}{6!1!}(0.75)^6(0.25)$$

$$+ \frac{7!}{5!2!}(0.75)^5(0.25)^2$$

$$+ \frac{7!}{4!3!}(0.75)^4(0.25)^3$$

$$+ \frac{7!}{3!4!}(0.75)^3(0.25)^4$$

$$+ \frac{7!}{2!5!}(0.75)^2(0.25)^5$$

$$+ \frac{7!}{1!6!}(0.75)(0.25)^6$$

$$+ (0.25)^7$$

$$= 0.1335 + 0.3115 + 0.3115 + 0.1730$$

$$+ 0.0577 + 0.0115 + 0.0013 + 0.0001$$

Listing these probabilities by their respective outcomes, we have

Outcome	Probability
all pass	0.1335
6 pass	0.3115
5 pass	0.3115
4 pass	0.1730
3 pass	0.0577
2 pass	0.0115
1 pass	0.0013
0 pass	0.0001

Note that the sum of the probabilities is 1.0, within rounding error. These probabilities represent an underlying distribution of possible outcomes for this example.

The binomial theorem provides an efficient method of expanding the binomial. The theorem is given in formula 8.6.

The Normal Distribution as an Underlying Distribution

The normal curve, as an underlying distribution, can also be used to determine the probability of a particular outcome. Consider the distribution of mathematics achievement scores for 15,000 third-grade and fourth-grade students in a large metropolitan school district. For the purposes of this example, it is known—not assumed—that the scores are normally distributed, that the mean is 60, and that the standard deviation is 10. The distribution is illustrated in Figure 8.4.

Suppose each of the 15,000 scores is recorded on a small slip of paper and placed in a container. Since the scores are normally distributed, it is possible to use the properties of the normal curve to calculate the probabilities associated with this distribution of scores. For example, we can determine the probability of drawing from the container a slip of paper bearing a number between 50 and 60. Using the normal distribution as the underlying distribution, we find that 34.13% of all the scores on that mathematics achievement test are between 50 and 60. Thus the probability of drawing a number between 50 and 60 from the container is 0.3413.

Now consider the probability of drawing a number that is greater than 80 or less than 40. To determine this probability, we use the normal curve table (Table 1 in the Appendix) as discussed in Chapter 5. The first step is to convert the raw scores 80 and 40 to standard scores.

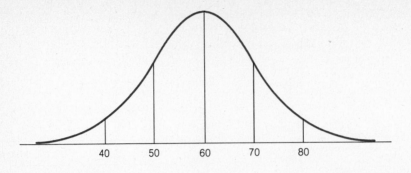

Figure 8.4   Normal distribution of 15,000 scores with μ = 60 and σ = 10

$$z_{80} = \frac{80 - 60}{10} = +2.00$$

$$z_{40} = \frac{40 - 60}{10} = -2.00$$

Now, using Table 1, we find that 0.0228 or 2.28% of the area is above a *z* score of 2.00 and that the same amount is below a *z* score of −2.00. Adding these two areas, we find a probability of 0.0228 + 0.0228 = 0.0456 that a score selected at random will be greater than 80 or less than 40. In other words, it is highly unlikely (there is a low probability) that a number less than 40 or greater than 80 will be drawn from the container. On the other hand, it is very likely (highly probable) that a number between 50 and 70 will be drawn. The actual probability is 0.6826; it is determined in the same manner by using the standard normal curve. Note again that the sum of the probabilities of all possible outcomes is 1.00, the total area under the curve.

Although the foregoing example is somewhat contrived, the normal distribution is used extensively in inferential statistics. It is a commonly used underlying distribution, because many statistics are normally distributed or nearly so.

The underlying distribution is defined as the distribution of all possible outcomes of a particular event. The normal distribution is one of the more commonly used underlying distributions in inferential statistics.

Sampling

Much of the research in the behavioral sciences is conducted by studying populations through the use of samples selected from those populations. This is because it is not feasible to gather information for all members of the population. Such data collection may be too costly or too time-consuming. The alternative is to select a sample and collect data on members of the sample only. It is the sample that provides the empirical information or data on which the conclusions of the research are based.

Before discussing the role of sampling in inferential statistics, we present a brief overview of sampling procedures. There are numerous sampling procedures; some are relatively complex.[1] However, it is sufficient for our purposes to discuss several methods that are widely used by researchers in the behavioral sciences. No attempt is made to deal comprehensively with these procedures. They are simply defined and illustrated.

Simple Random Sampling

When making inferences about the characteristics of a population from the characteristics of a sample, the researcher generally asserts that the sample selected for the investigation is random. The researcher is usually referring to one kind of probability sample, the simple random sample. A *probability sample* is a sample selected in such a way that each member or element of the population has a known and nonzero probability of being included in the sample. Note that, in a probability sample, it is not necessary for all population members to have the same probability of selection. It is only necessary that no population members be completely excluded from the possibility of entering into the sample.

In order to define a *simple random sample* for theoretical purposes, it is necessary to distinguish between sampling with replacement and sampling without replacement. Sampling with replacement means that, as a members of the population are selected for the sample, they are "replaced" into the population and can be selected again. In *sampling with replacement,* a simple random sample is one selected in such a way that all members of the population have the same probability of selection.

In behavioral science research, however, the more common practice is to sample without replacement. Once selected for the sample, a population member is not replaced into the population and cannot be selected a second time. *In sampling without replacement,* a simple random sample is one in which all possible samples of a given size have the same probability of selection. This is only a slight adjustment. In either case, all members of the sample are selected independently of one another.

There are several common procedures for selecting a simple random sample (with or without replacement) from a finite population. Suppose the reseacher has a population of 500 members and wants to select a 10% sample, or 50 members of the population. In this case the *sampling fraction,* or the ratio of the size of the sample to the size of the population, is $50/500 = 0.10$. Conceptually, simple random sampling involves recording all names (or some identification code) on small slips of paper and placing them in a container. Then 50 of these slips of paper are drawn without bias from the container, either with or without replacement; the names drawn constitute the simple random sample. This procedure for selecting a simple random sample is cumbersome, and there are more practical approaches. One common

[1]See Seymour Sudman, *Applied Sampling* (New York: Academic Press, 1976).

approach involves the use of a table of random numbers. Another involves a computer or programmable hand calculator that will generate a list of random numbers. In both these procedures, each member of the population is initially assigned an identification number from 1 to N and is included in the sample if its identification number is generated in the list of random numbers.

A simple random sample is a sample in which all population members have the same probability of being selected and the selection of each member is independent of the selection of all other members.

Systematic Sampling

Systematic sampling is a procedure for selecting a probability sample that is commonly used in research studies when a listing of the members of the population is readily available. In *systematic sampling* from a list, the investigator chooses every kth member of the list for the sample. The procedure initially involves determining the sample size and the sampling fraction. Suppose a researcher wants to select a sample of size 300 from a population of 15,000. The sampling fraction is

$$\frac{n}{N} = \frac{300}{15,000} = \frac{1}{50}$$

Assuming that a list containing the names of the 15,000 population members is available, systematic sampling involves the following steps:

1. Randomly select a number between 1 and 50, inclusive—say 37.
2. Individual number 37 is the first sample member selected. All subsequent members are identified by adding multiples of 50 to 37. So the second member selected is number 87, the third is number 137, and so on.
3. Continue this procedure until the list is exhausted, at which point the 300 sample members will have been selected.

Systematic sampling is more convenient than simple random sampling when a list of the population members is available. It also provides sampling throughout the population by spacing the selections over the entire population list. It should be noted that the foregoing example is most elementary in that the denominator of the sampling fraction is the whole number 50. The systematic sampling procedure becomes slightly more complex when the denominator is a decimal.[2]

A systematic sample is a probability sample in which every kth member of the population is selected and in which $1/k$ is the sampling fraction. The first member of the sample is determined by randomly selecting an integer between 1 and k.

---

[2]Sudman, *Applied Sampling,* pp. 54–55.

Cluster Sampling

Cluster sampling can be used when the individual members of a population are found in naturally formed groups called clusters and when it is impractical to select individual members for the sample. In *cluster sampling,* the clusters are randomly selected from the population of clusters. Once a cluster is selected, all members of that cluster are included in the sample. Each member of the population must belong to only one cluster, but it is not necessary for all the clusters to have the same number of members.

Suppose the research director of a large school system is asked to conduct a survey of parents regarding the counseling services that are available in the elementary schools. Whereas it would be difficult logistically to select either a simple random sample or a systematic sample, it would be relatively easy to randomly select classes of third-, fourth-, and fifth-grade students and mail the survey to the parents of these students. In this case the class is the cluster, and the parents of *all* the students in the classes selected are included in the sample.

For this survey, the research director decides that the total number in the sample should be about 750. Because there are an average of 26 students in each of these classes, the research director decides to randomly select 29 classes (750 ÷ 26 = 28.85, or 29). After the classes are selected, the survey is mailed to the parents.

Cluster sampling involves the random selection of clusters (groups of population members) rather than individual population members. When a cluster is selected for the sample, all members of that cluster are included in the sample.

Stratified Sampling

In simple random sampling, systematic sampling, and cluster sampling, the population is assumed to be generally homogeneous. However, this is not always the case. The population may be heterogeneous and consist of several subpopulations. For example, consider the political party affiliation of all registered voters in a small midwestern town. Assume that the population has four subpopulations called *strata* with 40% of the registered voters belonging to the Republican Party (stratum A), 30% belonging to the Democratic Party (stratum B), 15% belonging to the Conservative Party (stratum C), and 15% registered as independent voters (stratum D).

Stratified sampling yields a highly representative sample. By providing for sampling throughout the entire population, it ensures that members from each stratum definitely will be included in the sample. In the stratification process, the researcher not only determines the strata but also decides how many members of each stratum to include in the sample. The procedure most commonly used in determining the number of members from each stratum is *proportional allocation.* With proportional allocation, each stratum contributes to the sample a number of members that is proportional to its size relative to other strata in the population.

In stratified random sampling, the population is divided into subpopulations called strata. All strata are represented in the sample, often via proportional allocation according to size.

The Role of Sampling in Inferential Statistics

In Chapter 1, we defined the mean of a sample ($\overline{X}$) as a *statistic* (a descriptive measure of a sample), whereas the mean of a population (μ) was defined as a *parameter* (a descriptive measure of a population). Up to this point in the book, there has been no need to distinguish between a statistic and a parameter. But, as we have seen, financial and time constraints often make it impossible or not feasible to gather data on all members of a population. The alternative is to collect data on a sample drawn from the population and to compute the appropriate statistics, such as the sample mean ($\overline{X}$). Subsequently, conclusions about the corresponding parameter in the population (μ) are based on the sample statistic ($\overline{X}$).

Consider an example. Suppose we have a distribution of IQ scores for a sample of 144 junior high school students. Using the methods introduced in Chapter 3, we find the mean IQ for this sample of students ($\overline{X}$) to be 106. What does this sample mean reflect? Certainly we expect it to reflect the mean of the population. But can we expect it to be exactly equal to the mean of the population? No, we cannot, due to sampling fluctuation. When a sample is drawn, the statistics of the sample reflect the respective parameters in the population, within fluctuations due to sampling. Subject to certain limitations, the researcher can make inferences about the population parameters from knowledge of the sample statistics. This chain of reasoning is part of inferential statistics, and it is illustrated in Figure 8.5. To understand this chain of reasoning, we must consider the relationship of sampling to probability theory and the concept of the underlying distribution of all possible outcomes.

The Sampling Distribution of the Mean

Earlier in the chapter, we discussed underlying distributions of events such as the sum of the roll of two dice and the outcome of a toss of several coins. However, in the example involving the IQ scores for junior high students, the statistic under consideration is the sample mean ($\overline{X} = 106$). Whereas the underlying distributions of the outcomes of the roll of two dice and the toss of several coins are relatively easy to generate, a different approach is necessary to define the underlying distribution of the sample mean. In the example, we have a sample of only 144 taken from a large population. The only way to plot all values of the sample mean would be to draw all possible samples of size 144, to compute the mean for each, and to develop the frequency distribution of these means. This frequency distribution is referred to as the

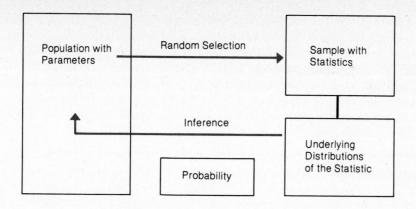

(a) We have a population and we want to make decisions about measures of the population, namely parameters
(b) We select a random sample and compute measures of the sample, which are statistics.
(c) The statistics reflect the corresponding parameters and sampling fluctuation.
(d) We observe the statistics, which are the facts that we have, and infer back to the parameters in the light of the underlying distributions and probability.

Figure 8.5 Chain of reasoning for inferential statistics

sampling distribution of the mean. In practice, statisticians do not spend their time plotting actual sampling distributions. Rather, they concern themselves with the characteristics of *theoretical sampling distributions*. Theoretical sampling distributions are based on information from a single sample in conjunction with mathematical theory. In this example, the mathematical theory we need is the central limit theorem. This theorem, along with information derived from the single sample of 144 students, is sufficient for describing the theoretical sampling distribution of the mean.

As with any distribution, theoretical or actual, three kinds of information are required to describe its nature: shape, location, and variability. The *central limit theorem* provides this information for the sampling distribution of the mean. It is as follows:

As the sample size (n) increases, the sampling distribution of the mean ($\bar{X}$) of simple random samples of n cases taken from any population with a mean equal to μ and a finite variance equal to σ^2 has the following properties.

1. The distribution of sample means approaches a normal distribution.   (shape)
2. The mean of the distribution of sample means equals μ.   (location)
3. The variance of the distribution of sample means equals σ^2/n. The standard deviation of the distribution of sample means equals $\sigma/\sqrt{n}$.   (variability)

The Sampling Distribution of the Mean

You will note that this theorem does not require the population to be infinite or to be normally distributed, and it does not assume that an infinite number of samples of a specified size have been selected. The only stipulation is that the population must have a finite variance. Almost every population of interest in the behavioral sciences meets this criterion, so the central limit theorem is readily applicable to almost any research setting that requires knowledge of the sampling distribution of the sample means. Note that the central limit theorem provides for us the shape, location, and variability of the distribution.

The central limit theorem provides the mathematical basis for using the normal distribution as the sampling distribution of all sample means of a given sample size. The theorem states that this distribution of sample means (1) is normally distributed, (2) has a mean equal to μ and (3) has a variance equal to σ^2/n.

Problems arise in applying the central limit theorem when it is used with distributions that are far from normal and with samples that are relatively small. If the distribution of the population is nearly normal, the theorem holds even for sample sizes that are quite small. However, as the distribution of the population departs substantially from normal, a larger sample size is necessary for it to be the appropriate sampling distribution of the mean. Statisticians have been concerned with the problem of determining the sample size necessary for legitimately applying the central limit theorem. They have shown that, for most populations, a sample size greater than 30 is generally sufficient. That is, even when the original population is not normally distributed, the distribution of sample means is normal if n is greater than 30.

Using the Normal Distribution as the
Sampling Distribution of the Mean

For our example of the IQ scores, we have only one of all possible samples of size 144 that could have been drawn from the population of junior high school students. By the central limit theorem, the sampling distribution of the mean, which is basically a theoretical distribution, contains the means of all possible samples of 144 students. This distribution is normally distributed with a mean equal to μ and a standard deviation equal to $\sigma/\sqrt{n}$. The standard deviation of the sampling distribution is called the *standard error of the mean* and is denoted as follows:

$$\sigma_{\overline{X}} = \sigma/\sqrt{n} \qquad (8.7)$$

The commonly used IQ tests have a mean of 100 and a standard deviation of 15. Therefore, for this example, the theoretical sampling distribution of the mean would be normally distributed with a mean equal to 100 and a standard error of the mean equal to $15/\sqrt{144} = 1.25$. This distribution is illustrated in Figure 8.6.

Using the properties of the normal distribution, we can determine the probability of randomly selecting a sample of size 144 and having the mean ($\overline{X}$) be, say, less than 98 or greater than 103. To begin, we note that the distribution shown in Figure 8.6 is normal, so we can use the normal distribution as the sampling distribution. In

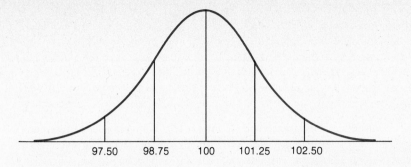

| 97.50 | 98.75 | 100 | 101.25 | 102.50 |

Figure 8.6   Sampling distribution of the mean for $\mu = 100$ and $\sigma_{\bar{X}} = 1.25$

this example, it is the sampling distribution of all possible means from samples of size 144. Thus, by using Table 1, we can determine the probability of selecting a sample of size 144 and finding the mean to be between 98.75 and 101.25. Because these two points are -1.0 and $+1.0$ standard errors below and above the mean, respectively, the probability is 0.6826. Similarly, the probability of finding a mean between 97.50 and 102.50 (-2.0 and $+2.0$ standard errors from the mean) is 0.9544. In order to determine the probability of finding a mean less than 98.0, we need to use the concept of the standard score. In this example, the standard score would be computed as follows:

$$z = \frac{\bar{X} - \mu}{\sigma_{\bar{X}}}$$

Next we find the probability of obtaining a sample mean less than 98.

$$z = \frac{98 - 100}{1.25}$$

$$= \frac{-2}{1.25}$$

$$= -1.60$$

Using Table 1, we find the associated probability to be 0.0548. Similarly, we can find the probability of obtaining a sample mean greater than 103.

$$z = \frac{103 - 100}{1.25}$$

$$= \frac{3}{1.25}$$

$$= 2.40$$

The associated probability of obtaining a sample mean greater than 103 is 0.0082, if the population mean is 100. Note that in these calculations the statistic is the sample mean, $\bar{X}$, and the standard deviation of the distribution is the standard error of the mean, $\sigma_{\bar{X}}$.

> The sampling distribution of the mean consists of the sample means ($\overline{X}$) of all possible samples of a given size.

Summary

This chapter introduces probability and sampling, two key concepts of inferential statistics. Probability is defined as the ratio of the number of favorable outcomes to the number of possible outcomes. Primary emphasis falls on probability as it is related to the underlying distribution of all possible outcomes. Several underlying distributions are presented and discussed, including the binomial distribution and the normal distribution. As we will see in subsequent chapters, once the underlying or sampling distribution has been identified, determining the probability of a given event occurring is relatively straightforward.

Inferential statistics involves studying a population by using a sample drawn from the population. For example, when a sample is selected and a sample mean is calculated on some variable, this sample mean reflects not only the mean of the population but also sampling fluctuation. That is, if all possible samples of a given size were selected and the means computed, we would not expect them all to be the same value. There would be a distribution of sample means.

The central limit theorem is used to describe the nature of this distribution of means, which is called the sampling distribution of the mean. Using the central limit theorem, we know that the sampling distribution is normal with a mean equal to μ and a standard deviation, called the standard error of the mean, equal to $\sigma_{\overline{X}} = \sigma/\sqrt{n}$.

Near the close of the chapter, we saw how we can use the properties of the unit normal curve to determine the probability of obtaining a given sample mean when μ and σ are known.

Chapter 9 will introduce hypothesis testing. The concept of the sampling distribution and the associated probability of a given event will be applied to test whether a hypothesized value for μ is tenable, given the corresponding value of $\overline{X}$.

Key Concepts

Inferential statistics
Probability
Laws of probability
Underlying distribution
Binomial distribution
Binomial theorem
Factorial
Normal distribution
Simple random sample
Probability sample

Sampling fraction
Systematic sampling
Cluster sampling
Stratified sampling
Parameter
Statistic
Sampling distribution of the mean
Theoretical sampling distribution
Central limit theorem
Standard error of the mean

Exercises

8.1 Consider the Euler diagram shown in Figure 8.1. Use the addition rules to determine the following probabilities.
   a. P(selecting the number 6)
   b. P(selecting the number 6 *or* 4)
   c. P(selecting an odd number)
   d. P(selecting a number that is a multiple of 3; that is, 3, 6, 9, and so on)

8.2 Consider the Euler diagram shown in Figure 8.1. Define the following compound events.
   Event A = selecting an odd number
   Event B = selecting an even number
   Event C = selecting a number that is a multiple of 3

   Use the addition rules to determine the following probabilities.
   a. $P(A\ or\ B)$
   b. $P(A\ or\ C)$
   c. $P(B\ or\ C)$

8.3 Consider the compound events defined in Exercise 8.2. Use the multiplication rules for statistically independent events to determine the following probabilities.
   a. $P(A\ and\ B)$
   b. $P(A\ and\ C)$
   c. $P(B\ and\ C)$

8.4 Consider the compound events defined in Exercise 8.2. Use the multiplication rules for statistically dependent events to determine the following probabilities.
   a. $P(A\ and\ B)$
   b. $P(A\ and\ C)$
   c. $P(B\ and\ C)$

8.5 Consider a deck of playing cards as the sample space (that is, the sample space contains 52 sample points). Use the rules of probability to determine the following.
   a. P(drawing the ace of spades)
   b. P(drawing an ace)
   c. P(drawing a heart)
   d. P(drawing the ace of spades *and* the ace of hearts)—with and without replacement.
   e. P(drawing three aces)—with and without replacement
   f. P(drawing ace, king, and queen)—with and without replacement

8.6 Consider the simultaneous toss of 5 coins. Use the binomial theorem to determine the probabilities of all possible outcomes (no heads, 1 head, . . . , 5 heads).

8.7 Beer drinkers claim that they can identify their favorite beer blindfolded. Suppose a sample of 10 known beer-drinking experts are blindfolded and given 4 different beers to drink, one of which is their favorite. Use the Binomial Theorem to determine the following probabilities.
   a. P(none identify favorite beer)
   b. P(1 identifies favorite beer)
   c. P(2 identify favorite beer)
   d. P(fewer than 5 identify favorite beer)

8.8 Given a distribution of 3000 numbers that are assumed to be normally distributed with $\mu = 120$ and $\sigma = 25$, use the properties of the normal distribution to determine the following probabilities.
   a. P(selecting a number less than 100)
   b. P(selecting a number greater than 150)
   c. P(selecting a number between 110 and 125)
   d. P(selecting a number less than 75 or greater than 160)

8.9 Over the past 10 years, a high school Spanish teacher has been giving the same comprehensive final examination to all students in first-year Spanish. The mean on this examination is 78.4 and the standard deviation is 14.8. To pass the course, the student must score at least 55. To be placed in the honors section, the student must score at least 99. Use the properties of the normal distribution to determine the following probabilities.
   a. P(student will pass the course)
   b. P(student will be placed in the honors section)
   c. P(student will score greater than 80)
   d. P(student will score less than 85)

8.10 Consider the distribution of scores given in Exercise 8.8. Use the central limit theorem and describe the sampling distribution of the mean for samples of size 144 and 400.

8.11 Using the sampling distribution of the mean for the samples of size 144 in Exercise 8.10 and the properties of the normal distribution, determine the following probabilities.
   a. $P(\overline{X} > 121.4)$
   b. $P(\overline{X} < 118.2)$
   c. $P(\overline{X} < 120.8)$
   d. $P(\overline{X} > 119.4)$

8.12 Using the sampling distribution of the mean for the samples of size 400 in Exercise 8.10 and the properties of the normal distribution, determine the following probabilities.
   a. $P(\overline{X} > 121.4)$
   b. $P(\overline{X} < 118.2)$
   c. $P(\overline{X} < 120.8)$
   d. $P(\overline{X} > 119.4)$

8.13 On the basis of the latest census data, the average family income for a family of four is $\mu = \$18,465$ and the standard deviation is $\sigma = \$7438$. Suppose a random sample of 400 families is selected. Use the central limit theorem to develop the sampling distribution of the mean, and then use the properties of the normal distribution to determine the following probabilities.
   a. $P(\overline{X} < \$19,000)$
   b. $P(\overline{X} > \$20,000)$
   c. $P(\$18,000 < \overline{X} < \$19,000)$
   d. $P(\$17,000 > \overline{X} > \$20,000)$

⊞ Key Strokes for Selected Exercises

		Value	Key Stroke	Display
8.6	Determine $P(0$ heads$)$.			
	$(.5)^5$	.5 (p)	×	.5
		.5	×	.25
		.5	×	.125
		.5	×	.0625
		.5	=	.03125
	Determine $P(1$ head$)$.			
	$5(.5)^4(.5)$	5	×	5
		.5 (p)	×	2.5
		.5	×	1.25
		.5	×	.625
		.5	×	.3125
		.5 (q)	=	.15625
	Determine $P(2$ heads$)$.			
	$10(.5)^3(.5)^2$	10	×	10.
		.5 (p)	×	5.
		.5	×	2.5
		.5	×	1.25
		.5 (q)	×	.625
		.5	=	.3125
8.7a	Determine $P(\text{none})$.			
	$(.75)^{10}$	.75 (q)	×	.75
		.75	×	.5625
		.75	×	.421875
		.75	×	.316406
		.75	×	.237305
		.75	×	.177978

.75	×	.133484
.75	×	.100113
.75	×	.075085
.75	=	.056314

b. Determine $P(1)$.

$$\frac{10!}{1!\ 9!}(.25)(.75)^9$$

10	×	10
.25 (p)	×	2.5
.75 (q)	×	1.875
.75	×	1.40625
.75	×	1.054688
.75	×	.791017
.75	×	.593262
.75	×	.444946
.75	×	.333709
.75	×	.250282
.75	=	.187712

c. Determine $P(2)$.

$$\frac{10!}{2!\ 8!}(.25)^2(.75)^8$$

Step 1: Determine $\dfrac{10!}{2!\ 8!}$.

10	×	10
9	÷	90
2	=	45
	STO	45

Step 2: Determine $P(2)$.

	RCL	45
	×	45
.25	×	11.25
.25	×	2.8125
.75	×	2.109375
.75	×	1.582031
.75	×	1.186523
.75	×	.889893
.75	×	.667419
.75	×	.500565
.75	×	.375423
.75	=	.281567

9

Hypothesis Testing: One-Sample Case for the Mean

In Chapter 8, we introduced the concept of probability and related it to the sampling distribution of the mean. In this chapter, we will apply these two concepts to testing hypotheses about the value of the population mean. For example, suppose a large city school district has 14,000 junior high school students and the central office needs more information on the academic aptitudes of this junior high school population to supplement the results of the existing standardized testing program. The director of research and evaluation is assigned the task of selecting the appropriate academic aptitude test and then carrying out the project under the constraints of a somewhat restricted budget. After selecting the ABC Test of Academic Aptitude, the director realizes that the complexity and cost of purchasing enough tests for each student, administering them to everyone, and scoring the tests would be prohibitive. The problem becomes one of identifying a sample of junior high school students via the methods described in Chapter 8 and then administering the test to the sample.

Studying the nature of a population through the use of samples was introduced in the previous chapter. The process involves testing each member of the sample on the relevant measure (for example, the ABC Test of Academic Aptitude) and computing the appropriate statistics (for example, the sample mean, $\overline{X}$). As we have indicated, this sample mean not only reflects the mean of the population but also is affected by sampling fluctuation. That is, we cannot expect the sample mean to be exactly equal to the population mean, even though we have conducted the sampling procedure very meticulously. However, we can use this sample mean to make inferences about the value of the population mean. Hence the term ''inferential statistics.'' In general, we use inferential statistics in making inferences about the nature of the population on the basis of observations made on the sample.

This chapter is concerned with *hypothesis testing:* determining whether some hypothesized value for an unknown population parameter is tenable, or justifiable. For example, we could test the hypothesis that the mean academic aptitude for the junior high school students is 90. If our test indicates that the hypothesized value is tenable, we *retain* the hypothesis; if not, we *reject* it.

Inferential statistics involves making inferences about the nature of the population on the basis of observations made on a sample drawn from the population.

The logic of hypothesis testing follows the chain of reasoning for inferential statistics that was illustrated in Figure 8.5. Researchers use this logic when formulating and testing theories about the effects of an experimental treatment on some criterion variable and generalizing these theories to a given population. We begin the process of hypothesis testing by stating a hypothesis about the nature of the population. For example, consider a health and physical education teacher in a senior high school who wants to determine whether the female students in the school district have a level of physical fitness that differs from the national average. After reviewing the literature, the teacher selects the most appropriate fitness test. The mean performance, nationally, on this test for senior high school female students is 200 and the standard deviation is 25. The process of testing the hypothesis for this example is outlined in the following steps.

Step 1: Stating the Hypothesis

You may well have an intuitive understanding of the terms ''hypothesis'' and ''hypothesis testing,'' but a formal definition and a brief discussion are appropriate here. In the context of inferential statistics, a *hypothesis* is a conjecture about one or more population parameters. In our example, the health and physical education teacher wants to test the hypothesis that the mean level of physical fitness of the senior high school females in the school district equals 200.

In research projects, a hypothesis usually provides the general framework for the investigation and further delineates the statement of the problem and the variables under investigation. It also has implications for the research procedures to be used and the data to be collected and analyzed. The researcher then uses the data collected from the sample to determine whether the hypothesis is tenable. Testing the specific hypothesis does *not* prove or disprove the conjecture. Rather, the outcome of hypothesis testing either supports or refutes the *tenability* of the hypothesis.

The usual research convention in hypothesis testing is to test the *null hypothesis* which is ordinarily a statement of no difference or no relationship. We will deal with hypotheses such as these in later chapters of the book. For this specific example, the null hypothesis is that the mean fitness level in the population equals 200. In symbols,

$$H_0: \quad \mu = 200 \quad \text{or} \quad H_0: \quad \mu - 200 = 0$$

where

H_0 = symbol for the null hypothesis

μ = population mean

200 = hypothesized value to be tested

It is also necessary to state an *alternative hypothesis*, which includes the possible outcomes that were not covered by the null hypothesis. We will discuss the different types of alternative hypotheses in a later section. For our present example, we will use the alternative hypothesis that the mean fitness level in the population is *not* 200. In symbols:

$$H_a: \quad \mu \neq 200$$

To summarize this first step in the process of hypothesis testing, the health and physical education teacher must test the null hypothesis (the mean fitness level for the population of senior high school females in the school district equals 200) against the alternative hypothesis (the mean fitness level is not 200). In symbols,

$$H_0: \quad \mu = 200$$

$$H_a: \quad \mu \neq 200$$

A hypothesis is a conjecture about the nature of the parameters in a population.

Step 2: Computing the Test Statistic

After stating the hypothesis, the second step in the process of hypothesis testing is to analyze the data collected. In our example, suppose the health and physical education teacher randomly selected 81 female students in the senior high school and administered the fitness test to each. Suppose further that the mean fitness level for the 81 students in the sample was 192.50. Note that this sample mean does not equal 200, the hypothesized value for μ. Recall that, even though we have a random sample from the population, we do not expect the sample mean to be exactly equal to the hypothesized value of the population mean. So the question becomes "How different can the observed sample mean ($\overline{X}$) be from the hypothesized population mean (μ) before the null hypothesis is rejected?"

To answer this question, we must use the central limit theorem. Recall from Chapter 8 that statisticians and researchers do not spend time developing sampling distributions of means by randomly selecting all samples of a specified size and computing the sample means. They work instead with *theoretical sampling distributions,* some of which are based on the central limit theorem. The theoretical sampling distribution for our example would be the distribution of mean fitness scores of all possible samples of 81 female students selected randomly from the population. Because the population mean and the standard deviation are specified, this sampling distribution is normal with mean (μ) equal to 200 and standard error ($\sigma_{\overline{X}} = \sigma/\sqrt{n}$) equal to $25/\sqrt{81} = 2.78$. This distribution is illustrated in Figure 9.1. As we did in Chapter 8, we refer to this distribution as the underlying distribution of all possible outcomes—that is, all possible sample means of sample size 81. Thus we can use the properties of the normal curve to determine the probability of observing a sample mean less than a value such as 192.50 when the null hypothesis is true.

First we use the concept of z scores to determine the number of standard errors the observed sample mean is from the hypothesized value. In symbols,

$$z = \frac{\overline{X} - \mu}{\sigma_{\overline{X}}} \qquad (9.1)$$

For this example,

$$z = \frac{192.50 - 200}{2.78}$$

$$= -2.70$$

We find that the observed sample mean ($\overline{X} = 192.50$) is 2.70 standard errors below the hypothesized value of the population mean ($\mu = 200$). Using Table 1 in the Appendix, we find that the probability of observing a sample mean less than or equal to 192.50 when the null hypothesis is true is 0.0035. In other words, such an occurrence is not likely if the H_0 is true.

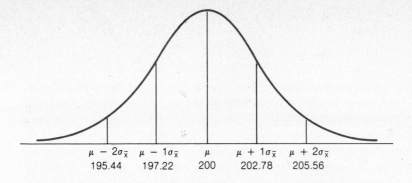

Figure 9.1 **Sampling distribution of the mean for the hypothesis H_0: $\mu = 200$ and $\sigma_{\overline{X}} = 2.78$**

Calculating the z score is referred to as calculating the *test statistic*. We use the value of the test statistic to decide whether to retain or reject the null hypothesis. The rationale for rejecting the H_0 in our example is that the value of the test statistic is -2.70. This value indicates that the difference between the observed sample mean ($\overline{X}$) and the hypothesized value of the population mean (μ) is too great to attribute only to sampling fluctuation and that, consequently, we should reject the null hypothesis. In the next section we will discuss the strategy for determining the criterion for rejecting the H_0.

The test statistic for testing the null hypothesis about the population mean is a standard score indicating the difference between the observed sample mean and the hypothesized value of the population mean.

Step 3: Setting the Criterion for Rejecting H_0

When computing the test statistic, the researcher determines the number of standard errors the sample mean ($\overline{X}$) is from the hypothesized value of the population mean (μ). Then the researcher must decide how different $\overline{X}$ can be from μ before it indicates that the null hypothesis should be rejected. If the sample mean is close to the hypothesized value, the hypothesized value may still be tenable and the difference can be attributed to sampling fluctuation. On the other hand, if the sample mean is quite different from the hypothesized value, the researcher would probably conclude that this hypothesized value for the population is not tenable. That is, the researcher would reject the null hypothesis.

Errors in Hypothesis Testing

In deciding to either retain or reject a null hypothesis, we must be aware of the possible errors inherent in this decision. One of two types of errors may be made: (1) rejecting the null hypothesis when it is true or (2) retaining the null hypothesis when it is false. The four possible outcomes, in terms of the relationship between the researcher's decision and the actual truth or falsity of the hypothesis, are illustrated in Figure 9.2. As indicated, the hypothesis can be either true or false and the hypothesis can be either rejected or retained. If the hypothesis is true and we retain it, we have made the proper decision. If the hypothesis is false and we reject it, we have made the proper decision. On the other hand, if we reject a true hypothesis or retain a false hypothesis, our decision is in error. We make a *Type I error* when we reject a true hypothesis; we make a *Type II error* when we retain a false hypothesis.

Which of the two errors is more serious? Consider an example that involves the use of a newly developed drug to treat a very serious disease. Suppose that a group of patients with this disease are randomly assigned to one of two treatment groups. One group is given the new drug, which is considerably more expensive, and the second group receives the traditional drug treatment. The hypothesis tested in this investigation is that there is no difference between the effects of the two drug treatments.

Now consider the consequence of making either a Type I or a Type II error. If the new drug is not more effective than the existing drug but the hypothesis is rejected anyway (Type I error), the new drug will begin to be used to treat the disease even though it is considerably more expensive and no more helpful. On the other hand, if the hypothesis is not rejected even though it is false (Type II error), the new drug will not be used. The consequence of the Type I error is that the patients will incur additional cost for the new drug even though it is not more effective. The consequence of the Type II error is that the patients will not incur the additional expense for the new drug but will not receive its additional benefits.

This example illustrates that a value judgment is required to determine which of the errors is more serious. The general approach to hypothesis testing is to show that the alternative hypothesis is true by rejecting the null hypothesis. This approach focuses on the Type I error, rejecting the null hypothesis when in fact it is true. This is not to say that the Type I error is more serious. The specific conditions of the research situation determine whether the Type I error or the Type II error is more serious. Ideally, both types of errors should be minimized in any research investigation.

The two types of errors that arise in hypothesis testing are Type I error (rejecting a hypothesis when in fact it is true) and Type II error (retaining a hypothesis when in fact it is false).

State of Nature (Reality of the Parameters)

Hypothesis	Hypothesis Is False
	Proper Decision
…ion	Type II Error

…utcomes in hypothesis testing

…a Type I error when the null hypothesis is rejected is,
…nificance. If the H_0 is not rejected, the probability of
…not so easily determined.[1] Nevertheless, this probability
…cance. Holding other factors constant, raising the level
…) (increasing the probability of making a Type I error)
…aking a Type II error. On the other hand, lowering the
…to .01 increases the probability of making a Type II
…ng the level of significance for an investigation, a re-
…er the consequences of making either type of error in
…e on an appropriate level of significance.
…α level, is defined as the probability below which we
…ur physical fitness example, we found that the proba-
…an less than 192.50 when the null hypothesis was true
…Such an occurrence is not likely, so we indicated that
…jected.
…ual probability of making a Type I error, researchers
…l of significance beforehand. The two most frequently
…evels) are .05 and .01. If the probability of making a
…ual to .05 or .01 (whichever level of significance is
…jected. In deciding to reject the hypothesis at one of
…ws that the decision to reject the hypothesis may be
…, respectively. However, the researcher is willing to
…process. When the null hypothesis is rejected at the
…sult is said to be ''statistically significant at the .05
…example, the result is that there is a ''significant''

…obability of making a Type II error, the value of the alternative
…ed value for the parameter and not an expression such as ''greater
…'' For a thorough discussion of the Type II error, see William L.
…(New York: Holt, 1973), pp. 353–75.

175

difference between the observed sample value ($\overline{X}$ = 192.5) and the hypothesized value for the population (μ = 200) at the .05 level. In other words, *the probability that a sample mean ($\overline{X}$) of 192.5 or less would have occurred by chance, when in fact μ was equal to 200, is less than .05.*

The level of significance is the probability of making a Type I error if H_0 is rejected.

Whereas the actual selection of the level of significance is somewhat arbitrary, the .05 and .01 levels are the most frequently used. In some research settings, however, smaller α levels, such as .005 and .001, are used. Reducing these levels is referred to as "going in a more conservative direction." In other research settings less conservative α levels, such as .10 and .20, are used. To some extent, the researcher should be influenced by the conditions of the research in selecting the level of significance. The criterion for selecting the appropriate level of significance is the specific consequence of making either a Type I error or a Type II error. In some research settings, failing to reject a false hypothesis might be considered a very serious error. In other settings, indicators of direction or a trend might be important and would be legitimately evidenced by a less substantial departure from the null hypothesis. In these cases, a less conservative level of significance (.10 or .20) might be used. On the other hand, if major expenditures or changes would result from the rejection of H_0, the researcher would probably use a very conservative level of significance (.005 or .001). In this case, the researcher wants to reduce the probability of making a Type I error (rejecting H_0 when it is true).

The commonly used α levels are .05 and .01. Other α levels may also be appropriate, depending on the consequences of making either a Type I error or a Type II error.

Region of Rejection

In the physical fitness example, we used the normal distribution as the underlying distribution. Thus we were able to use the properties of the normal curve in our initial consideration of whether to retain or reject the null hypothesis on the basis of the value of the test statistic. The properties of the normal curve will now be used to determine those values of the test statistic for which we will reject the H_0. Suppose the health and physical education teacher decides to test H_0: μ = 200 at the .05 level of significance. The sampling distribution of the mean for this example, first illustrated in Figure 9.1, is illustrated again in Figure 9.3. The mean of this distribution equals the hypothesized value (μ = 200), and the standard error equals 2.78. Note the shaded areas under the curve below the point $\mu - 1.96\ \sigma_{\overline{X}}$ (194.55) and above

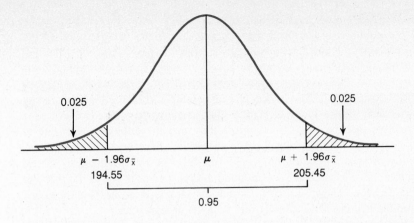

Figure 9.3 **Sampling distribution of the mean for the hypothesis H_0: $\mu = 200$ and $\sigma_{\bar{X}} = 2.78$**

the point $\mu + 1.96\,\sigma_{\bar{X}}$ (205.45). These areas represent a total of 5% of the area under the normal curve, or 2.5% in each tail of the distribution. When we are testing the null hypothesis against the alternative hypothesis, H_a: $\mu \neq 200$, these areas represent values of the sample mean that are highly improbable if the null hypothesis is true. That is, the probability of obtaining a sample mean less than 194.55 or greater than 205.45 is less than .05. The shaded areas in Figure 9.3 are referred to as the *region of rejection*. The remainder of the area in the sampling distribution is called the region of "nonrejection" or the region of retention. The values of the sample mean that fall in this area are more probable if the null hypothesis is true. In Figure 9.3, 5% of the area in the sampling distribution of the mean is in the region of rejection and 95% is in the region of retention.

Note that the region of rejection is equally divided between the two tails of the distribution. The reason is that the null hypothesis is to be tested against the nondirectional alternative, H_a: $\mu \neq 200$. Therefore, if a sample mean is observed that is *either* much greater than or much less than the hypothesized value, the null hypothesis will be rejected. Such a procedure is called a *two-tailed test*, or a *nondirectional test*, of the null hypothesis.

The region of rejection is the area of the underlying distribution that represents those values of the sample statistic that are highly improbable if the null hypothesis is true.

If we assume that the null hypothesis (H_0: $\mu = 200$) is true, the region of rejection contains sample means less than or equal to 194.55 and greater than or equal to 205.45; the region of retention contains all sample means between these two values.

These values, 194.55 and 205.45, are called the *critical values* used for testing the hypothesis (H_0: μ = 200). The observed sample mean of 192.5 falls in the region of rejection, and such a value has less than a .05 probability of appearing if the null hypothesis is true. Therefore, the researcher rejects the null hypothesis in favor of the alternative hypothesis. By rejecting the null hypothesis, the researcher is saying that the difference between the observed sample mean and the hypothesized population mean is too great to be attributed to chance fluctuation in sampling. However, the researcher realizes that there is a small probability that the difference was due to chance fluctuation and that rejecting the hypothesis may have been a Type I error. The probability of such an occurrence is .05 (the level of significance), so the researcher decides that 1 chance in 20 of making a Type I error is worth the risk and stands by the decision. The result, then, is that the sample mean is "significantly" different from the hypothesized value at the .05 level of significance.

Suppose, however, that the sample mean for the 81 students was found to be 199.3. In this case, the observed value does *not* fall in the region of rejection, the researcher would not reject the null hypothesis, and the hypothesized value would be retained as a probable value. This decision is based on the properties of the sampling distribution and on the fact that this observed sample mean is not sufficiently different from the hypothesized value of the population to warrant rejection of the null hypothesis. The researcher is willing to attribute this "nonsignificant" difference to sampling fluctuation.

Note that, in using the properties of the normal curve and the concept of standard scores, we converted the sampling distribution of the mean with mean equal to μ and standard error equal to $\sigma_{\overline{X}}$ to a normal distribution with mean equal to 0 and standard error equal to 1. Thus, rather than compute the critical values, μ + 1.96 $\sigma_{\overline{X}}$ and μ − 1.96 $\sigma_{\overline{X}}$, we use the *generalized critical values* −1.96 and +1.96. In this way, we compare the observed value of the test statistic to these generalized critical values in deciding whether to reject the null hypothesis. At the .05 level of significance, if the test statistic is less than −1.96 or greater than +1.96, the null hypothesis is rejected. Figure 9.4 can be considered a generalization of the concept of the sampling distribution for all two-tailed tests of hypotheses about a population mean, and for any standard error of the mean, at the .05 level of significance.

In the foregoing example, the test statistic was found to be

$$z = \frac{192.5 - 200}{2.78} = -2.70$$

Thus the observed sample mean ($\overline{X}$ = 192.50) is 2.70 standard errors below the hypothesized mean. This is greater than 1.96 standard errors below the mean, indicating that the value of the sample mean falls in the region of rejection for this particular theoretical sampling distribution. On the other hand, if the sample mean were found to be 199.3, the corresponding test statistic would be

$$z = \frac{199.3 - 200}{2.78} = -0.25$$

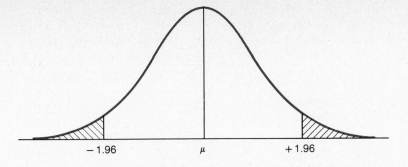

-1.96 μ $+1.96$

Figure 9.4   Theoretical sampling distribution with generalized critical values of the test statistic for $\alpha = .05$

In this case, the sample mean is only 0.25 standard errors below the hypothesized value for the population mean and thus is in the region of retention. Figure 9.5 illustrates these two test statistics on the theoretical sampling distribution.

The critical values of the test statistic are those values of the underlying distribution that represent the beginning of the region of rejection. When the observed value of the test statistic exceeds the critical value, the null hypothesis is rejected. For the nondirectional alternative hypothesis, the region of rejection is located in both tails of the underlying distribution. The test of the null hypothesis against this nondirectional alternative is referred to as a two-tailed test.

Region of Rejection for Directional Alternative Hypotheses

The previous tests of hypotheses were two-tailed, because the null hypothesis was tested against a nondirectional alternative hypothesis:

$$H_0: \quad \mu = 200$$

$$H_a: \quad \mu \neq 200$$

To the researcher who is more experienced and knowledgeable about the variables under investigation, testing the null hypothesis against a *directional* alternative may often seem more appropriate. In the physical fitness example, the health and physical education teacher might want to show that the average fitness level is less than 200. In symbols,

$$H_0: \quad \mu = 200$$

$$H_a: \quad \mu < 200$$

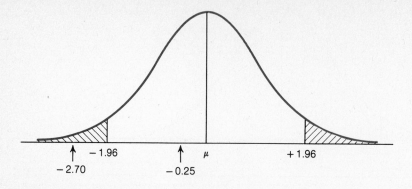

Figure 9.5 Theoretical sampling distribution for the hypothesis H_0: $\mu = 200$, illustrating the values of the test statistics when $\overline{X} = 192.50$ and when $\overline{X} = 199.3$

Testing the null hypothesis against a directional (as opposed to a nondirectional) alternative is the more powerful test, because the null hypothesis is rejected more readily when the difference between the hypothesized value of the parameter and the sample value is in the appropriate direction.

An alternative hypothesis can be either nondirectional or directional. A directional alternative hypothesis states that the parameter is greater than or less than the hypothesized value. A nondirectional alternative hypothesis merely states that the parameter is different from (not equal to) the hypothesized value.

When the alternative hypothesis is directional, the strategy for testing the null hypothesis at a particular level of significance is essentially the same. However, the critical values for the test statistic differ. For example, if the concern of the research is to show that the average fitness level is less than 200, only the left-hand tail of the theoretical sampling distribution needs to be considered. Under these circumstances, rather than dividing the 5% of the area between the two tails, we put the entire 5% into the left-hand tail. For the .05 level of significance, the sampling distribution and the region of rejection are illustrated in Figure 9.6. Using the properties of the normal curve, we find that a value of -1.645 corresponds to the beginning of the region of rejection and thus is the critical value of the test statistic for the null hypothesis. When the sample mean was 192.5, the test statistic was -2.70, which was less than the critical value of -1.645. Because -2.70 falls in the region of rejection, the null hypothesis is rejected in favor of the directional alternative. On the other hand, if the value of the test statistic fell in the region of retention, the null hypothesis would be retained as a probable value of the population parameter. Such a procedure with a directional alternative hypothesis is called a *one-tailed test*, or a *directional test*.

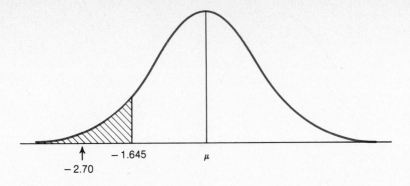

−1.645   μ

↑
−2.70

Figure 9.6   Theoretical sampling distribution for the hypothesis H_0: $\mu = 200$ and the directional alternative H_a: $\mu < 200$; $\alpha = .05$

The region of rejection for a directional alternative hypothesis is located in one of the two tails of the underlying distribution. The specific tail of the distribution is determined by the direction of the alternative hypothesis. The test of the null hypothesis against this directional alternative is referred to as a one-tailed test.

Now suppose the alternative hypothesis stated that the mean fitness level was greater than 200. The corresponding null and alternative hypotheses would be

$$H_0: \quad \mu = 200$$

$$H_a: \quad \mu > 200$$

In this case, the region of rejection would be established in the right-hand tail, because the researcher would reject H_0 and retain H_a only if the sample mean were substantially greater than 200. The critical value of the test statistic for this set of hypotheses is again 1.645, but it is in the right-hand tail—that is, at +1.645 (see Figure 9.7). In other words, the observed sample mean would have to be greater than 1.645 standard errors above the value 200 for the researcher to reject the null hypothesis at the .05 level of significance. Using the same test statistic, $z = -2.70$, the researcher would fail to reject the null hypothesis because the test statistic falls in the region of retention. Thus, when the alternative hypothesis is directional, there is only one critical value for the test statistic. When the alternative hypothesis is nondirectional, there are two critical values. The null hypothesis is used to locate the sampling distribution, in either a one-tailed or a two-tailed test, and the test statistic is calculated in the same way. However, we must be careful in locating the rejection region and in identifying the correct critical value of the test statistic, especially when dealing with directional alternatives. An easy way to remember the tail in which to establish the region of rejection and the corresponding critical value is to make an arrow out of the inequality

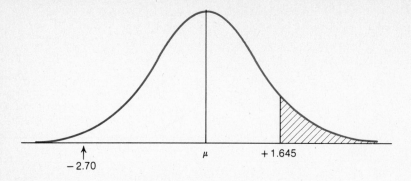

Figure 9.7   Theoretical sampling distribution for the hypothesis H_0: $\mu = 200$ and directional alternative H_a: $\mu > 200$; $\alpha = .05$

sign ($<$ or $>$) that appears in the alternative hypothesis. This arrow will point to the appropriate tail. For example, consider the first set of directional hypotheses we dealt with.

$$H_0: \quad \mu = 200$$

$$H_a: \quad \mu < 200 \qquad \text{becomes} \qquad H_a: \quad \mu \leftarrow 200$$

The arrow that we substitute for $<$ points to the left. This tells us that the critical value for testing this hypothesis, -1.645, is in the left-hand tail.

If the significance level is changed, the critical value(s) of the test statistic also change. For example, if the researcher established the level of significance at .01 for a two-tailed test, the critical values would be -2.576 and $+2.576$. For a one-tailed test at the .01 level, the critical value is either -2.326 or $+2.326$. The critical values for both two-tailed and one-tailed tests for the most commonly used levels of significance are presented in Table 9.1.

Step 4: Deciding Whether to Reject or Retain the Null Hypothesis

In our physical fitness example, the health and physical education teacher was trying to decide whether the mean performance of the population of the local female senior high school students was the same as the national average ($\mu = 200$). On the basis of a sample of 81 students, the mean physical fitness level was 192.5. Following the procedures outlined earlier in the chapter, we used the central limit theorem to find the sampling distribution of the mean and then the properties of the normal curve to determine the extent of the difference between the observed sample mean ($\overline{X} = 192.50$) and the value of the hypothesized population mean ($\mu = 200$). The

Level of Significance, Two-Tailed Test	Level of Significance, One-Tailed Test	Critical Value of Test Statistic
.20	.10	1.282
.10	.05	1.645
.05	.025	1.960
.02	.01	2.326
.01	.005	2.576
.001	.0005	3.291

Table 9.1 Critical values of the test statistic, using the normal curve as the sampling distribution

test statistic was equal to -2.70, which indicated that the observed sample mean was 2.70 standard errors below the hypothesized value. At the .05 level of significance, this observed value of the test statistic exceeded the critical value. (In this case, the observed value was less than the critical value.) The critical values for this example were -1.96 and $+1.96$, because the null hypothesis was tested against the nondirectional alternative hypothesis at the .05 level of significance ($\alpha = .05$).

This means that, if the null hypothesis were true, it would be very unlikely that the health and physical education teacher would have found a sample mean equal to 192.50. That is, *the probability*[2] *is less than .05 that the observed sample mean* ($\overline{X} = 192.50$) *would have occurred by chance if the null hypothesis were true* ($\mu = 200$). Therefore, we reject the null hypothesis and conclude that the mean fitness level for the population of female senior high school students in the teacher's school district is *not* equal to 200, which is the national average. By rejecting the null hypothesis, we are saying that the value 200 is not tenable. In the next chapter, we will illustrate and discuss procedures for identifying a range of tenable values for the population mean.

The steps in testing a hypothesis are as follows:

1. State the hypothesis.
2. Compute the test statistic.
3. Set the criterion for rejecting H_0.
4. Decide whether to reject or retain the null hypothesis.

---

[2]Because the sample mean is only a point located on the line (see Figure 9.5) and a point in a distribution has no area, this probability statement has a slight conceptual flaw. The probability is less than .05 that an observed sample mean this small *or smaller* would have occurred by chance if the null hypothesis were true. For all practical purposes we are concerned with the observed sample mean, so we will continue to use the less cumbersome probability statement.

Hypothesis Testing When σ^2 Is Unknown

In the previous example, the normal distribution was used as the sampling distribution for testing the null hypothesis H_0: $\mu = 200$. We could use the normal distribution because we assumed that the population variance (and therefore the standard deviation) was known. When σ^2 is known, the central limit theorem is directly applicable. The sampling distribution of the mean is normally distributed with μ assumed to be equal to the hypothesized value, and the standard deviation (the standard error of the mean) is equal to $\sigma/\sqrt{n}$. However, when σ is unknown, the researcher estimates the standard deviation of the population (σ) by using the standard deviation of the sample (s). This estimate is then used in conjunction with the central limit theorem to test the hypothesis about the value of the population mean. In other words, when s is used to estimate σ, $s/\sqrt{n}$ is the estimate of the *standard error of the mean*, or the standard deviation of the theoretical sampling distribution. The notation for the estimate of the standard error of the mean is

$$s_{\overline{X}} = s/\sqrt{n} \qquad (9.2)$$

For testing the hypothesis about a population mean, we estimate the standard deviation of the population (σ) by using the standard deviation of the sample (s). Thus the estimated standard error of the sampling distribution of sample means is given by $s/\sqrt{n}$.

Student's *t* Distributions

In hypothesis testing, when σ is unknown and s is used to estimate it, the normal curve is inappropriate for describing the sampling distribution of the mean. This was discovered shortly after the beginning of the twentieth century by William S. Gossett, a young chemist employed at a brewery in Dublin, Ireland. His work, which was concerned with quality control, led to the realization that traditional statistical procedures using the normal distribution as the sampling distribution are inadequate for small samples. Lacking mathematical sophistication, Gossett began to develop sampling distributions for various sample sizes. During these empirical investigations he found that, for small samples, the sampling distributions departed substantially from the normal distribution and that, as sample sizes changed, the distributions changed. This gave rise to not one distribution but a family of distributions. Gossett also noted that, as the sample size increased, the distributions increasingly approximated the normal distribution. Later, with the assistance of a mathematician, he developed a general formula for these sampling distributions and published the results in 1908 under the pen name "Student." Thus these sampling distributions are referred to as Student's *t* distributions.

Student's t distributions are a family of symmetric distributions. As sample size increases, the specific t distribution increasingly approximates the normal distribution.

The use of Student's t distributions in hypothesis testing is analogous to the use of the normal distribution. As we have indicated the sampling distribution of the mean changes as the sample size changes. Thus, for testing hypotheses, there is a specific t distribution for every sample of a given size. In order to determine the appropriate t distribution to use in these procedures, we must define the concept of degrees of freedom.

Degrees of Freedom

The concept of degrees of freedom as it applies to a statistic is fundamentally mathematical in nature. For our discussion, it is sufficient to say that the number of degrees of freedom of a statistic depends on the number of sample observations (n). Specifically, the number of *degrees of freedom* (df) can be defined as the number of observations less the number of restrictions placed on them. For example, suppose the sum of two numbers is *known* to be 50. Then, for a sample of two numbers, once the first number is known, so is the second. The restriction is the known total and, in this case, only one of the numbers is free to vary.

Suppose the mean of a population is being estimated from a sample of 5 and that the five sample values are 21, 23, 24, 27, and 30. The restriction in this case is that the sum of the deviations from the mean is equal to zero, $\Sigma(X - \overline{X}) = 0$. There are five scores, so there are five deviations. But only 4 of the 5 deviations are free to vary. In other words, once the mean is specified (in this case $\overline{X} = 25$), the fifth value is not free to vary, because the last deviation score must be such that the sum of the deviations equals zero. In this example there are $n - 1$, or 4, degrees of freedom.

Number of degrees of freedom is a mathematical concept defined as the number of observations less the number of restrictions placed on them.

In order to relate the concept of degrees of freedom to Student's t distributions, three of the t distributions for degrees of freedom equal to 5, 15, and ∞ are illustrated in Figure 9.8. Like the normal curve, the various t distributions are symmetric and bell-shaped. The table of t distributions appears as Table 2 in the Appendix. The distributions in this table are in standard-score form with the mean equal to 0 and the standard deviation equal to 1. Note that the t distribution for infinite degrees of freedom is identical to the normal distribution. Note also that this table differs from Table 1, the normal curve table. Each row of Table 2 represents a different t distribution, and each distribution is associated with a unique number of degrees of freedom. The column headings in the table (.10, .05, .025, and so on) represent the proportion of

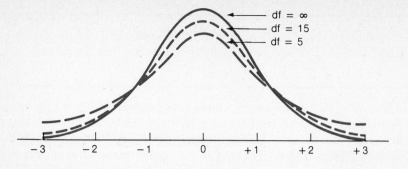

Figure 9.8 Student's *t* distribution for 5, 15, and ∞ degrees of freedom

the area remaining in the tails of the distribution. To illustrate the use of this table, consider the *t* distribution with 15 degrees of freedom (Figure 9.9). For this distribution, 2.5% of the area is in the right-hand tail beyond the point that is 2.131 standard deviations above the mean. Similarly, 5% of the area is in the left-hand tail below a point corresponding to −1.753 standard deviations below the mean. The values 2.131 and −1.753 are actually standard scores required for the corresponding areas of 2.5% and 5% in the tails of the distribution. Note that, for degrees of freedom equal to ∞, these points would be 1.960 and −1.645, respectively, the same points that were shown before for the normal distribution.

In hypothesis testing, if we were to use the normal distribution as an approximation of a *t* distribution, this approximation would become increasingly accurate as the number of degrees of freedom (or sample size) was increased. When the approximation is adequate is, of course, somewhat subjective. From a practical standpoint, the normal distribution can *almost always* be used as an adequate approximation for the *t* distribution when the number of degrees of freedom exceeds 120.

As sample size increases, the difference between the normal distribution and the corresponding *t* distribution decreases. From a practical standpoint, the normal distribution is an adequate approximation of the *t* distribution when the sample size exceeds 120.

An Example Using the *t* Distribution

Suppose the athletic director at a large university was asked to investigate the claim that the student athletes were doing less well academically than the average student in the university. Data supplied by the Office of Institutional Research revealed that the overall grade point average (GPA) for the university was 2.50 on a 4-point scale. To investigate the claim, the athletic director randomly selected 20 student athletes

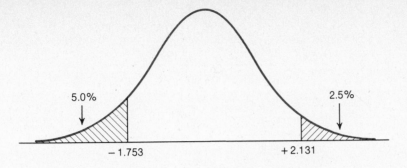

Figure 9.9 Student's *t* distribution for 15 degrees of freedom

and obtained the current GPA for each. These data are given in Table 9.2. As can be seen, the sample mean for these 20 student athletes was 2.45 and the sample standard deviation was 0.54. In completing the investigation, the athletic director would follow the four steps for hypothesis testing.

Step 1: Stating the Hypothesis

The claim made in this example was that the average GPA of the student athletes was less than the average of the population of students at the university. Thus the null hypothesis would be tested against the directional alternative; that is, a one-tailed test of the hypothesis would be conducted. In symbols, the hypotheses would be written

$$H_0: \quad \mu = 2.50$$

$$H_a: \quad \mu < 2.50$$

We will assume that the level of significance was established beforehand at .05 ($\alpha = .05$).

Step 2: Computing the Test Statistic

In this example, we make the assumption that the population variance and standard deviation are unknown, and we will take the sample variance (or standard deviation) as an estimate. Thus the sampling distribution of the mean would be the *t* distribution with $n - 1$, or 19, degrees of freedom. The mean of this sampling distribution would be 2.50, the hypothesized value of μ; the standard deviation (standard error of the mean) would be computed as follows:

$$s_{\overline{X}} = s/\sqrt{n}$$

$$= 0.54/\sqrt{20}$$

$$= 0.12$$

1.8	2.0	1.2	3.0	$\Sigma X = 48.9$
3.1	3.2	2.5	2.9	
2.6	2.8	2.3	2.0	$\Sigma X^2 = 125.15$
2.4	2.7	2.0	1.9	
2.2	3.3	2.8	2.2	

1 sample case

2 sample P 213

$$\overline{X} = \frac{\Sigma X}{n} \qquad s^2 = \frac{\Sigma X^2 - \dfrac{(\Sigma X)^2}{n}}{n-1} \qquad s = \sqrt{0.29}$$

$$= \frac{48.9}{20} \qquad = \frac{125.15 - \dfrac{(48.9)^2}{20}}{19} \qquad = 0.54$$

$$= 2.45 \qquad\qquad = 0.29$$

Table 9.2 GPA data for the 20 student athletes

We compute the test statistic in exactly the same way as we did when the variance was known. The only difference is that we now call the test statistic a *t* score. This reflects the fact that we are using the *t* distribution as the sampling distribution rather than the normal distribution. For this example,

$$t = \frac{\overline{X} - \mu}{s_{\overline{X}}} \tag{9.3}$$

$$= \frac{2.45 - 2.50}{0.12} = -0.42$$

This *t* value indicates that the sample mean is 0.42 standard errors below the hypothesized value of the population mean.

Step 3: Setting the Criterion for Rejecting H_0

The purpose of computing the *t* value (test statistic) is to determine the magnitude of the difference, in standard-score form, between the observed sample mean ($\overline{X}$) and the hypothesized value of the population mean (μ). The next step is to determine how great this difference must be before it causes us to reject the null hypothesis. Specifically, this step involves determining the critical value(s) of the test statistic. The critical value depends on (1) the distribution used as the sampling distribution of the mean (normal distribution versus *t* distributions), (2) the nature of the alternative hypothesis (nondirectional versus directional) and (3) the level of statistical significance (α level). For this example, the *t* distribution for $n - 1$, or 19, degrees of freedom would be used. Second, because the alternative hypothesis was directional (H_a: $\mu < 2.50$), the region of rejection would be located only in the left-hand tail of the distribution. Finally, with the level of significance set at .05, the critical value of the test statistic from Table 2 would be -1.73:

$$t_{cv} = -1.73$$

Step 4: Deciding Whether to Reject or Retain the Null Hypothesis   In deciding whether to retain or reject the null hypothesis, we compare the observed value of the test statistic (t value) to the critical value of the test statistic (t_{cv}). Because $t = -0.42$ does *not* exceed $t_{cv} = -1.73$ (in absolute value), the null hypothesis is retained as tenable. The associated probability statement would be: *The probability is greater than .05 that the observed sample mean* ($\overline{X} = 2.45$) *would have occurred by chance if the null hypothesis were true* ($\mu = 2.50$). That is, the difference between the sample mean and the hypothesized value of the population was not sufficient to attribute it to anything other than sampling fluctuation. Thus the athletic director would conclude that the academic performance of the student athletes was *not* different from the average academic performance of the student body of the university.

Statistical Significance versus Practical Importance

In the physical fitness example, the null hypothesis (H_0) was $\mu = 200$ and the conclusion was that the observed sample mean ($\overline{X} = 192.50$) was "significantly" different from the hypothesized value at the .05 level of significance. But what is meant by *statistical significance?* In this example, and in general, the difference between the hypothesized population parameter and the corresponding sample statistic is said to be statistically significant when the probability that the difference occurred by chance is less than the significance level (α level). But how large must the difference be? And, just because the difference is statistically significant, is it necessarily of practical importance?

One consideration in answering these questions is the level of significance established for testing the hypothesis. The selection of the level of significance in behavioral science research does not often receive the attention it merits. The .05 and .01 levels are commonly used, but sometimes there seems to be no rationale for their use other than the fact that R. A. Fisher used them in his agricultural experiments several decades ago. In previous sections of this chapter, we indicated that the criterion for selecting the appropriate level of significance for a specific research study is the gravity of the consequences of making either a Type I error (rejecting a true hypothesis) or a Type II error (retaining a false hypothesis). There has been a tendency in behavioral science research to guard against the Type I error, but some critics assert that this tendency has resulted in setting α levels that are too conservative. Sometimes this results in retaining the null hypothesis even though the probability that the statistic would occur if the hypothesis were true is relatively small, say .07 or .08. Thus the area of the research and the characteristics of the specific study must be considered. In the final analysis, the level of significance should not be arbitrarily set. It should be given very careful attention by the researcher.

A second consideration in answering these questions is related to the concept of statistical precision. In testing hypotheses such as the one for this example, we define *statistical precision* as the inverse of the standard error of the mean. In other words, the smaller the standard error, the greater the statistical precision. (Statistical precision is discussed further in Chapter 10.) For example, consider formula 9.1. A smaller value for the standard error of the mean will result in a larger absolute value of the test statistic and a greater likelihood of rejecting the null hypothesis.

But how can the researcher increase statistical precision? One way is to increase the sample size. Increasing the sample size yields a more precise estimate of the population parameter (μ), because the sample statistic ($\overline{X}$) is based on more observations. Consider the formula for the standard error of the mean (formula 9.2). Note that the standard deviation (s) is divided by $\sqrt{n}$. Thus, as n increases, $s_{\overline{X}}$ decreases and the statistical precision increases.

What statistical precision is adequate for research in the behavioral sciences? The observant reader may have noted that a sufficiently large sample size leads to the rejection of any null hypothesis based on a *fixed* difference between the hypothesized parameter and the observed sample statistic. On the other hand, a researcher might *not* reject a null hypothesis even though there is a seemingly large difference between the parameter and the corresponding statistic. It is possible that the sample size is simply too small. Critics of inferential statistics have charged that this makes statistically based research a ''numbers game.'' They are also concerned about the reporting and publishing of only ''statistically significant'' results (that is, when the null hypothesis has been rejected). This charge is unfounded if the researcher has been thorough in the design and conduct of the investigation and if the results are interpreted in light of practical considerations as well as theoretical, statistical implications.

Finally, it is important to note that the matter of statistical significance versus practical importance is *not* settled by statistical tests. As we have discussed, increasing the statistical precision can render almost any difference between the observed statistic and the hypothesized parameter statistically significant. But what does statistical significance mean in the practical context of the research situation? Further, when does a difference become large enough to warrant corrective action? In the physical fitness example, does this difference of 7.5 points merit the institution of a remedial fitness program? These kinds of questions are not fully answered by inferential statistics. They can be answered only on the basis of a thorough knowledge and understanding of the research area and the variables under consideration. Inferential statistics are only tools for analyzing data. They are not substitutes for knowledgeable interpretation of those data.

Summary

This chapter introduces inferential statistics. Specifically, it discusses the process of hypothesis testing. Hypothesis testing involves determining whether an *a priori* value of a parameter (for example, μ) is tenable. This *a priori* value is established by the researcher on the basis of the research literature and/or previous research experience in the particular area of inquiry. A sample is drawn, and the appropriate sample statistic ($\overline{X}$) is computed. The decision to retain or to reject the hypothesis, in terms of probability, is based on the magnitude of the difference between the observed statistic ($\overline{X}$) and the hypothesized parameter (μ).

The discussion of hypothesis testing includes (1) the directional nature of the alternative hypothesis, (2) the errors that arise in hypothesis testing, (3) the level of significance, (4) the region of rejection, and (5) the critical value(s) of the test statistic. Specifically, when drawing inferences about the population mean, we develop the theoretical sampling distribution on the basis of the sample size and the standard deviation of the sample. In testing hypotheses about the population mean, we use the hypothesized value of μ as the mean of this sampling distribution.

The general formula for calculating the test statistic is

$$\text{Test statistic} = \frac{\text{statistic} - \text{parameter}}{\text{standard error of the statistic}}$$

The critical value of this test statistic is found by using the tables for either the normal distribution or Student's t distributions. In subsequent chapters the same basic logic of hypothesis testing will be applied to testing hypotheses about other population parameters. The general formula for the test statistic will also remain the same, but additional test statistics will be introduced.

Chapter 10 will discuss the important process of estimation. Specifically, the method of estimating μ from $\overline{X}$ will be presented.

Key Concepts

Hypothesis testing

Null hypothesis

Alternative hypothesis

Sampling distribution

Test statistic

Type I error

Type II error

Level of significance

Region of rejection

Nondirectional or two-tailed test

Directional or one-tailed test

Critical value

Standard error of the mean

t distribution

Degrees of freedom

Statistical significance

Statistical precision

Exercises

9.1 An almanac listed the income per household in a particular city as $19,000. Five years later, a researcher surveyed a random sample of 60 households in that city and found the average income for this sample to be $21,200; the standard deviation was $2400. Test the hypothesis that the population mean is unchanged from the number cited in the almanac against the alternative that the average increase has increased. Use $\alpha = .05$.

a. State the hypothesis.

b. Compute the test statistic.

c. Set the criterion for rejecting the null hypothesis.

d. Decide whether to reject or to retain the null hypothesis.

9.2 One would expect a student to be able to guess correctly the answers to 10 items on a 20-item true–false test. The scores for 14 students on a 20-item true–false test follow. Test the appropriate null hypothesis against the alternative that the average is above that which could be expected by chance. Use $\alpha = .01$.

11.64

12	10	9	13	13
8	11	7	14	11
15	17	11	12	

a. State the hypothesis.

b. Compute the test statistic.

c. Set the criterion for rejecting the null hypothesis.

d. Decide whether to reject or to retain the null hypothesis.

9.3 Which of the following is more likely to lead to the rejection of the null hypothesis?

a. one-tailed or two-tailed test

b. $\alpha = .05$ or $\alpha = .01$

c. $n = 144$ or $n = 400$

9.4 The average reading level for the norm group on a certain standardized test was 5.4. A test coordinator for a local school district was interested in whether the students in the school district were equivalent to the norm group. A random sample of size 26 was selected; the average reading level for the sample was 5.2 with a standard deviation of 0.4. Test the hypothesis that the students in this school district are equivalent to the norm group in reading level. Use $\alpha = .05$.

a. State the hypothesis

b. Compute the test statistic.

c. Set the criterion for rejecting the null hypothesis.

d. Decide whether to reject or to retain the null hypothesis.

9.5 A study on the reaction time of children with cerebral palsy reported an average of 1.6 seconds on a particular task. A researcher believed that reaction time could be reduced by using a motivating set of directions. An equivalent set of children were located, and they completed the same task under the condition of the motivating set of directions. The reaction times for the 12 children follow. Determine whether the reaction time was improved when the motivating directions were used ($\alpha = .01$).

Child	Reaction Time	Child	Reaction Time
A	1.4	G	1.5
B	1.8	H	2.0
C	1.1	I	1.4
D	1.3	J	1.9
E	1.6	K	1.8
F	0.8	L	1.3

a. State the hypothesis.
b. Compute the test statistic.
c. Set the criterion for rejecting the null hypothesis.
d. Decide whether to reject or to retain the null hypothesis.

9.6 a. If the null hypothesis H_0: $\mu = a$ is rejected in favor of the alternative hypothesis H_a: $\mu > a$, what can be said about retaining or rejecting the same null hypothesis in favor of the alternative hypothesis H_a: $\mu \neq a$?

b. If the null hypothesis H_0: $\mu = a$ is rejected in favor of the alternative hypothesis H_a: $\mu \neq a$, what can be said about retaining or rejecting the same null hypothesis in favor of the alternative hypothesis H_a: $\mu > a$?

9.7 The director of research in a large school system was charged with running a pilot study to see whether decreasing class size to 15 or fewer would result in increased levels of achievement. The implication was that, if the achievement levels were significantly higher with smaller class sizes, the school system would reduce class sizes throughout the system.

a. Which type of error (Type I or Type II) would be more serious in this example?
b. How might you reduce this error?

▦ Key Strokes for Selected Exercises

9.2 Use Key Strokes described in the Appendix to determine ΣX and ΣX^2

	Value	Key Stroke	Display
Step 1: Determine $\overline{X}$.			
	163 (ΣX)	÷	163
	14 (n)	=	11.64
			(save)
Step 2: Determine $s_{\overline{X}}$.			
	163 (ΣX)	X^2	26569
		÷	26569
	14 (n)	=	1897.7857
		STO	1897.7857
	1993 (ΣX^2)	−	1993
		RCL	1897.7857
		=	95.2143
		÷	95.2143
	13 ($n - 1$)	=	7.3242
		$\sqrt{}$	2.7063 = s
		÷	2.7063
	14	$\sqrt{}$	3.7417
		=	0.7233 = $s_{\overline{X}}$
		STO	0.7233

Step 3: Determine t.

$$
\begin{array}{llll}
11.64\ (\overline{X}) & - & 11.64 \\
10\ (\mu) & = & 1.64 \\
& \div & 1.64 \\
& \text{RCL} & 0.7233 \\
& = & \underline{2.27} \quad = t
\end{array}
$$

9.4 Step 1: Determine $s_{\overline{X}}$.

$$
\begin{array}{llll}
0.4\ (s) & \div & 0.4 \\
26\ (n) & \sqrt{} & 5.099 \\
& = & 0.0784 \\
& \text{STO} & \underline{0.0784} = s_{\overline{X}}
\end{array}
$$

Step 2: Determine t.

$$
\begin{array}{llll}
5.2\ (\overline{X}) & - & 5.2 \\
5.4\ (\mu) & = & -0.2 \\
& \div & -0.2 \\
& \text{RCL} & 0.0784 \\
& = & \underline{2.765} \ = t
\end{array}
$$

10

Estimation: One-Sample Case for the Mean

In the previous chapter, we discussed the process of hypothesis testing. The process is based on the assumption that the researcher can establish logically an *a priori* value for the population parameter. This value is usually established through knowledge of the research literature and previous experience with the variables under investigation. In the two examples used in the last chapter to illustrate the process of hypothesis testing, we assumed that the health and physical education teacher and the athletic director were interested in testing the respective null hypotheses. On the other hand, rather than testing the tenability of a hypothesized value of the population mean, they could have tried to answer the question "What is the average fitness level for the female senior high school students in the school district?'' or ''What is the average level of academic performance for the student athletes?'' In such a case, the research strategy shifts from testing hypotheses about specific values to estimating the value of the population mean. This process is referred to as *statistical estimation.* It is important to note that, because we never know the population mean for sure unless we measure every member of the population, we are still in an inferential situation. That is, we are making inferences about the population on the basis of what we observe in the sample.

Statistical estimation is the process of estimating a parameter from the corresponding sample statistic.

In both of the cases we considered in Chapter 9, the sample mean $(\overline{X})$ is referred to as a *point estimate* and represents the ''best'' estimate of the population value (μ). Although this is the best point estimate available to the researcher, there is no basis for arguing conclusively that, for the physical fitness example, the population mean is 192.50 or that, for the athletes' academic performance, the population mean is 2.45. (These values were the observed sample means in the two examples.) The reason is that the sample mean reflects not only the population mean but also sampling fluctuation. Therefore, even though the sample mean is the single best estimate, we would not expect it to be exactly equal to the population mean.

Confidence Intervals

When the purpose of a research investigation is to estimate the unknown population mean on the basis of the mean of the corresponding sample, the estimate is enhanced if we can determine a range of values that we are confident contains the population mean. This range of values is called a *confidence interval* and is founded in probability theory. Consider the example of the academic performance of the student athletes, and assume that the population mean $\mu = 2.50$. The sampling distribution for the mean is the t distribution with 19 degrees of freedom; the standard error of the mean $(s_{\overline{X}})$ was calculated and found to be 0.12. This distribution is illustrated in Figure 10.1. Note that, because we are assuming that $\mu = 2.50$, 95% of all possible sample means of sample size 20 would be between $\mu - 2.09 s_{\overline{X}}$ and $\mu + 2.09 s_{\overline{X}}$ (2.25 to 2.75). Similarly, 99% of all possible sample means would be between $\mu - 2.861$ $(s_{\overline{X}})$ and $\mu + 2.861(s_{\overline{X}})$ (2.16 to 2.84).

Using this property of the t distributions and the standard error of the mean $(s_{\overline{X}})$, we can determine the 95% confidence interval (denoted CI_{95}) for any or all possible sample means by using the formula

$$CI_{95} = \overline{X} \pm (t_{cv})(s_{\overline{X}})$$

(10.1)

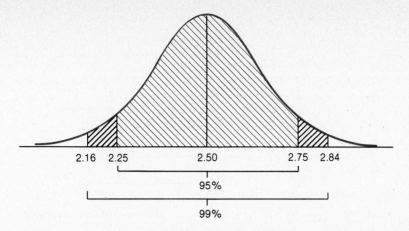

Figure 10.1 Sampling distribution of the mean for the hypothesis H_0: μ = 2.50 and $s_{\overline{X}}$ = 0.12

where

$\overline{X}$ = sample mean

t_{cv} = critical value of t using the appropriate t distribution

$s_{\overline{X}}$ = standard error of the mean

For this example, the appropriate critical value of t is 2.09 and the interval is calculated as follows:

$$CI_{95} = 2.45 \pm (2.09)(0.12)$$
$$= 2.45 \pm 0.25$$
$$= (2.20, 2.70)$$

The confidence interval is a range of values that we are confident contains the population parameter.

Level of Confidence

Level of confidence is defined as the researcher's degree of confidence that the computed interval contains the parameter being estimated. In the foregoing example, we computed the 95% confidence interval (CI_{95}) using $\overline{X}$ = 2.45. Suppose we computed the CI_{95} for 29 other sample means of sample size 20. These intervals are found in Table 10.1. Note that 2 of the 30 confidence intervals do *not* contain the point

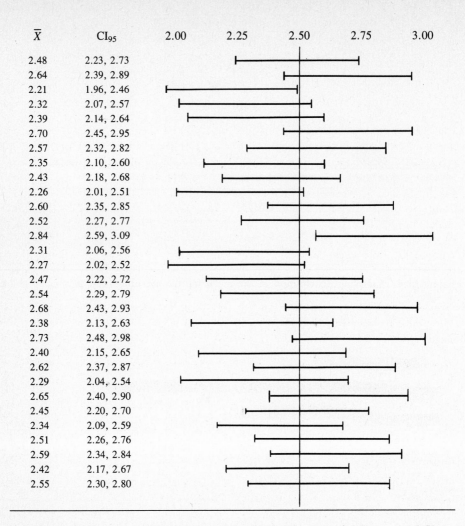

$\overline{X}$	CI_{95}
2.48	2.23, 2.73
2.64	2.39, 2.89
2.21	1.96, 2.46
2.32	2.07, 2.57
2.39	2.14, 2.64
2.70	2.45, 2.95
2.57	2.32, 2.82
2.35	2.10, 2.60
2.43	2.18, 2.68
2.26	2.01, 2.51
2.60	2.35, 2.85
2.52	2.27, 2.77
2.84	2.59, 3.09
2.31	2.06, 2.56
2.27	2.02, 2.52
2.47	2.22, 2.72
2.54	2.29, 2.79
2.68	2.43, 2.93
2.38	2.13, 2.63
2.73	2.48, 2.98
2.40	2.15, 2.65
2.62	2.37, 2.87
2.29	2.04, 2.54
2.65	2.40, 2.90
2.45	2.20, 2.70
2.34	2.09, 2.59
2.51	2.26, 2.76
2.59	2.34, 2.84
2.42	2.17, 2.67
2.55	2.30, 2.80

Table 10.1   95 Percent confidence intervals for 30 sample means for H_0: $\mu = 2.50$, with $s_{\overline{X}} = 0.12$

2.50, which was assumed to be the value of μ. These are the 95% confidence intervals for $\overline{X} = 2.21$ and $\overline{X} = 2.81$. As indicated in the sampling distribution in Figure 10.1, 95% of all possible sample means will be between 2.25 and 2.75 under the assumption that $\mu = 2.50$. However, it is possible to have a sample mean less than 2.25 or greater than 2.75. The two sample means $\overline{X} = 2.21$ and $\overline{X} = 2.84$ are examples. Therefore, we can expect only 95% of the CI_{95} constructed around the respective sample means to contain $\mu = 2.50$. In other words, we are 95% confident that $\overline{X}$

$\pm\ 2.09 s_{\overline{X}}$ will contain μ, where $\overline{X}$ is *one* of all possible sample means—for example, $\overline{X} = 2.45$.

If the CI_{95} were developed for all possible sample means, 5% of these intervals would not contain μ. In this case, the level of confidence is the complement of the level of significance. That is, if the level of significance is .05, the corresponding level of confidence is $1 - .05$ or .95. As with the α level, the level of confidence selected is based on the seriousness of making an error by concluding that a particular confidence interval contains μ when in fact it does not. If the consequences are not serious, the researcher may use a less conservative level such as .90 or .80. If the consequences are serious, the more conservative .99 level may be used.

It is important to note that we *cannot* say that the probability is .95 that the interval from 2.20 to 2.70 contains μ. Either it does or it does not. The probability statement relates to all possible sample means for which the CI_{95} can be computed. For this reason, these intervals are called confidence intervals rather than probability intervals. To emphasize this point, we interpret the confidence interval in terms of the confidence we have that the interval contains μ. In our example, the athletic director would conclude with 95% confidence that the interval from 2.20 to 2.70 contains the mean academic performance of the student athletes.

If the null hypothesis is true, 95% of all possible sample means will be between $\mu\ -\ (t_{cv})(s_{\overline{X}})$ and $\mu\ +\ (t_{cv})(s_{\overline{X}})$. Correspondingly, 95% of all CI_{95} computed for each of these sample means will contain μ.

Statistical estimation is an extremely valuable tool in research investigations, because it is not limited to situations in which an *a priori* value for the population parameter can be logically established. In the physical fitness example introduced in Chapter 9, the null hypothesis (H_0: $\mu = 200$) was rejected when the sample mean ($\overline{X} = 192.50$) was found to be significantly different from the hypothesized value of the population mean. In rejecting the null hypothesis, we concluded that the probability was too low to attribute the difference between $\overline{X}$ and the hypothesized value of μ to sampling fluctuation and therefore that the population mean was probably different from 200. The conclusion does not specify an estimate for μ. To do so, it would be necessary to develop an appropriate confidence interval. In a more general sense, rejecting the null hypothesis yields little information about the variable under investigation. Therefore, regardless of whether the researcher decides to test a hypothesis about the value of the population parameter, statistical estimation is especially important in research studies wherein one of the major purposes is to continue the development of a theory.

The Relationship Between
Hypothesis Testing and Estimation

In both hypothesis testing and estimation, the computation involves using the sample mean ($\overline{X}$) and the standard error of the mean ($s_{\overline{X}}$). (See formulas 9.2 and 9.3.) In addition, the critical value of the t distribution (t_{cv}) is used in interpreting the test statistic and computing the confidence interval. Hence it is not surprising that the two procedures are related.

To illustrate this relationship, consider the physical fitness example. Recall that the following null hypothesis was tested at the .05 level of significance:

$$H_0: \quad \mu = 200$$

$$H_a: \quad \mu \neq 200$$

The value of the test statistic was -2.70 and, because the critical values were -1.96 and $+1.96$, the null hypothesis was rejected. For this example, the 95% confidence interval could have been computed as follows:

$$CI_{95} = 192.50 \pm (1.96)(2.78)$$

$$= 192.50 \pm 5.45$$

$$= (187.05, 197.95)$$

The conclusion for this confidence interval would be that the health and physical education teacher can be 95% confident that this interval contains the average physical fitness score for the female senior high school students in the school district.

Note that the hypothesized population value ($\mu = 200$) is *not* contained in the interval. It should not be in the interval, because the interval contains only tenable values for μ based on the observed sample mean ($\overline{X} = 192.50$). In other words, rejecting the null hypothesis means that we do not consider $\mu = 200$ a tenable value and thus would *not* expect it to be contained in the interval.

Constructing a confidence interval can be considered a means of testing a large number of hypotheses simultaneously. In this example, any value in the range of the confidence interval would not be rejected if it were stated in the form of a null hypothesis. Examples include

$$H_0: \quad \mu = 189 \qquad H_0: \quad \mu = 194 \qquad H_0: \quad \mu = 196$$

On the other hand, any value outside the range of the confidence interval would be rejected if it were stated in the form of a null hypothesis. Examples include

$$H_0: \quad \mu = 200 \qquad H_0: \quad \mu = 185 \qquad H_0: \quad \mu = 180$$

In summary, any values between 187.05 and 197.95 are tenable values for the population mean and all other values are not.

> Developing a confidence interval can be viewed as testing many hypotheses simultaneously. Any value within the range of the interval is a tenable value for the population parameter. All values outside the interval are not tenable.

Statistical Precision in Estimation

Much of the preliminary planning of studies in which estimates of population parameters are determined is concerned with enhancing the accuracy of the estimates. The accuracy with which a confidence interval can be used to estimate a population parameter can be thought of as the *statistical precision* of the estimate. Generally speaking, the narrower the width of the confidence interval, the more precise the estimate.

The major concerns in planning studies for which confidence intervals are used to estimate parameters are the level of confidence to be used and the appropriate sample size. Both are important considerations, but the effect of the size of sample is more directly related to the concept of statistical precision. Consider the example of the grade point average for the student athletes. On the basis of the sample size of 20, the 95% confidence interval was

$$CI_{95} = 2.45 \pm (2.09)(0.12)$$

$$= (2.20, 2.70)$$

As can be seen, the width of this interval is 0.50.

For this example, the statistical precision could be enhanced by increasing the sample size. Suppose that the sample size had been 121 rather than 20, and assume that both the sample mean and the standard deviation were the same ($\overline{X} = 2.45$, $s = 0.54$). How much would this affect the CI_{95}? First, the standard error would be smaller:

$$s_{\overline{X}} = 0.54/\sqrt{121}$$

$$= 0.049$$

Second, the t_{cv} would be based on 120 degrees of freedom rather than 19, so $t_{cv} = 1.98$. The CI_{95} would then be

$$CI_{95} = 2.45 \pm (1.98)(0.049)$$

$$= (2.35, 2.55)$$

This confidence interval has a width of 0.20. Thus, by increasing the sample size, we make the standard error of the mean smaller and the width of the confidence interval narrower. In other words, the width of the confidence interval, which we have defined as the accuracy of the estimation, is directly related to the size of the sample.

Statistical precision in interval estimation is concerned with the width of the interval; as the width decreases, the precision is enhanced. The width of the interval can be decreased by increasing the sample size.

The other factor that the researcher can manipulate to affect the width of the confidence interval is the level of confidence. The choice of the level of confidence is analogous to the selection of the level of significance in hypothesis testing. The most commonly used levels are .95 and .99; the use of these levels is based primarily on tradition.

The effect of choice of the level of confidence on the width of the confidence interval is easy to illustrate. Consider again the student athlete example with the sample size to 20. The CI_{95} was found as follows:

$$CI_{95} = \overline{X} \pm (t_{cv})(s_{\overline{X}})$$

$$= 2.45 \pm (2.09)(0.12)$$

$$= (2.20, 2.70)$$

Now consider the CI_{99} for the same example. The only change will be in t_{cv}. From Table 2, we find $t_{cv} = 2.861$. Thus the CI_{99} would be

$$CI_{99} = 2.45 \pm (2.86)(0.12)$$

$$= (2.11, 2.79)$$

Note that the width of the confidence has increased from 0.50 to 0.68. This will always be true. That is, the greater the level of confidence, the wider the interval.

As the level of confidence is increased, say from .95 to .99, the width of the interval is also increased.

Summary

The first step in estimating a population parameter, such as the population mean (μ), involves determining the best point estimate. In estimating μ, the sample mean ($\overline{X}$) is the point estimate. The second step involves determining the confidence interval, which is defined as a range of values that we have a specified degree of confidence will contain μ. The development of this interval is based on the sample mean ($\overline{X}$), the standard error of the mean ($s_{\overline{X}}$), and the sampling distribution of the mean. In general,

Estimation: One-Sample Case for the Mean

the formula for the confidence interval in the estimation of any population parameter is given by

$$CI = \text{statistic} \pm (\text{critical value})(\text{standard error of the statistic})$$

The width of the confidence interval in a given situation depends on the sample size and the level of confidence that is chosen. As the interval becomes narrower, statistical precision is increased. The narrower the confidence interval, the more accurate the estimate of the parameter.

Chapters 9 and 10 discuss hypothesis testing and estimation, respectively. These are the two basic procedures of inferential statistics. The examples are presented in the context of testing hypotheses about means and estimating means. But the basic reasoning underlying these procedures is the same regardless of the parameters and statistics involved. It is the chain of reasoning in inferential statistics. As you proceed with the following chapters, be sure to keep in mind this underlying logic.

Key Concepts

Statistical estimation
Confidence interval
Point estimate

Level of confidence
Statistical precision

Exercises

10.1 Using the data given in Exercise 9.1, develop the 95% confidence interval (CI_{95}) for estimating the average household income.

10.2 Using the data given in Exercise 9.2, develop the 99% confidence interval (CI_{99}) for estimating the average number of correct responses on the 20-item true–false test.

10.3 What sample size do we need to estimate the average IQ in a given population to within ± 3 points when the CI_{95} is used? Assume $s = 15$.

10.4 Using the data given in Exercise 9.4, develop the 95% confidence interval (CI_{95}) for estimating the average reading level.

10.5 Using the data given in Exercise 9.5, develop the 99% confidence interval (CI_{99}) for estimating the average reaction time.

10.6 Given that the 95% confidence interval (CI_{95}) for the mean of a certain population is (21.84, 40.56), find CI_{90} and CI_{99}. Assume $n = 22$. (*Hint*: Find $\overline{X}$, t_{cv} and $s_{\overline{X}}$)

10.7 Given that $\overline{X} = 22.6$ and $s = 2.34$, compare the 95% confidence intervals for $n = 150$ and $n = 10$.

⌨ Key Strokes for Selected Exercises

	Value	Key Stroke	Display
10.1 Determine CI_{95}.			
	2.0 (t_{cv})	×	2.0
	309.84 ($s_{\overline{X}}$)	=	619.68
		STO	619.68
	21,200 ($\overline{X}$)	−	21,200
		RCL	619.68
		=	20,580.32
	21,200 ($\overline{X}$)	+	21,200
		RCL	619.68
		=	21,819.68

10.7 Step 1: Find $s_{\overline{X}}$ for $n = 150$.

	Value	Key Stroke	Display
	2.34 (s)	÷	2.34
	150 (n)	$\sqrt{}$	12.2474
		=	0.191

Step 2: Find CI_{95}.

	Value	Key Stroke	Display
	.191	×	0.191
	1.96 (t_{cv})	=	0.374
		STO	0.374
	22.6 ($\overline{X}$)	−	22.6
		RCL	0.374
		=	22.226
	22.6 ($\overline{X}$)	+	22.6
		RCL	0.374
		=	22.974

Step 3: Find $s_{\overline{X}}$ for $n = 10$.

	Value	Key Stroke	Display
	2.34 (s)	÷	2.34
	10 (n)	$\sqrt{}$	3.162
		=	0.740

Step 4: Find CI_{95}.

	Value	Key Stroke	Display
	0.740	×	0.740
	2.262 (t_{cv})	=	1.674
		STO	1.674
	22.6 ($\overline{X}$)	−	22.6
		RCL	1.674
		=	20.926
	22.6 ($\overline{X}$)	+	22.6
		RCL	1.674
		=	24.274

11

Hypothesis Testing and Estimation: Two-Sample Case for the Mean

In Chapter 9, we introduced the example of the administrator who was assigned the task of obtaining information about the academic ability of a junior high school population. Rather than test the entire population, the evaluation director selected a random sample. Knowing the techniques introduced in Chapters 9 and 10, the director could test hypotheses about the population mean or estimate the population mean.

But suppose the central office staff is not satisfied with inferring only to the population mean. All the data generated by selecting the random sample and testing the students selected should yield considerably more information. There are a number of subgroups or subpopulations that may have different means, for instance. What about the means of boys and girls, those of the eighth and ninth grades, those of school A and school B? Are all these population means equal? We know how to use inferential statistics when one mean is involved. What if we have pairs of means, the means from two samples?

Testing hypotheses about two means can lead to interesting hypotheses dealing with such issues as the difference between means of experimental and control groups, the difference between means of men and women, or the difference between the pretest mean and the posttest mean for the same group of individuals in an experiment. In the latter case, the means would come from the same individuals tested at different times.

As we will see, the logic and procedures of inferential statistics (hypothesis testing and confidence intervals) that were developed in the previous chapters also apply when two means are involved. The only difference is that here, rather than comparing the sample mean to some hypothesized value, we will compare two sample means. We will use the same general formula for the test statistic:

$$\text{Test statistic} = \frac{\text{statistic} - \text{hypothesized parameter}}{\text{standard error of the statistic}} \qquad (11.1)$$

and for developing a confidence interval for the difference between the two sample means:

$$\text{CI} = \text{statistic} \pm (t_{cv})(\text{standard error of the statistic}) \qquad (11.2)$$

The basic steps in hypothesis testing will remain unchanged.

1. Stating the hypothesis.
2. Computing the test statistic.
3. Setting the criterion for rejecting H_0.
4. Deciding whether to reject or retain the null hypothesis.

The basic reasoning of inferential statistics is the same whether we are testing hypotheses about a single mean or about the difference between two means.

Testing H_0: $\mu_1 - \mu_2 = 0$, Independent Samples

First we will consider the situation in which samples are selected from two separate and distinct populations. Such samples are said to be *independent samples*. For such samples, no one individual can belong to both populations or subsequently to both samples. Following the selection of the samples, we compute the mean for each sample, $\bar{X}_1$ and $\bar{X}_2$, and we will consider the statistic $(\bar{X}_1 - \bar{X}_2)$ as the difference between the sample means.

The Null Hypothesis

In the general case, the null hypothesis that we test when we are comparing the population means of the two groups is

$$H_0: \quad \mu_1 - \mu_2 = a$$

That is, the difference between two population means equals some value, a. The difference a can be any hypothesized value, but the usual null hypothesis tested is that there is *no* difference between the population means:

$$H_0: \quad \mu_1 - \mu_2 = 0$$

This hypothesis could also be written

$$H_0: \quad \mu_1 = \mu_2$$

That is, the mean of population 1 equals the mean of population 2. For example, we might want to determine whether there is a difference between ninth-grade boys and ninth-grade girls in their "attitude toward mathematics." Using the foregoing null hypothesis, we would test whether the difference between the population mean of the boys (μ_1) and the population mean of the girls (μ_2) equals zero.

The Alternative Hypothesis

As we discussed in Chapter 9, the alternative hypothesis can be either nondirectional or directional. For our example, the nondirectional alternative hypothesis would be

$$H_a: \quad \mu_1 - \mu_2 \neq 0 \qquad \text{or} \qquad H_a: \quad \mu_1 \neq \mu_2$$

That is, the difference between the mean "attitude toward mathematics" for the population of boys (μ_1) is *not* equal to the mean for the population of girls (μ_2).

A directional alternative hypothesis for the same example would be that the mean attitude of the population of boys is greater than the mean attitude for the population of girls:

$$H_a: \quad \mu_1 - \mu_2 > 0 \qquad \text{or} \qquad H_a: \quad \mu_1 > \mu_2$$

The Sampling Distribution of $(\overline{X}_1 - \overline{X}_2)$ for Independent Samples

Suppose the two populations from which the samples were selected have distributions such as those shown in Figure 11.1. These two distributions have means μ_1 and μ_2, respectively. In addition, the variances of these two populations, σ_1^2 and σ_2^2, are assumed to be equal to each other. In other words, the variances are equal to some common value:

$$\sigma_1^2 = \sigma_2^2 = \sigma^2$$

This does not mean that, in order to be independent, the population distributions cannot overlap. Indeed, if H_0 is true, $\mu_1 = \mu_2$, the distributions would exhibit

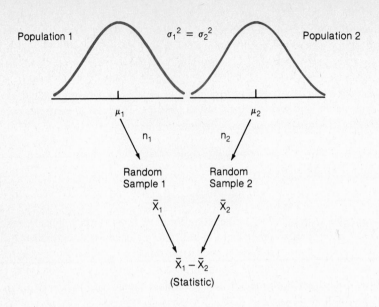

Figure 11.1 Illustration of a pair of independent random samples from two populations

considerable overlap or possibly be identical. The distributions are separated in the figure simply to set them off.

Suppose we select a random sample of size n_1 from the first population and another sample of size n_2 from the second population. (It is not necessary that $n_1 = n_2$). For both samples we could compute the mean and then determine the difference between the two sample means $(\overline{X}_1 - \overline{X}_2)$. We could repeat this procedure for a second pair of means and then a third pair of means, and so on, using samples of size n_1 and n_2, respectively. If we continued this process indefinitely, we would generate a distribution of differences between two sample means. This distribution would be the *sampling distribution for the differences between two sample means*. What is this distribution like? Remember that to know a distribution is to know its shape, location, and dispersion. This sampling distribution of differences is analogous to the sampling distribution of the mean discussed in Chapter 9 and can also be defined by using the central limit theorem. Specifically, the sampling distribution for the difference between two sample means has the following characteristics:

1. The distribution approximates a normal distribution.
2. The mean of the distribution is $\mu_1 - \mu_2$.
3. The standard deviation (standard error of the difference) is given by

$$\sigma_{\overline{X}_1 - \overline{X}_2} = \sqrt{\sigma^2 \left(\frac{1}{n_1} + \frac{1}{n_2} \right)}$$

Hypothesis Testing and Estimation: Two-Sample Case for the Mean

As discussed in Chapter 9, when the variance (σ^2) is known, the normal distribution is the appropriate sampling distribution as defined by the central limit theorem. However, when σ^2 is unknown, s^2 is used as the estimate of the population variance, $s_{\overline{X}}$ is used as the estimate of the standard error of the mean, and the t distribution for $n - 1$ degrees of freedom is the appropriate sampling distribution. For the sampling distribution for the difference between the two sample means when $\sigma_1^2 = \sigma_2^2 = \sigma^2$ and σ^2 is unknown, we estimate the standard error of the difference by using the variances from the two samples. The formula is

$$s_{\overline{X}_1 - \overline{X}_2} = \sqrt{s^2 \left(\frac{1}{n_1} + \frac{1}{n_2} \right)} \tag{11.3}$$

where

$$s^2 = \frac{(n_1 - 1)s_1^2 + (n_2 - 1)s_2^2}{n_1 + n_2 - 2} \tag{11.4}$$

Two sample case

In formulas 11.3 and 11.4, s^2 is called the *pooled estimate* of σ^2. It takes into consideration not only the variances of the two samples, s_1^2 and s_2^2, but also the respective samples sizes n_1 and n_2. In addition, when σ^2 is unknown, the sampling distribution of the differences is not a normal distribution but rather a t distribution with $n_1 + n_2 - 2$ degrees of freedom. Thus, when σ^2 is unknown, the sampling distribution of the differences is defined as follows:

1. The distribution is a t distribution with $n_1 + n_2 - 2$ degrees of freedom.
2. The mean of the distribution is $\mu_1 - \mu_2$.
3. The standard deviation of the distribution is estimated by

$$s_{\overline{X}_1 - \overline{X}_2} = \sqrt{s^2 \left(\frac{1}{n_1} + \frac{1}{n_2} \right)}$$

Note that, by combining formulas 11.3 and 11.4, we can use the following formula for the estimated *standard error of the difference:*

$$s_{\overline{X}_1 - \overline{X}_2} = \sqrt{\frac{(n_1 - 1)s_1^2 + (n_2 - 1)s_2^2}{n_1 + n_2 - 2} \left(\frac{1}{n_1} + \frac{1}{n_2} \right)} \tag{11.5}$$

When σ^2 is unknown and is estimated with the sample variances, the sampling distribution of $\overline{X}_1 - \overline{X}_2$ is the t distribution with $n_1 + n_2 - 2$ degrees of freedom.

It is important to note that the samples in the preceding development have been *independent*. That means that no individual's score in one sample can be paired with any individual's score in the other sample. Similarly, no score in sample 2 can be linked or correlated with any score from sample 1. The concept of independence will become more apparent when the t test for dependent samples is developed later in this chapter.

A Computational Example

Returning to our example of the difference between ninth-grade boys and ninth-grade girls in their "attitude toward mathematics," suppose that 20 girls and 15 boys are randomly selected and administered the "attitude toward mathematics" inventory. The summary statistics are as follows:

Boys	Girls
$\overline{X}_B = 25$	$\overline{X}_G = 21$
$s_B^2 = 18$	$s_G^2 = 16$
$n_B = 15$	$n_G = 20$

Now consider this example in terms of the four basic steps in hypothesis testing.

Step 1: Stating the Hypothesis   The null and alternative hypotheses are

$$H_0: \quad \mu_B - \mu_G = 0$$

$$H_a: \quad \mu_B - \mu_G \neq 0$$

The null hypothesis is that there is no difference between the mean of the boys and the mean of the girls. The parameter is $(\mu_B - \mu_G)$, and the corresponding statistic is $(\overline{X}_B - \overline{X}_G)$. The level of significance to be used is .05.

Step 2: Computing the Test Statistic   Because we do not know σ^2, the sampling distribution is the t distribution with $n_B + n_G - 2$, or 33, degrees of freedom. We can use formula 11.5 to compute $s_{\overline{X}_1 - \overline{X}_2}$. At this point we compute the test statistic, t, using the general formula (formula 11.1). For the two-sample case for the mean, the formula is

$$t = \frac{(\overline{X}_1 - \overline{X}_2) - (\mu_1 - \mu_2)}{\sqrt{\dfrac{(n_1 - 1)s_1^2 + (n_2 - 1)s_2^2}{n_1 + n_2 - 2}\left(\dfrac{1}{n_1} + \dfrac{1}{n_2}\right)}} \tag{11.6}$$

For our example, let the boys be group 1 and the girls be group 2; then

$$t = \frac{(25 - 21) - 0}{\sqrt{\dfrac{(14)(18) + (19)(16)}{15 + 20 - 2}\left(\dfrac{1}{15} + \dfrac{1}{20}\right)}}$$

$$= \frac{4}{\sqrt{\dfrac{252 + 304}{33}(.1166)}}$$

$$= \frac{4}{1.40}$$

$$= 2.85$$

Step 3: Setting the Criterion for Rejecting H_0

The sampling distribution for $(\overline{X}_B - \overline{X}_G)$ under the null hypothesis is illustrated in Figure 11.2. This is the t distribution for 33 degrees of freedom. From Table 2 in the Appendix, we find that the critical values of the test statistic for a two-tailed test at the .05 level of significance are ± 2.04.

Step 4: Deciding Whether to Reject or Retain the Null Hypothesis

The calculated t (2.85) exceeds the critical value of t (2.04), so our decision is to reject the null hypothesis. The probability statement after the statistical test is: *The probability is less than .05 that the observed difference in sample means occurred by chance if the difference in population means is zero*.

We conclude that the population means for boys and girls on the measure of "attitude toward mathematics" are not equal. The result of the statistical test is said to be *statistically significant* at $\alpha = .05$.

A Second Computational Example

In order to illustrate the test of H_0: $\mu_1 - \mu_2 = a$ where a is not equal to zero, we use the hypothesis $\mu_B - \mu_G = 5$ instead of the hypothesis $\mu_B - \mu_G = 0$. In symbols,

$$H_0: \quad \mu_B - \mu_G = 5$$

and the alternate hypothesis would be

$$H_a: \quad \mu_B - \mu_G \neq 5$$

The remaining conditions are the same, so the sampling distribution of $(\overline{X}_B - \overline{X}_G)$ is still the t distribution with 33 degrees of freedom, and $s_{\overline{X}_1 - \overline{X}_2}$ does not change. The change in the computation occurs in the test statistic. Instead of hypothesizing $\mu_B - \mu_G = 0$, we hypothesize $\mu_B - \mu_G = 5$. Using formula 11.6, we find that

$$t = \frac{(25 - 21) - (5)}{\sqrt{\dfrac{(14)(18) + (19)(16)}{15 + 20 - 2}\left(\dfrac{1}{15} + \dfrac{1}{20}\right)}}$$

$$= \frac{-1}{1.40}$$

$$= -0.71$$

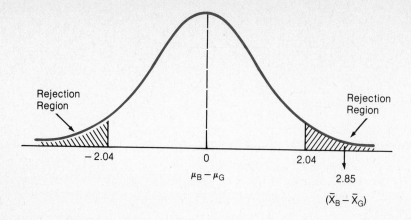

Figure 11.2   Sampling distribution of $\overline{X}_B - \overline{X}_G$; t distribution with 33 df.

The critical values from Table 2 are still ± 2.04. Because -0.71 falls between the two critical values, we retain the hypothesis that $\mu_B - \mu_G = 5$. The probability statement after the statistical test is: *The probability is greater than .05 that the observed difference in sample means occurred by chance if the difference in population means is 5.*

The sampling distribution of $(\overline{X}_B - \overline{X}_G)$ has the same shape (t distribution with 33 degrees of freedom) and standard deviation as the distribution shown in Figure 11.2. The only difference is that the change in hypothesis from $\mu_B - \mu_G = 0$ to $\mu_B - \mu_G = 5$ has shifted the distribution 5 points to the right on the scale of measurement.

Confidence Interval for $\mu_1 - \mu_2$

We find the confidence interval for estimating $(\mu_1 - \mu_2)$, the difference between the population means, by using the general formula for the confidence interval, which was discussed earlier in this chapter as well as in the preceding chapter. For this example, because we used $\alpha = .05$ in the test of the null hypothesis, we will compute the 95% confidence interval. The sampling distribution is the t distribution with 33 degrees of freedom, so the critical value for the confidence interval (from Table 2) is 2.04. Using the standard error of the difference $(s_{\overline{X}_B - \overline{X}_G} = 1.40)$ derived from computing the test statistic for testing the hypothesis, we find the 95% confidence interval to be

$$CI_{95} = (\overline{X}_1 - \overline{X}_2) \pm (t_{cv})(s_{\overline{X}_1 - \overline{X}_2}) \tag{11.7}$$

$$= (25 - 21) \pm (2.04)(1.40)$$

$$= 4 \pm 2.86$$

$$= (1.14, 6.86)$$

The conclusion is that the researcher is 95% confident that the interval from 1.14 to 6.86 contains the difference between μ_1 and μ_2. Note that 5 is in the interval but that 0 is not. These results are consistent with the earlier tests of the two hypotheses. They further illustrate the relationship between hypothesis testing and interval estimation.

Parametric Assumptions

The mathematical concepts underlying tests of hypotheses and confidence intervals require several assumptions.

1. The two samples have been randomly selected.
2. The samples are independent.
3. The distributions of scores in the populations are normal.
4. The population variances are equal; that is, $\sigma_1^2 = \sigma_2^2$.

These assumptions are referred to as the *parametric assumptions;* procedures that require these assumptions are called *parametric statistics*.

When these assumptions are not met, various consequences can occur. For example, if the samples were not randomly selected, we do not have a population or populations to which the results can be generalized, and a statistical inference is not appropriate. If the samples are not independent, the sampling distribution for the differences between means cannot be used. If the population(s) is(are) not normally distributed, the probability statements associated with tests of hypotheses and confidence intervals may be slightly incorrect. For example, an α level of .05 may actually be .047 or .051. The probability statements are also affected when $\sigma_1^2 \neq \sigma_2^2$.

The method of sampling determines whether assumptions 1 and 2 have been met. If the sample data appear to be non-normal, it is possible that the scores in the population are not normally distributed. The consequences of non-normality are so slight that assumption 3 can be ignored if the sample size exceeds 30. Assumption 4 is referred to as the assumption of *homogeneity of variance*. The statistical test of this assumption will be explained in Chapter 12. If $\sigma_1^2 \neq \sigma_2^2$, the probability statements may be greatly affected, especially when $n_1 \neq n_2$.

Parametric assumptions are required for testing hypotheses about the difference between two population means.

Testing H_0: $\mu_1 - \mu_2 = 0$, Dependent Samples

In the previous example, we tested the null hypothesis, H_0: $\mu_1 - \mu_2 = 0$, under the assumption that we had independent samples. That is, the subjects were randomly selected from two populations and measurements were taken, or the subjects were randomly assigned to the two treatment conditions. Due to this random selection/

assignment, the two groups can be assumed to be equivalent before the beginning of the experiment. This assures that whatever treatment effect emerges will not be confounded with initial differences between the groups.

An alternative procedure is to have the subjects act as their own controls and to study them under both the treatment and the control conditions. For example, subjects in a drug experiment are measured before and after administration of the drug. The same subjects are measured under both conditions, and the data are *correlated data* because a subject scoring high on the first measurement tends to score high on the second, and vice versa. The scores for each subject are *not* independent; the scores under one condition are *dependent* on the scores under the other condition.

The steps in testing the null hypothesis and developing the confidence interval for *dependent samples* are the same as those we use with independent samples. However, because the assumption of independent samples has been violated, a different sampling distribution is needed. The sampling distribution of the difference of sample means $(\bar{X}_1 - \bar{X}_2)$ for dependent samples is defined by the central limit theorem as follows:

1. The sampling distribution approximates the normal distribution.
2. The mean of the distribution is $\mu_1 - \mu_2$.
3. The standard deviation of the distribution is given by

$$\sigma_{\bar{d}} = \sqrt{\frac{\sigma_d^2}{n}}$$

where

σ_d^2 = variance of the differences in paired scores

n = number of pairs of scores

When σ_d^2 is unknown, s_d^2 is used as the estimate, and the sampling distribution is the t distribution with $n - 1$ degrees of freedom. The standard error of the difference is thus estimated by

$$s_{\bar{d}} = \sqrt{\frac{s_d^2}{n}} \tag{11.8}$$

where

$$s_d^2 = \frac{\Sigma(d - \bar{d})^2}{n - 1} = \frac{\Sigma d^2 - \dfrac{(\Sigma d)^2}{n}}{n - 1} \tag{11.9}$$

A Computational Example

Suppose a psychologist is to investigate the effect of a drug on a psychomotor task. Ten introductory psychology students are randomly selected and complete the task without the drug and under the influence of the drug. Table 11.1 gives the data collected. Let us test the null hypothesis, using the four steps that have been used previously.

Without Drug	With Drug	d	d^2
126	144	18	324
113	116	3	9
146	152	6	36
135	133	-2	4
118	128	10	100
129	135	6	36
121	126	5	25
125	120	-5	25
154	161	7	49
138	145	7	49
		55	657

Table 11.1   Data for the hypothesis test for the difference between the means of dependent samples

Step 1: Stating the Hypothesis

For this example, the null hypothesis will be tested against the nondirectional alternative. In symbols,

$$H_0: \quad \mu_{\text{with}} - \mu_{\text{without}} = 0$$

$$H_a: \quad \mu_{\text{with}} - \mu_{\text{without}} \neq 0$$

In testing the null hypothesis that the mean performance under the influence of the drug is the same as the mean performance without the drug, we will use the .01 level of significance.

Step 2: Computing the Test Statistic

For the foregoing null hypothesis with dependent samples, the statistic corresponding to the hypothesized population parameter is the mean of the differences between the paired scores—that is,

$$\bar{d} = \frac{\Sigma d}{n} \tag{11.10}$$

for this example,

$$\bar{d} = \frac{55}{10}$$

$$= 5.5$$

The variance of these difference scores is calculated as follows:

$$s_d^2 = \frac{657 - \dfrac{(55)^2}{10}}{9}$$

$$= 39.39$$

Thus the standard error of the difference for dependent samples is

$$s_{\overline{d}} = \sqrt{\frac{s_{\overline{d}}^2}{n}}$$

$$= \sqrt{\frac{39.39}{10}}$$

$$= 1.98$$

and the test statistic is computed via the general formula (formula 11.1):

$$t = \frac{\overline{d} - 0}{s_{\overline{d}}}$$

$$= \frac{5.5 - 0}{1.98}$$

$$= 2.78$$

Step 3: Setting the Criterion for Rejecting H_0
The sampling distribution of $\overline{d}$ under the null hypothesis is the t distribution with $n - 1$ degrees of freedom. For this example, the sampling distribution is the t distribution with 9 degrees of freedom. From Table 2 in the Appendix, we learn that the critical values of t for the two-tailed test at the .01 level of significance are ± 3.250.

Step 4: Deciding Whether to Reject or Retain the Null Hypothesis
The calculated t (2.78) does not exceed the critical value (3.250), so we decide *not* to reject the null hypothesis. The probability statement after the statistical test is: *The probability is greater than .01 that the mean of the paired differences occurred by chance if the mean difference in the population were equal to zero.*

Thus the conclusion is that there is no difference between mean performance on the psychomotor skill under the influence of the drug and mean performance on that skill without the drug.

Confidence Interval for the Mean Difference with Dependent Samples

We construct the confidence interval for the mean difference with dependent samples by using the general formula for confidence intervals (formula 11.2):

$$\text{CI} = \overline{d} \pm (t_{cv})(s_{\overline{d}})$$

For this example, assume that the psychologist wants to construct CI_{99}.

$$\text{CI}_{99} = 5.5 \pm 3.250(1.98)$$

$$= 5.5 \pm 6.44$$

$$= (-0.94, 11.94)$$

The psychologist could conclude with 99% confidence that the interval from -0.94 to 11.94 contains the mean difference. Note that the interval contains 0, which was the hypothesized value for the mean difference. Thus these results again illustrate the relationship between the test of the hypothesis and the confidence interval.

Summary

This chapter introduces the hypothesis test and the corresponding confidence interval that should be used when making inferences about the difference between two population means. As in Chapters 9 and 10, the central limit theorem is used to define the sampling distribution, and the four basic steps in hypothesis testing are applied. The general formulas for both the test statistic and the confidence interval are also used again.

In this chapter, we differentiate between independent and dependent samples when testing the null hypothesis, H_0: $\mu_1 - \mu_2 = 0$. For testing the null hypothesis with independent samples when the population variance is unknown, the sampling distribution is the t distribution with $n_1 + n_2 - 2$ degrees of freedom. The estimate of the population variance is called the pooled estimate, because it takes into consideration the variances of the two independent samples. For dependent samples, the variance under consideration is the variance of the differences between the paired scores.

The parametric assumptions, along with the consequences of violating one or more of them, are introduced in this chapter. The assumptions of randomness and independence are controlled by the researcher through the use of adequate sampling procedures. However, the researcher does not have the same kind of control over the assumptions of normality and homogeneity of variance. The best way to exercise limited control over them is to use samples that are approximately equal in size and greater than 30.

In the next chapter, we will continue examining procedures for testing hypotheses and developing confidence intervals for other parameters and their corresponding statistics. Even though there will be certain technical differences between these tests of hypotheses and confidence intervals, such as different sampling distributions, the logic and general formulas for testing hypotheses and developing confidence intervals will remain the same.

Key Concepts

Independent samples
Pooled estimate
Standard error of the difference
Parametric assumptions
Parametric statistics

Equal variances (homogeneity of variance)
Correlated data
Dependent samples

Exercises

11.1 In a study of the effects of class size on reading achievement, 50 second-grade students are assigned to either classes with fifteen or fewer students or classes with twenty-five or more students. Summary statistics for the two groups follow. Determine whether the students in the smaller classes had significantly higher achievement (Use $\alpha = .05$).

Classes ≤ 15 (Group 1)	Classes ≥ 25 (Group 2)
$\bar{X} = 2.8$	$\bar{X} = 2.4$
$s = 0.52$	$s = 0.49$
$n = 25$	$n = 25$

a. State the hypothesis.
b. Compute the test statistic
c. Set the criterion for rejecting the null hypothesis.
d. Decide whether to reject or to retain the null hypothesis.
e. Develop the CI_{95}.

11.2 A researcher is interested in whether children would learn better with just positive illustrations of the concept or with both positive and negative illustrations. Twenty children are randomly assigned to each of the two experimental conditions; their scores on the concept attainment task follow. Determine whether there is a difference between the two methods. Use $\alpha = .01$.

Positive Illustrations (Group 1)	Positive and Negative Illustrations (Group 2)
8	14
10	8
7	7
12	10
6	12
9	6
10	15
11	11
6	9
13	8
92	100

a. State the hypothesis.
b. Compute the test statistic.
c. Set the criterion for rejecting the null hypothesis.
d. Decide whether to reject or to retain the null hypothesis.
e. Develop the CI_{99}.

11.3 College students were measured on a finger dexterity task under two conditions, a neutral condition and a stressful condition. Their scores follow. Determine whether the scores were significantly lower in the stressful condition. Use $\alpha = .05$.

Neutral	Stressful	d	d²
16	13	3	9
19	15	4	16
14	14	0	0
20	16	4	16
18	15	3	9
11	13	-2	4
15	15	0	0
17	14	3	9
		17	63

n = 8

a. State the hypothesis.
b. Compute the test statistic.
c. Set the criterion for rejecting the null hypothesis.
d. Decide whether to reject or to retain the hypothesis.
e. Develop the CI_{95}.

11.4 An automobile manufacturer wanted to know if the new model had better gasoline performance (number of miles per gallon) than the old model. Thirty new-model automobiles were selected and the gasoline performance was determined. This year's data were then compared to data from 25 older-model automobiles. The data follow. Determine whether the performance is better for the newer model. Use $\alpha = .05$.

Older Model	Newer Model
$\overline{X} = 52$	$\overline{X} = 56$
$s^2 = 220$	$s^2 = 210$
$n = 25$	$n = 30$

a. State the hypothesis.
b. Compute the test statistic.
c. Set the criterion for rejecting the null hypothesis.
d. Decide whether to reject or to retain the null hypothesis.
e. Develop the CI_{95}.

11.5 Test the hypothesis that there is no difference between the average scores of husbands and wives on an attitude-toward-smoking scale. Use $\alpha = .05$.

Husbands	Wives
16	15
20	18
10	13
15	10
8	12
19	16
14	11
15	12

a. State the hypothesis.
b. Compute the test statistic.
c. Set the criterion for rejecting the null hypothesis.
d. Decide whether to reject or to retain the null hypothesis.
e. Develop the CI_{95}.

▦ Key Strokes for Selected Exercises

	Value	Key Stroke	Display
11.1 Step 1: Determine s^2.			
	24 ($n_1 - 1$)	×	24
	0.52 (s_1)	X^2	0.2704
		=	6.4896
		+	6.4896
	24 ($n_2 - 1$)	×	24
	0.49 (s_2)	X^2	0.2401
		=	12.252
		÷	12.252
	48 (df)	=	0.255 $= s^2$
		STO	0.255

Step 2: Determine $S_{\overline{X}_1 - \overline{X}_2}$.

	Value	Key Stroke	Display
	25 (n_1)	1/X	0.04
		+	0.04
	25 (n_2)	1/X	0.04
		=	0.08
		×	0.08
		RCL	0.255
		=	0.0204
		√	0.143 $= s_{\overline{X}_1 - \overline{X}_2}$
		STO	0.143

Step 3: Determine t.

	Value	Key Stroke	Display
	2.8 ($\overline{X}_1$)	−	2.8
	2.4 ($\overline{X}_2$)	÷	0.4
		RCL	0.143
		=	2.797 $= t$

Step 4: Determine CI_{95}.

	Value	Key Stroke	Display
		RCL	0.143
		×	0.143
	2.01 (t_{cv})	=	0.287
		STO	0.287

Value	Key Stroke	Display
$0.4\ (\overline{X}_1 - \overline{X}_2)$	−	0.4
	RCL	0.287
	=	0.113
0.4	+	0.4
	RCL	0.287
	=	0.687

11.3 Step 1: Determine $\overline{d}$.

Value	Key Stroke	Display
15 (Σd)	÷	15
8 (n)	=	1.875 = $\overline{d}$
		(save)

Step 2: Determine $s_{\overline{d}}$.

Value	Key Stroke	Display
15 (Σd)	X^2	225
	÷	225
8 (n)	=	28.125
	STO	28.125
63 (Σd^2)	−	63
	RCL	28.125
	=	34.875
	÷	34.875
7 (df)	=	4.982
	$\sqrt{\ }$	2.232 = s_d
	÷	2.232
8 (n)	$\sqrt{\ }$	2.828
	=	0.789 = $s_{\overline{d}}$
	STO	0.789

Step 3: Determine t.

Value	Key Stroke	Display
1.875 ($\overline{d}$)	÷	1.875
	RCL	0.789
	=	2.376 = t

Step 4: Determine CI_{95}.

Value	Key Stroke	Display
	RCL	0.789
	×	0.789
2.365 (t_{cv})	=	1.866
	STO	1.866
1.875 ($\overline{d}$)	−	1.875
	RCL	1.866
	=	0.009
1.875	+	1.875
	RCL	1.866
	=	3.741

12

Hypothesis Testing for Other Statistics

In Chapters 9, 10, and 11, we were concerned about tests of hypotheses and confidence intervals for a single population mean or for the difference between two population means. However, there are many research situations in which other statistics are of primary interest. For example, consider a national survey designed to answer the following questions.

1. What proportion of the nonwhite high school graduates are attending community/junior colleges?

2. Is the proportion of white high school graduates attending community/junior colleges different from the proportion of nonwhite graduates attending such colleges?

3. What is the relationship between family income and father's level of education?
4. Is there higher variability of family income for those students attending community/junior colleges than for those attending colleges/universities?

Clearly, answering these questions calls for the use of statistics other than means.

In this chapter, we will extend the concepts of hypothesis testing and interval estimation to hypotheses relating to population values of proportions (P), correlation coefficients (ρ) and variances (σ^2). Even though the statistic of primary interest will be different, we will use the same four steps of hypothesis testing. The only difference is that a different statistic will be computed and that the subsequent statistical inference will be made relative to a different population parameter.

Hypotheses About Proportions

A *proportion* is the fractional part of a group that possesses some specific characteristic. For example, if 35 of the 60 students in an introductory statistics class are female, we would say that the proportion of students in the class who are female is 35/60 or 0.583. Up to this point, we have used Greek letters to indicate population parameters. However, for proportions, we will use P rather than the Greek letter π so that it will not be confused with the mathematical constant π, which is the ratio of the circumference of a circle to its diameter.

Testing H_0: $P = a$

Proportions can take on values from 0 to 1.0, inclusive, and we can test hypotheses about any value of a population proportion in this range. For example, suppose a political scientist hypothesizes that only half of the voters in city A will favor a certain tax reform proposal. A random sample of 400 voters is surveyed, and 220 of these voters indicate that they are in favor of the proposal. That is, the proportion of the sample that favors the proposal is 220/400 or 0.55. We now consider the four steps of hypothesis testing.

Step 1: Stating the Hypothesis
The null hypothesis for this example is that the population proportion equals 0.50, or

$$H_0: P = 0.50$$

Because there is no information given about whether the political scientist expects the proportion to be either greater than or less than 0.50, a two-tailed test is appropriate. The nondirectional alternative is

$$H_a: P \neq 0.50$$

For this example, we will use the .05 level of significance.

Step 2: Computing the Test Statistic

In this example, the statistic is the sample proportion, which was found to be 0.55. As was the case for the sample mean, we could develop the sampling distribution for the proportion (the sampling distribution is the binomial distribution, which was discussed in Chapter 8). However, when the sample size is reasonably large, say 50 or greater, the normal distribution is an adequate approximation to the binomial distribution, and the central limit theorem can be used to define the sampling distribution. Specifically, the sampling distribution of a proportion has the following characteristics:

1. The distribution is approximately normal.
2. The mean of the distribution equals P, the population proportion.
3. The standard deviation (standard error of a proportion) is given by

$$\sigma_P = \sqrt{\frac{PQ}{n}}$$

where

P = proportion of the population possessing the characteristic

$Q = (1 - P)$ = proportion of the population *not* possessing the characteristic

P and Q are unknown, so σ_P is estimated by

$$s_p = \sqrt{\frac{pq}{n}} \tag{12.1}$$

where

p = proportion of the *sample* possessing the characteristic

$q = (1 - p)$ = proportion of the *sample* not possessing the characteristic

For the data given in the example, we find that

$$s_p = \sqrt{\frac{(0.55)(0.45)}{400}}$$

$$= 0.025$$

Because the sampling distribution is the normal distribution, the test statistic is z rather than t. We will use the general formula for the test statistic:

$$z = \frac{\text{statistic} - \text{hypothesized parameter}}{\text{standard error of the statistic}}$$

For the test of H_0: $P = 0.50$, the formula is

$$z = \frac{p - P}{s_p} \tag{12.2}$$

$$= \frac{0.55 - 0.50}{0.025}$$

$$= 2.00$$

Step 3: Determining the Criterion for Rejecting H_0

The sampling distribution of p is normally distributed, so we use Table 1 in the Appendix to find the critical value of z. For a two-tailed test at the .05 level of significance, the critical values are ± 1.96.

Step 4: Deciding Whether to Reject or Retain the Null Hypothesis

The calculated statistic of 2.0 exceeds the critical value of 1.96, so our decision is to reject the null hypothesis. The probability is less than .05 that the observed sample proportion occurred by chance if the population proportion is .50.

The sampling distribution of p is the normal distribution (if n is large), with mean equal to P and standard deviation estimated by

$$s_p = \sqrt{\frac{pq}{n}}$$

Confidence Interval[1] for P As with the population mean, μ, we can construct an interval estimate for P. The general formula for constructing a confidence interval is used. It is

$$\text{CI} = \text{statistic} \pm (\text{critical value})(\text{standard error of the statistic})$$

For a proportion, this formula becomes

$$\text{CI} = p \pm (z_{cv})(s_p) \tag{12.3}$$

Suppose a 95% confidence interval is constructed for the example. The interval would be computed as follows

$$
\begin{aligned}
\text{CI}_{95} &= 0.55 \pm (1.96)(0.025) \\
&= 0.55 \pm 0.049 \\
&= (0.501, 0.599)
\end{aligned}
$$

Thus, we are 95% confident that the interval 0.501 to 0.599 includes the population proportion. Note that the interval does not include 0.50, a value of P rejected in the hypothesis test using $\alpha = .05$.

Testing H_0: $P_1 - P_2 = 0$

As was the case for the mean, it is possible to test the hypothesis that two population proportions are equal. We will limit the discussion to testing H_0: $P_1 - P_2 = 0$ for

---

[1]In this chapter, confidence intervals are discussed only for the one-sample statistics for proportions and correlation coefficients. However, they can be constructed for two-sample statistics; the reasoning and procedures are the same as for the one-sample case.

independent samples. (The case for dependent samples will be discussed in Chapter 15 as the McNemar test for significance of change.)

Consider again the example of the survey of registered voters in city A, and suppose the political scientist wants to test whether there is a difference between the proportion of men and the proportion of women who support the tax reform proposal. Suppose further that 60 of the 240 women ($60/240 = 0.25$) and 56 of the 160 men ($56/160 = 0.35$) surveyed indicated that they supported the proposal. Now apply the four steps of hypothesis testing.

Step 1: Stating the Hypothesis
The null hypothesis for this example is that there is no difference between the proportion of men who support the tax reform proposal (P_1) and the proportion of women who support the proposal (P_2). In symbols,

$$H_0: \ P_1 - P_2 = 0 \quad \text{or} \quad H_0: \ P_1 = P_2$$

The nondirectional alternative hypothesis for a two-tailed test would be

$$H_a: \ P_1 - P_2 \neq 0 \quad \text{or} \quad H_a: \ P_1 \neq P_2$$

The .01 level of significance is used for testing the null hypothesis against the nondirectional alternative in this example.

Step 2: Computing the Test Statistic
The statistic of interest for this example is the difference between the two sample proportions ($p_1 - p_2$). This statistic has a sampling distribution that is normally distributed if the sample sizes are reasonably large. The mean of the sampling distribution is the hypothesized value of the difference between the two population proportions ($P_1 - P_2$). The standard deviation of the sampling distribution is called the *standard error of the difference between proportions*. It is estimated by

$$s_{p_1 - p_2} = \sqrt{pq\left(\frac{1}{n_1} + \frac{1}{n_2}\right)} \tag{12.4}$$

where

p = proportion from both samples exhibiting the characteristic; that is, the common proportion across samples

$\quad = \dfrac{f_1 + f_2}{n_1 + n_2}$

$q = 1 - p$

f_1 = frequency of occurrence in the first sample

f_2 = frequency of occurrence in the second sample

Continuing with the example, we compute

$$p_1 = p_{\text{men}} = \frac{56}{160} = 0.35$$

$$p_2 = p_{women} = \frac{60}{240} = 0.25$$

$$p = \frac{56 + 60}{160 + 240} = \frac{116}{400} = 0.29$$

$$s_{p_1 - p_2} = \sqrt{(0.29)(0.71)\left(\frac{1}{160} + \frac{1}{240}\right)} = 0.046$$

The sampling distribution is normal, so the test statistic is again z and is defined as follows:

$$z = \frac{(p_1 - p_2) - (P_1 - P_2)}{s_{p_1 - p_2}} \qquad (12.5)$$

for the data in this example,

$$z = \frac{(0.35 - 0.25) - (0)}{0.046}$$

$$= \frac{0.10}{0.046}$$

$$= 2.159$$

Step 3: Setting the Criterion for Rejecting H_0: This is a two-tailed hypothesis test using the normal distribution. The .01 level of significance is being used, so the critical values for rejecting the null hypothesis are ±2.58.

Step 4: Deciding Whether to Reject or Retain the Null Hypothesis Because the calculated statistic, 2.159, is less than the critical value of +2.58, our decision is not to reject the null hypothesis. The probability is greater than .01 that the observed difference in sample proportions occurred by chance if the population proportions are equal.

The sampling distribution for the difference between two proportions is normally distributed with a mean of $P_1 - P_2$ and a standard deviation estimated by

$$\sqrt{pq\left(\frac{1}{n_1} + \frac{1}{n_2}\right)}$$

Summary of Hypothesis Testing for Proportions

The hypothesis tests for the value of a single proportion and the difference between two independent proportions follow the same logic and procedures that are used in hypothesis tests for means. Because the hypotheses are now about different parameters (proportions), we must use the standard error for the corresponding statistics and the appropriate sampling distributions. The normal distribution is used as an adequate approximation for the binomial distribution as the sampling distribution for proportions. The general form of the hypotheses and the process used in deciding to reject or not to reject the null hypothesis is the same as before. If the calculated test statistic is greater than the critical value from the table, the null hypothesis is rejected; if not, the null hypothesis is retained.

Hypotheses About Correlation Coefficients

The correlation coefficient was introduced in Chapter 6 as a descriptive statistic, specifically the index of the relationship between two variables. In this section, we use correlation coefficients in inferential tests of hypotheses. We will test hypotheses about the magnitude of a correlation coefficient (one-sample case) and about the difference between two correlation coefficients (two-sample case).

Testing H_0: $\rho = a$

Suppose a technical manual from a test publisher reported that the correlation between scores on the XYZ Aptitude Test and grade point average in technical school was .45. The school counselors wanted to test whether the correlation for students at the local technical school was different from the correlation for the general population of students for whom data were reported in the manual. A random sample of 50 students was selected from the local school population and tested, using the XYZ Aptitude Test. The scores on the test were correlated with those students' grade point averages after one year in technical school, and the correlation coefficient was .38.

Step 1: Stating the Hypothesis
The null hypothesis to be tested in this example is that the correlation between scores on the XYZ Aptitude Test and grade point average is .45. Suppose that this null hypothesis is tested against the nondirectional alternative. In symbols, these hypotheses are

$$H_0: \quad \rho = .45$$

$$H_a: \quad \rho \neq .45$$

The significance level is set at .05.

Step 2: Computing the Test Statistic
The sampling distributions we have used for testing hypotheses about means and proportions are the normal distribution and the t distributions. However, neither of these distributions is appropriate for testing the foregoing null hypothesis (except when the population correlation coefficient (ρ) equals zero; see the next section). This is because the shape of the sampling

distribution of the correlation coefficient actually differs depending on the value of ρ. The sampling distribution is symmetric when $\rho = 0.0$ and gets more and more skewed as ρ nears its limits of $+ 1.0$ or $- 1.0$. Figure 12.1 shows the sampling distributions for $\rho = -.90, 0$, and $+.90$.

The problem of the varying shapes of the sampling distribution of the correlation coefficient was solved by Sir R. A. Fisher. He developed a formula for transforming the correlation coefficient such that the sampling distribution of the transformed correlation coefficient is the normal distribution. This transformed statistic, z_r, is called the *Fisher z transformation*. It is given by the formula

$$z_r = \frac{1}{2} \log_e (1 + r) - \frac{1}{2} \log_e (1 - r) \qquad (12.6)$$

Fortunately, it will not be necessary to use this formula for computing each z transformation. The values of z_r that correspond to the range of possible values of r are given in Table 6 in the Appendix. We will illustrate the use of Table 6 in computing the test statistic for this example.

Fisher also showed that the standard deviation of the sampling distribution of z_r, called the *standard error of the transformed correlation coefficient*, is given by

$$s_{z_r} = \sqrt{\frac{1}{n - 3}} \qquad (12.7)$$

where $n =$ sample size.

The first step in computing the test statistic for this hypothesis is to convert the observed sample correlation coefficient ($r = .38$) to a z_r. Using Table 6 in the Appendix, we find $z_r = 0.400$ corresponding to $r = .38$. It is also necessary to transform the hypothesized value of ρ. Again using Table 6, we find $z_\rho = 0.485$ corresponding to $\rho = .45$. We compute the standard error of z_r by using formula 12.7.

$$s_{z_r} = \sqrt{\frac{1}{n - 3}}$$

$$= \sqrt{\frac{1}{47}}$$

$$= 0.145$$

The general formula for the test statistic can now be applied to testing the hypothesis for this example.

$$z = \frac{z_r - z_\rho}{s_{z_r}} \qquad (12.8)$$

where

$z_r =$ Fisher z transformation of the observed r

$z_\rho =$ Fisher z transformation of the hypothesized ρ

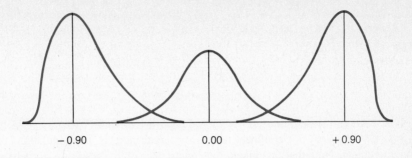

$$-0.90 \qquad 0.00 \qquad +0.90$$

Figure 12.1 **Sampling distribution for the correlation coefficient when** $\rho = -0.90$, **0,** and $+0.90$

For the data given in the example,

$$z = \frac{0.400 - 0.485}{0.145}$$

$$= -0.586$$

Step 3: Setting the Criterion for Rejecting H_0 In this example, the null hypothesis was tested against the nondirectional alternative at the .05 level of significance. Because the sampling distribution of the transformed correlation coefficient is the normal distribution, the critical values of the test statistic are ± 1.96.

Step 4: Deciding Whether to Reject or Retain the Null Hypothesis

The computed value of the test statistic was -0.586 and is between the critical values of ± 1.96. Therefore, the decision is *not* to reject the null hypothesis. The probability is greater than .05 that the observed value of the correlation coefficient occurred by chance if the population correlation coefficient is .45.

Confidence Intervals for ρ Confidence intervals can also be computed for estimating the value of ρ. We first construct the confidence interval around the transformed value of the sample correlation coefficient, using the general formula for confidence intervals. For example, the 95% confidence interval for this example would be computed as follows:

$$CI_{95} = z_r \pm (z_{cv})(s_{z_r}) \qquad (12.9)$$

$$= 0.400 \pm (1.96)(.145)$$

$$= 0.400 \pm .284$$

$$= (0.116, 0.684)$$

This is the 95% confidence interval for z_ρ. In order to find the 95% confidence interval for ρ, we must transform these values of z_r back to r values by using Table 6 in reverse.

$$z_r = 0.116 \rightarrow r = .115$$

$$z_r = 0.684 \rightarrow r = .594$$

Therefore, the confidence interval for the population correlation coefficient is from .115 to .594. We are 95% confident that this interval includes the population correlation coefficient. Note that this interval is consistent with the result we got when we tested H_0: $\rho = .45$. The statistical test was nonsignificant and the interval includes the hypothesized value, .45.

Testing H_0: $\rho = 0$

Behavioral science research sometimes involves an attempt to determine whether a relationship exists between two variables. Exploratory studies often search for relationships regardless of their magnitude. This gives rise to the null hypothesis that the correlation in the population is zero (H_0: $\rho = 0$).

When the correlation coefficient in the population (ρ) is zero, the sampling distribution of r is the t distribution with $n - 2$ degrees of freedom, and the formula for the test statistic is

$$t = r \sqrt{\frac{n - 2}{1 - r^2}} \tag{12.10}$$

However, it is unnecessary to compute this test statistic, because Table 7 in the Appendix contains the values of r that are necessary for rejecting the null hypothesis, H_0: $\rho = 0$. This table contains the minimum-value r for both one-tailed and two-tailed tests, for the different levels of significance, and for the varying degrees of freedom.

For example, how large must the correlation coefficient from a sample of 25 individuals be before we can reject H_0: $\rho = 0$ at the .01 level of significance for a two-tailed test? In Table 7 we go down the .01 column, two-tailed test, until we come to the row with df = 23. The value in the table is .505. Therefore, if the sample correlation coefficient is .505 or greater or $-.505$ or less, we will reject H_0: $\rho = 0$. In other words, if the absolute value of the sample coefficient equals or exceeds the tabled value, H_0: $\rho = 0$ is rejected.

For testing H_0: $\rho = 0$, the sampling distribution of r is the t distribution with $n - 2$ degrees of freedom. Tabled values enable us to test H_0: $\rho = 0$ directly by comparing the sample correlation coefficient with values in the table.

Testing H_0: $\rho_1 = \rho_2$ or $\rho_1 - \rho_2 = 0$

We can also test the hypothesis about the equality of two population correlation coefficients. As in the one-sample case for the correlation coefficient, we use the Fisher z transformation along with the basic procedures and steps of hypothesis testing. Suppose the counselors in a junior high school wanted to test whether the correlation between mathematics achievement and attitude toward mathematics was the same for eighth-grade boys and eighth-grade girls. Random samples of 20 boys and 20 girls were selected and administered an appropriate achievement test and an attitude inventory. A correlation coefficient was computed between the scores on these measures for both the boys and the girls. The correlation coefficient for the boys (r_1) was .48; for the girls (r_2) it was .29. The counselors have hypothesized that the correlation (in the population) is higher for boys than for girls.

Step 1: Stating the Hypothesis
The null hypothesis is that there is no difference between the correlations in the population—that the difference between the population correlations is zero.

$$H_0: \quad \rho_1 - \rho_2 = 0$$

where

ρ_1 = population correlation for boys

ρ_2 = population correlation for girls

The alternative hypothesis in this case is directional, because the counselors hypothesized that the correlation is higher for boys than for girls.

$$H_a: \quad \rho_1 - \rho_2 > 0 \qquad \text{or} \qquad H_a: \quad \rho_1 > \rho_2$$

The significance level is set at .01.

Step 2: Computing the Test Statistic
The statistic of interest is $r_1 - r_2$, the difference between two correlation coefficients from independent samples. However, before computing the test statistic, we must transform each of the sample correlation coefficients using the z transformation. The statistic of interest then becomes $z_{r_1} - z_{r_2}$, the difference between the two transformed correlation coefficients. In this example, the Fisher z transformations are

$$r_1 = .48 \rightarrow z_{r_1} = 0.523$$

$$r_2 = .29 \rightarrow z_{r_2} = 0.299$$

And the difference $(0.523 - 0.299)$ equals 0.224.

The sampling distribution of $(z_{r_1} - z_{r_2})$ is normally distributed, and the standard deviation is given by

$$s_{z_{r_1} - z_{r_2}} = \sqrt{\frac{1}{n_1 - 3} + \frac{1}{n_2 - 3}} \qquad (12.11)$$

This standard deviation is called the *standard error of* $z_{r_1} - z_{r_2}$. For the data of the example;

$$s_{z_{r_1} - z_{r_2}} = \sqrt{\frac{1}{17} + \frac{1}{17}} = 0.343$$

The general formula for the test statistic can now be applied and computed.

$$z = \frac{(z_{r_1} - z_{r_2}) - (0)}{s_{z_{r_1} - z_{r_2}}}$$

(12.12)

For this example, the computed value of the test statistic is

$$z = \frac{(0.523 - 0.229) - (0)}{0.343}$$

$$= 0.64$$

Step 3: Setting the Criterion for Rejecting H_0

This is an example of a one-tailed test at the .01 level of significance, and the normal distribution is the appropriate sampling distribution. Using the table of normal curve areas, Table 1 in the Appendix, we find that the critical value is $+2.33$.

Step 4: Deciding Whether to Reject or Retain the Null Hypothesis

The null hypothesis should be rejected if the computed statistic is greater than $+2.33$, the critical value. The computed statistic was only 0.64, so the null hypothesis is not rejected. The probability is greater than .01 that the observed difference in correlation coefficients occurred by chance, if the population correlation coefficients are equal. We cannot reject the null hypothesis, so the conclusion is that, in these eighth-grade populations, the correlations between mathematics achievement and attitude toward mathematics are the same for boys and girls.

In testing the hypothesis $\rho_1 = \rho_2$, we use the Fisher z transformation to generate the statistic $z_{r_1} - z_{r_2}$. This statistic is normally distributed with a standard error of

$$\sqrt{\frac{1}{n_1 - 3} + \frac{1}{n_2 - 3}}$$

Summary of Hypothesis Testing for Correlation Coefficients

Hypotheses can be tested that involve a single correlation coefficient or the difference between two correlation coefficients. The general procedure for testing hypotheses about correlation coefficients is the same as that for testing hypotheses about means and proportions. However, because the sampling distribution of r changes as ρ changes, a transformation is used converting r to z_r, a statistic with a stable sampling distribution. The one exception to this transformation is testing H_0: $\rho = 0$, in which case the sampling distribution is the t distribution with $n - 2$ degrees of freedom.

Hypotheses About Variances

The remainder of this chapter focuses on testing hypotheses about another statistic, the variance. These tests of hypotheses are different from those for means, proportions, and correlation coefficients in that they involve new sampling distributions: the chi square distributions and the F distributions. The two tests of hypothesis that we will discuss are the one-sample case, H_0: $\sigma^2 = a$, and the two-sample case, H_0: $\sigma_1^2 = \sigma_2^2$. For both of these hypotheses, the logic for hypothesis testing remains the same even though different sampling distributions are used.

Testing H_0: $\sigma^2 = a$

Suppose a school counselor believed that the children who were referred to the counseling center were more homogeneous with regard to their intelligence than the population in general. Stanford Binet Intelligence Test scores were available on the children; the variance in the population for this test is $(16)^2 = 256$. It was decided to test the hypothesis that the population of children referred to the center had a variance of 256 on the IQ test. Rather than collect the data from the files of all children, the counselor selected a random sample of 25 files and computed the variance of these 25 IQ scores. This variance of the sample, s^2, was 214. The test of the hypothesis would proceed as follows.

Step 1: Stating the Hypothesis   The null hypothesis in this example is that the variance for the population of referred children is the same as that of the population in general (the variance described in the manual). The alternative hypothesis is that the referred population's variance is less than that of the general population.

$$H_0: \quad \sigma^2 = 256$$

$$H_a: \quad \sigma^2 < 256$$

The level of significance is set at .05.

Step 2: Computing the Test Statistic   The statistic of interest in this example is the sample variance, s^2. Because the sampling distribution of the sample variance does not conform to either the normal distribution or the t distributions, we must use a different sampling distribution. This is the *chi square (χ^2) distribution*. As in

Student's t distributions, there is more than one χ^2 distribution; the shape of each varies according to the associated number of degrees of freedom. Unlike the t distributions, however, the χ^2 distributions are not symmetric.

The study of the development of the χ^2 distributions as the appropriate sampling distribution is somewhat complex, but it is enough to say at this point that the use of the χ^2 distributions for testing hypotheses about population variances (σ^2) is analogous to the use of either the normal distribution or Student's t distributions. The χ^2 distributions are discussed further in Chapter 15.

The test statistic for testing H_0: $\sigma^2 = a$ is

$$\chi^2 = \frac{(n-1)s^2}{a} \tag{12.13}$$

where

$$s^2 = \frac{\Sigma(X - \overline{X})^2}{n - 1}$$

a = hypothesized value of σ^2

For the data of this example, the calculated value of χ^2 is

$$\chi^2 = \frac{(25 - 1)(214)}{256} = 20.06$$

Step 3: Setting the Criterion for Rejecting H_0 Using the χ^2 distributions is essentially the same as using the t distributions; the appropriate sampling distribution depends on the degrees of freedom. As with the t distributions, if the calculated value of the test statistic falls outside the tabled value for the appropriate number of degrees of freedom and level of significance, the null hypothesis is rejected. However, the χ^2 distributions are *not* symmetric and Table 3 in the Appendix is constructed differently from the table for the t distributions. Table 3 contains the critical values for 30 distributions, those for degrees of freedom 1 through 30, inclusive.[2] (There is a distribution on each row of the table.) Across the top are provided the areas in the distribution to the right of the score indicated. Thus, for 1 degree of freedom, there is 5% of the area to the right of the χ^2 value of 3.841. Note that there is also 5% of the area to the left of 0.00393—that is, 95% of the area is to the right of 0.00393.

Returning to the example, the sample size was 25, so the number of degrees of freedom is $n - 1 = 24$. Recall that the null hypothesis, H_0: $\sigma^2 = 256$, was to be tested against the directional alternative, H_a: $\sigma^2 < 256$, at the .05 level of significance. Because the alternative hypothesis is directional, we are concerned only with the left-hand tail of the χ^2 distribution for 24 degrees of freedom. This distribution is illustrated in Figure 12.2. For this example, we need the χ^2 value that has 5% of the

[2]If the number of degrees of freedom exceeds 30 (seldom the case when using the χ^2 distribution), the statistic $\sqrt{2\chi^2} - \sqrt{2df} - 1$ is computed. This statistic has a sampling distribution that is approximately normal.

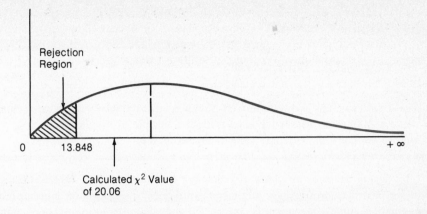

Rejection
Region

0   13.848   + ∞

Calculated χ^2 Value
of 20.06

Figure 12.2 **The relative positions of the critical value and calculated χ^2 value for the example; $\alpha = .05$, χ^2 distribution with df = 24. (One-tailed test)**

area to the left of it, or 95% of the area to the right. From Table 3, we find that the tabled value for 24 degrees of freedom and 95% of the area to the right is 13.848. Thus the critical value of the test statistic is 13.848.

Step 4: Deciding Whether to Reject or Retain the Null Hypothesis In this example the null hypothesis was tested against the directional alternative. Thus, in order for us to reject the null hypothesis, the computed value of χ^2 would have to be *less* than the critical value (13.848). The calculated χ^2 value is *not* less than the critical value, so we *cannot* reject the null hypothesis. The probability is greater than .05 that the observed sample variance occurred by chance if the population variance of the IQ scores of the population of children referred to the counseling center was not less than the variance of the scores in the general population.

> When testing a hypothesis about the value of a population variance (σ^2), we use the corresponding statistic s^2. The test statistic is defined by $(n - 1)s^2/a$ and is distributed as the χ^2 distribution with $n - 1$ degrees of freedom.

Testing H_0: $\sigma_1^2 = \sigma_2^2$

A frequently tested hypothesis about variances is that the variances of two populations are equal ($\sigma_1^2 = \sigma_2^2$). In Chapter 11, for example, we tested the null hypothesis (H_0: $\mu_1 = \mu_2$) under the assumption of homogeneity of variance (the variance of population 1 equals the variance of population 2). To test this hypothesis, we again use a different sampling distribution, the *F distribution*. Like the *t* distributions and χ^2 distributions, the *F* distributions are a family of distributions. These distributions, which are defined by two degrees-of-freedom values, are *not* symmetric and have a range of values from 0 to ∞. The development of the F distribution is beyond the scope of this book. However, using the F distributions in hypothesis testing follows the same logic as using the t distributions and the χ^2 distributions.

The F distributions are a family of sampling distributions that require two degrees-of-freedom values for identifying the specific distribution. The distributions are not symmetric and values range from 0 to $+\infty$.

Consider an example in which a sociologist hypothesizes that women are more homogeneous than men in their attitudes toward the liberalization of abortion laws. An attitudinal survey instrument is constructed and administered to a random sample of 31 men and 21 women. The variance of each sample is calculated; the variance among men's scores (s_1^2) is found to be 45 and the variance among women's scores (s_2^2) to be 36.

Step 1: Stating the Hypotheses

The null hypothesis in this case is that the population variances for men and women are equal. The alternative hypothesis is that the variance for men is greater than the variance for women. In symbols,

$$H_0: \quad \sigma_1^2 = \sigma_2^2 \quad \text{or} \quad \frac{\sigma_1^2}{\sigma_2^2} = 1$$

$$H_a: \quad \sigma_1^2 > \sigma_2^2 \quad \text{or} \quad \frac{\sigma_1^2}{\sigma_2^2} > 1$$

The level of significance is set at .05.

Step 2: Computing the Test Statistic

The statistics of interest for this example are the two sample variances, s_1^2 and s_2^2. The test statistic is the *ratio* of these variances:

$$F = \frac{s_1^2}{s_2^2} \tag{12.14}$$

where s_1^2 is the larger variance. For the example, the variance for the 31 men was larger than the variance for the 21 women. Thus

$$F = \frac{45}{36}$$

$$= 1.25$$

Step 3: Setting the Criteria for Rejecting H_0

The degrees of freedom associated with the sample variance for the men were $(n_1 - 1) = (31 - 1) = 30$. For the women, the degrees of freedom were $(n_2 - 1) = (21 - 1) = 20$. Thus the sampling distribution to be considered is the F distribution with 30 and 20 degrees of freedom. Because two degrees-of-freedom values are required to identify the specific F distribution, the table of critical values is constructed somewhat differently. Table 4 in the Appendix contains numerous F distributions, but it contains only four points in each distribution: those beyond which lie 25%, 10%, 5%, and 1% of the area in the right-hand tail. In general, we find the critical value of the test statistic by locating the value that corresponds to the degrees of freedom for the numerator, to the degrees of freedom for the denominator, and to the appropriate level of significance. For this example, the critical value of F for 30 and 20 degrees of freedom at $\alpha = .05$ is 2.04.

Step 4: Deciding Whether to Reject or Retain the Null Hypothesis

Because the calculated F ratio (1.25) is less than the critical value (2.04), the decision is *not* to reject the null hypothesis. The probability is greater than .05 that the observed ratio of variances occurred by chance if the null hypothesis was true. The sociologist was not able to show that women were more homogeneous than men in their attitudes toward the liberalization of abortion laws.

The null hypothesis of $\sigma_1^2 = \sigma_2^2$ or $\sigma_1^2/\sigma_2^2 = 1.0$ is tested using the F distribution with $(n_1 - 1)$ and $(n_2 - 1)$ degrees of freedom as the sampling distribution for s_1^2/s_2^2.

A Comment About One-Tailed and Two-Tailed Tests The critical values given in the F-distribution table are all for the right-hand tail of the distribution; there are no values less than 1.0. The designation of samples or groups as 1 or 2 is arbitrary, so the hypotheses should be tested by placing the larger sample (s_1^2) in the numerator. This avoids calculated F ratios less than 1.0 and allows us to use the F table directly.

The levels of significance in the table are appropriate for *one-tailed* tests of H_0: $\sigma_1^2 = \sigma_2^2$. For a two-tailed test, we merely double the α level. Had we used a two-tailed test at the .10 level of significance in the example, the critical value of F would have been 2.04, the same value we used for a one-tailed test at the .05 level of significance.

Summary of Hypothesis Testing for Variances

We have introduced two hypothesis tests for variances. The test of H_0: $\sigma^2 = a$ requires the use of the chi square (χ^2) distributions. The test of H_0: $\sigma_1^2 = \sigma_2^2$ requires the F distributions. However, even though two new sampling distributions were introduced, the logic and procedures were analogous to those of all the hypothesis tests that we have studied before.

In the preceding chapter, the parametric assumptions were discussed; one of these is the assumption of homogeneity of (population) variance. That is, the populations from which the two samples are selected have equal variances. This assumption can be tested using the procedure described here for testing H_0: $\sigma_1^2 = \sigma_2^2$. Of course, when this assumption is tested, the researcher is looking for a nonsignificant result. If the two sample variances produce a statistically significant F ratio, the assumption is violated.

Summary

This chapter focuses on a number of statistics for both one-sample and two-sample cases. Statistical tests of hypotheses involving proportions, correlation coefficients, and variances can be very useful and have wide application in the behavioral sciences.

The chain of reasoning for inferential statistics applies whenever we are testing hypotheses, regardless of the parameters and statistics involved. Hence the four steps of hypothesis testing are used throughout. However, in addition to the procedures for specific statistical tests, some new concepts are introduced in this chapter. One such concept is the transformation of a statistic with a sampling distribution that changes when the parameter changes to a statistic with a stable sampling distribution. The correlation coefficient is transformed by the Fisher z transformation to a statistic, z_r, that is normally distributed.

Two new sampling distributions are also introduced, the chi square (χ^2) distribution and the F distribution. Both are families of distributions; both are not symmetric. The chi square distribution requires one degrees-of-freedom value and the F distribution requires two such values to locate the specific distribution.

Key Concepts

Proportion
Standard error of the
   difference between proportions
Fisher z transformation

Standard error of the
   transformed correlation coefficient
Chi square (χ^2) distribution
F distribution

Exercises

12.1   A recent survey of a 10-county area showed that 7% of the adult population had sought professional help with psychological problems. A psychologist in one of the counties hypothesized the proportion seeking help in that county to be less than the proportion seeking help in the entire 10-county area. In a random sample of 100 adults, it was found that 4 had sought professional help. Determine whether the proportion of adults in that county who sought professional help was less than the proportion seeking help in the 10-county area (Use $\alpha = .01$.)

a. State the hypothesis.
b. Compute the test statistic.
c. Set the criteria for rejecting the null hypothesis.
d. Decide whether to reject or to retain the null hypothesis.
e. Develop CI_{99}.

12.2 A researcher wishes to test the hypothesis that more business majors in college hold part-time jobs than do engineering majors. The researcher selects a random sample of 10% of the business and engineering majors at a nearby college and finds that 83 of the 210 business students and 48 of the 160 engineering students have part-time jobs. Determine whether the difference between the two proportions is statistically significant. (Use $\alpha = .05$.)

a. State the hypothesis.
b. Compute the test statistic.
c. Set the criteria for rejecting the null hypothesis.
d. Decide whether to reject or to retain the null hypothesis.

12.3 Is a correlation coefficient of .35 based on a random sample of 40 people significantly larger than zero, using (1) the .05 level of significance and (2) the .01 level of significance? Assume that a researcher has a sample of size 72. (1) How large must r be in order for us to conclude that the population correlation is greater than zero? (2) How large must the absolute value of r be in order for us to conclude that the population correlation is different from zero? (Use the .01 level of significance.)

12.4 A research study reported that the correlation between test anxiety and test performance of senior high school students was $-.21$. A teacher correlated the scores on a six-week test with scores from a test anxiety measure for a class of 30 students. The correlation coefficient (r) was $-.10$. Assuming that the class was a random sample of the senior high school population, test the hypothesis that the correlation in the population was $-.21$. (Use $\alpha = .05$.)

a. State the hypothesis.
b. Compute the test statistic.
c. Set the criterion for rejecting the null hypothesis.
d. Decide whether to reject or to retain the null hypothesis.
e. Develop CI_{95}.

12.5 Test developers at a commercial publisher were concerned that an academic aptitude test being developed was not equally valid for bilingual and monolingual students. The correlation coefficient for a random sample of 125 bilingual students was $r = .38$, whereas that for a random sample of 200 monolingual students was .43. Determine whether the difference between the two correlation coefficients is statistically significant. (Use $\alpha = .05$.)

a. State the hypothesis.

b. Compute the test statistic.

c. Set the criterion for rejecting the null hypothesis.

d. Decide whether to reject or to retain the null hypothesis.

12.6 The manual of a widely used achievement test stated that the standard deviation for the norm group was 12. When the test was given to a random sample of 26 students from Parkside Elementary School, their standard deviation was 15. Determine whether the variance of the student population at Parkside School is the same as the variance of the norm group. (Use the .10 level of significance.)

a. State the hypothesis.

b. Compute the test statistic.

c. Set the criterion for rejecting the null hypothesis.

d. Decide whether to reject or to retain the null hypothesis.

12.7 From a random sample of 25 individuals, a researcher randomly assigned 12 to group A and 13 to group B and administered different treatments. In some preliminary measurements, it was found that the variance on the dependent variable was 262 for group A and 195 for group B. Determine whether $\sigma_1^2 = \sigma_2^2$, using a two-tailed test at the .10 level of significance.

a. State the hypothesis.

b. Compute the test statistic.

c. Set the criterion for rejecting the null hypothesis.

d. Decide whether to reject or to retain the null hypothesis.

12.8 For a sample of size 30, is a correlation coefficient of $r = .20$ significantly different from zero at the .05 level of significance?

a. State the hypothesis.

b. Compute the test statistic.

c. Set the criterion for rejecting the null hypothesis.

d. Decide whether to reject or to retain the null hypothesis.

e. Develop CI_{95}.

12.9 For the data given in Exercise 12.8, is a correlation coefficient of $r = .20$ significantly different from a hypothesized value of .35?

a. State the hypothesis.

b. Compute the test statistic.

c. Set the criterion for rejecting the null hypothesis.

d. Decide whether to reject or to retain the null hypothesis.

e. Is the confidence interval constructed in part e of Exercise 12.8 consistent with the results of these tests of hypothesis?

12.10 A congressman decided to vote for a certain bill that he personally opposed if more than 60% of his constituents in the district favored the bill. A survey of 200 randomly selected voters showed that 65% favored the bill. Test the appropriate hypothesis in order to determine how the congressman should vote. (Use $\alpha = .05$.)

a. State the hypothesis.
b. Compute the test statistic.
c. Set the criterion for rejecting the null hypothesis.
d. Decide whether to reject or to retain the null hypothesis.

12.11 Suppose that, in Exercise 12.10, 66% of the sample favored the bill. How should the congressman vote? (Use $\alpha = .05$.)
a. State the hypothesis.
b. Compute the test statistic.
c. Set the criterion for rejecting the null hypothesis.
d. Decide whether to reject or to retain the null hypothesis.

12.12 A researcher was about to test the hypothesis H_0: $\mu_1 - \mu_2 = 0$ using the t test for independent samples. One of the assumptions of this t test is that $\sigma_1^2 = \sigma_2^2$. The two samples had the following summary statistics.

	Group A	Group B
$\overline{X}$	64	72
s^2	48	22
n	15	10

Is the researcher violating the assumption of homogeneity of variance? (Use $\alpha = .10$.)
a. State the hypothesis.
b. Compute the test statistic.
c. Set the criterion for rejecting the null hypothesis.
d. Decide whether to reject or to retain the null hypothesis.

▦ Key Strokes for Selected Exercises

	Value	Key Stroke	Display

12.1 Step 1: Determine s_p.

	Value	Key Stroke	Display
	.04 (p)	×	0.04
	.96 (q)	÷	0.0384
	100 (n)	=	0.000384
		$\sqrt{}$	0.019595 = s_p
		STO	0.019595

Step 2: Determine z.

	Value	Key Stroke	Display
	.04 (p)	−	0.04
	.07 (P)	÷	−0.03
		RCL	0.019595
		=	−1.531 = z

Step 3: Determine CI_{95}.

	RCL	0.019595
	×	0.019595
2.576(z_{cv})	=	0.050479
	STO	0.050479
.04 (p)	−	0.04
	RCL	0.050479
	=	-0.010479
.04 (p)	+	0.04
	RCL	0.050479
	=	0.090479

12.2 Step 1: Determine p_1 and p_2.

83 (f_1)	÷	83	
210 (n_1)	=	0.3952	$= p_1$
		(save)	
48 (f_2)	÷	48	
160 (n_2)	=	0.3000	$= p_2$
		(save)	

Step 2: Determine $s_{p_1-p_2}$.

83 (f_1)	+	83	
48 (f_2)	÷	131	
370 ($n_1 + n_2$)	=	0.3541	$= p$
	STO	0.3541	
	+/−	-0.3541	
	+	-0.3541	
1	=	0.6359	$= q$
	×	0.6359	
	RCL	0.3541	
	=	0.2287	$= pq$
	STO	0.2287	
210 (n_1)	1/X	0.004762	
	+	0.004762	
160 (n_2)	1/X	0.00625	
	=	0.011012	
	×	0.011012	
	RCL	0.2287	
	=	0.002518	
	$\sqrt{}$	0.05018	$= s_{y_1 - p_2}$
	STO	0.05018	

Step 3: Determine z.

.3952 (p_1)	−	0.3952	
.3000 (p_2)	÷	0.0952	
	RCL	0.05018	
	=	1.897	$= z$

12.4 Step 1: Determine s_{z_r}.

27 $(n − 3)$	1/X	0.037037	$= s_{z_r}$
	$\sqrt{}$	0.19245	
	STO	0.19245	

Step 2: Determine z.

.10 (z_r)	+/−	−0.10	
	−	−0.10	
.213 (z_p)	+/−	−0.213	
	=	0.113	
	÷	0.113	
	RCL	0.19245	
	=	0.5872	$= z$

Step 3: Determine CI_{95}.

	RCL	0.19245
	×	0.19245
1.96	=	0.37720
	STO	0.37720
.10	+/−	−0.10
	−	−0.10
	RCL	0.37720
	=	−0.47720
.10	+/−	−0.10
	+	−0.10
	RCL	0.37720
	=	0.27720

Step 4: Convert to r values.

13

One-Way Analysis
of Variance

Suppose that an experiment is being conducted in an animal exercise laboratory to determine the effects of five different drug dosages on the performance of rats. Five groups of rats are randomly assigned to the drug dosages; they are administered the drugs, and their performance is measured. Mean performance can be computed for the groups and the effects of the different drug dosages are manifest in the differences between means. The researchers know that a difference between two sample means can be tested using a t test, but they want to compute a statistical test involving all five means simultaneously. What should they do?

An initial reaction might be to compute t tests for all combinations of two means. But such an approach would involve numerous t tests—10 in this situation. Using any one of the many computer programs presently available, the numerous t tests can be computed very easily and quickly.

However, such a strategy leads to a major problem; namely, that the *Type I error rate* (the probability of rejecting a true hypothesis) is increased drastically. In the foregoing example, if the researcher computed all 10 possible t tests and used the .05 level of significance for each, the Type I error rate would be determined by

$$1 - (1 - \alpha)^c$$

where

α = level of significance for each of the separate tests
c = number of comparisons

In the example, the Type I error rate would be

$$1 - (1 - .05)^{10} = .40$$

Thus the probability of making at least one Type I error in the comparisons of the five means is *not* .05, as one might expect. Instead the probability of making a Type I error is .40.

The problem with computing multiple t tests on data from several samples is that doing so increases the Type I error rate beyond the intended α level.

The statistical procedure that can be used to test simultaneously the equality of k sample means while maintaining the Type I error rate at the established α level is the Analysis of Variance (ANOVA). ANOVA is one of the more widely used statistical procedures in behavioral science research. In this chapter, we introduce ANOVA and the post hoc procedures that can be used following the ANOVA.

The Variables in ANOVA

In an ANOVA, there are two kinds of variables, independent and dependent. (These terms are defined in Chapter 1.) The *independent variables* are for the most part categorical variables and are used to form the groupings of observations. If the data from several different samples are used, say the five groups of rats assigned to the different drug dosages, the variable in terms of which the group is classified is the independent variable. In this example, drug dosage is the independent variable, and we say that there are five *levels* of the independent variable. When only one independent variable is considered in the ANOVA, we refer to it as a *one-way ANOVA*, regardless of the number of levels. On the other hand, when we have two or more independent variables, we refer to the ANOVA as a "factorial" ANOVA. Specifically, for two independent variables, we have two-way ANOVA, which will be discussed in the next chapter.

The *dependent variable* is the variable that is, or is presumed to be, affected by the manipulation of the independent variable. In our example, the performance scores for the five groups of rats are the dependent variables and the differences among the mean scores for the five groups are presumed to be due partially to the different drug dosages. In ANOVA, the hypothesis is that the mean performance in the population is the same for all groups (equality of population means). Differences among the observed sample means are tested for statistical significance.

Notation in ANOVA

Because data in an ANOVA consist of scores in k comparison groups, with $k \geq 2$, they can be described for the general case as shown in Table 13.1. Such a data array is called a *data layout*. The comparison groups are labeled group 1, group 2, and so forth, up to group k. Group j, or the jth column, represents the general case.

The scores have two subscripts. The first subscript indicates the individual observation in the group, and the second subscript indicates which group the observation is in. Thus X_{11} stands for score 1 in group 1, X_{23} stands for score 2 in group 3, and so forth. The symbol X_{ij} stands for the general case, or the ith score in the jth group. There are n_1 scores in group 1, n_2 scores in group 2, and so on, through n_k, which is the number of scores in group k. The numbers of scores for the different groups may or may not be equal, but $n_1 + n_2 + \cdots + n_k = N$, which is the total number of observations in the data set.

The mean of group 1 can be found by finding T_1, the sum of the scores in group 1, and then dividing T_1 by n_1, the number of scores in that group. Such a notational system may seem a little excessive, but it simplifies formulas and computations that we will encounter later.

Data Layout—An Example

Suppose research is being conducted concerning graduate students in four different disciplines with regard to their opinions about the need for grants to support student research. Thirty students are randomly selected from the graduate student populations in these disciplines: 7 from chemistry, 10 from sociology, 6 from engineering, and 7 from history. These students then respond to a 20-item scale that measures their opinions in such a way that a higher score on the scale reflects a greater perceived need for student grants. The data layout for this example is presented in Table 13.2.

			Group			
1	2	3	$\cdots$	j	$\cdots$	k
X_{11}	X_{12}	X_{13}	$\cdots$	X_{1j}	$\cdots$	X_{1k}
X_{21}	X_{22}	X_{23}	$\cdots$	X_{2j}	$\cdots$	X_{2k}
X_{31}	X_{32}	X_{33}	$\cdots$	X_{3j}	$\cdots$	X_{3k}
.	.	.		.		.
.	.	.		.		.
.	.	.		.		.
X_{i1}	X_{i2}	X_{i3}	$\cdots$	X_{ij}	$\cdots$	X_{ik}
.	.	.		.		.
.	.	.		.		.
X_{n_11}	X_{n_22}	X_{n_33}	$\cdots$	X_{n_jj}	$\cdots$	X_{n_kk}
T_1	T_2	T_3	$\cdots$	T_j	$\cdots$	T_k
$\overline{X}_1$	$\overline{X}_2$	$\overline{X}_3$	$\cdots$	$\overline{X}_j$	$\cdots$	$\overline{X}_k$

where

$$T_j = \sum_{i=1}^{n_j} X_{ij} \text{ (column sum)}$$

$$\overline{X}_j = \frac{T_j}{n_j} \text{ (column mean)}$$

Note: $T = \sum_{j=1}^{k} T_j$ and $\overline{X} = \frac{T}{N}$

where

$\overline{X}$ = grand mean

$$N = \sum_{j=1}^{k} n_j$$

Table 13.1   General data layout for ANOVA

The Logic of ANOVA

The null hypothesis tested in ANOVA is that the population means from which the k samples are selected are equal. Symbolically,

$$H_0: \quad \mu_1 = \mu_2 = \cdots \mu_k$$

where k is the number of levels of the independent variable. In our example of opinion regarding support for student research, the null hypothesis is

$$H_0: \quad \mu_1 = \mu_2 = \mu_3 = \mu_4$$

		Major		
Chemistry	Sociology	Engineering	History	
20	12	20	12	
18	14	20	10	
14	18	14	16	
12	10	18	11	
19	13	19	15	
18	12	17	11	
13	10		13	
	15			
	13			
	9			

n_j	7	10	6	7	$N = 30$
T_j	114	126	108	88	$T = 436$
$\overline{X}_j$	16.286	12.600	18.000	12.571	$\dfrac{T^2}{N} = 6336.53$
$\sum\limits_{i=1}^{n_j} X_{ij}^2$	1918	1652	1970	1136	$\sum\limits_{j=1}^{k}\sum\limits_{i=1}^{n_j} X_{ij}^2 = 6676$
$\dfrac{T_j^2}{n_j}$	1856.57	1587.60	1944.00	1106.29	$\sum\limits_{j=1}^{k}\left(\dfrac{T_j^2}{n_j}\right) = 6494.46$

Table 13.2   Scores of students on an opinion scale

The alternative hypothesis is that the population means differ; more specifically, that at least one population mean differs from the rest. Symbolically,

$$H_a: \quad \mu_i \neq \mu_j \text{ for some } i, j$$

The procedure for testing this null hypothesis against the alternative is precisely what the name implies, *the analysis of the variance of the scores on the dependent variable*. In this analysis, the total variation of the scores is partitioned into two components. The two component parts are (1) the variation of the scores *within* the k groups and (2) the variation *between* (among) the group means and the mean of the total group (referred to as the grand mean).

Suppose that the distribution of scores for the four samples is as illustrated in Figure 13.1. In order to illustrate graphically the partitioning of the total variation of the 30 scores, suppose (1) that the scores in each of the four groups are normally distributed; (2) that there is only minor overlapping of scores among the groups; and (3) that the variance of the scores for each group is the same.

Consider the total variation for the set of 30 scores. The computation of the variance would involve summing the squared deviations from the mean of the total group and

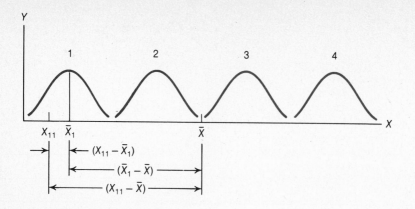

Figure 13.1 Partitioning the variation into between-groups and within-groups components

then dividing by 29, that is, $\Sigma(X - \bar{X})^2/29$. For subject 1 in group 1 (X_{11}), the deviation score $(X_{11} - X)$ would be this subject's contribution to the total variation of the scores. Note, in Figure 13.1, that this deviation score can be partitioned into two component parts. The first component is the deviation of X_{11} from the mean of group 1 $(X_{11} - \bar{X}_1)$. The second component is the deviation of the group 1 mean from the grand mean $(\bar{X}_1 - \bar{X})$. The first component reflects, in the dependent variable, the inherent or natural variation due to individual differences observed among the subjects exposed to the same treatment. This variation among all subjects within a group is referred to as the *within-groups* variation; we will denote this variation as s_w^2. It is assumed that within-group variation of a similar magnitude would exist in each of the four groups. This variation within any one group is a function of the specific subjects selected at random and included in the group. Therefore, we can attribute variation within groups to random sampling fluctuation.

The second component is concerned with the differences among group means. Even if there were absolutely no treatment effect, it would be unlikely that the sample means for the four groups would be identical. A more reasonable expectation is that the group means will differ, even without treatment, simply as a function of the individual differences of the subjects. For example, if we reassign the subjects to different groups at random *prior* to the treatment, we would expect somewhat different group means. Thus, we would expect the group means to vary somewhat due to the random selection (assignment) process in the formulation of the groups.

If, in addition, different treatments are applied to the different groups that do have an effect on the dependent variable, we can expect even larger differences among the group means. Thus, the *between-groups* variation, denoted s_B^2, would reflect variation due to the treatment *plus* the variation attributable to the random process by which the subjects were selected and assigned to groups.

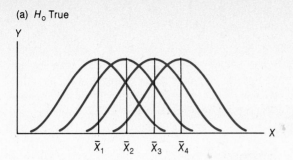

(a) H_o True

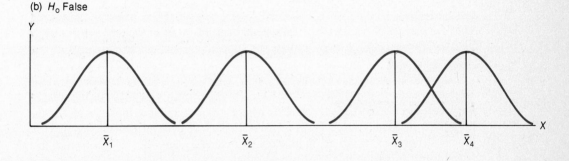

(b) H_o False

Figure 13.2 Illustration of sample group means when H_o is true (a) and H_o is false (b)

In one-way ANOVA, the total variance can be partitioned into two sources: (1) variation of scores "within" groups and (2) variation "between" the group means and the grand mean. Both of these sources reflect variation due to random sampling. In addition, the between-group variation reflects variation due to differential treatment effects.

Suppose the null hypothesis is true, that is, $\mu_1 = \mu_2 = \mu_3 = \mu_4$. This is graphically illustrated in Figure 13.2(a). Recall that the within-groups variation (s_W^2) is that source of natural variation attributable to individual differences and random sampling fluctuation. Thus we can consider this variance an estimate of the population variance that we will call *error variance*, or σ_e^2. It is error variance because it is not controlled.

Note that in Figure 13.2(a), there are slight differences among the group means even though the null hypothesis is true. Recall that the between-groups variation (s_B^2) contains variation due to differential treatment effects plus variation due to random sampling. Since we have assumed that the null hypothesis is true, there is no treatment effect and s_B^2 is also an estimate of σ_e^2. Therefore, the ratio of these two estimates

$(s_B{}^2/s_W{}^2)$ is approximately 1.00. (Due to random sampling, we would not expect this ratio to be exactly equal to 1.00.)

Now suppose the null hypothesis is false, or, $\mu_i \neq \mu_j$ for some i,j. This is graphically illustrated in Figure 13.2(b). The within-groups variation would again be an estimate of $\sigma_e{}^2$, but the between-groups variation would be an estimate of $\sigma_e{}^2$ *plus* the variation due to the difference of the population means as the result of differential treatment effects. We denote this additional source of variation[1] as $\sigma_t{}^2$. In this case, the ratio of $s_B{}^2/s_W{}^2$ would estimate:

$$\frac{\sigma_e{}^2 + \sigma_t{}^2}{\sigma_e{}^2}.$$

As can be seen when the null hypothesis is not true, this ratio is greater than 1.00. Thus we can ask the question, "How much can this ratio of variances depart from 1.00 before we can no longer attribute the between-groups variance to random sampling fluctuation?"

The statistical test of the null hypothesis in ANOVA uses this ratio of the between-groups variation to the within-groups variation and the test uses the F-distribution as the underlying distribution. The specific procedures are similar to those for testing the hypothesis of no difference between the variances of two independent samples. In ANOVA, however, if the null hypothesis is true (H_0: $\mu_1 = \mu_2 = \mu_3 = \mu_4$), this is equivalent to saying that the between-group variation is only an estimate of error variance. In other words, if the four population means are equal, then the variance between them is zero.

The null hypothesis in ANOVA is tested using the ratio of the two variance estimates $s_B{}^2/s_W{}^2$.

Partitioning the Sum of Squares

Although ANOVA has been described as a process of partitioning the total variation, the procedure actually involves partitioning the sum of squared deviations around the grand mean, $\Sigma (X - \overline{X})^2$. Consider again the deviation score for subject 1 in group 1 and the partitioning of that deviation score into two components:

$$(X_{11} - \overline{X}) = (X_{11} - \overline{X}_1) + (\overline{X}_1 - \overline{X})$$

To generalize this partitioning, we can consider the partitioning of the deviation score for the ith subject in the jth group,

$$(X_{ij} - \overline{X}) = (X_{ij} - \overline{X}_j) + (\overline{X}_j - \overline{X}) \qquad (13.1)$$

[1]The symbol σ_t^2 is used here only in the conceptual development and is without specific algebraic or statistical meaning. The elements of the numerator of the ratio will be discussed later in the chapter.

In order to obtain the total sum of squared deviations (SS$_T$), we must square each deviation score, $(X_{ij} - \overline{X})^2$, and then sum across all subjects and across all groups[2]; symbolically,

$$\left(SS_T = \sum_{j=1}^{k} \sum_{i=1}^{n_j} (X_{ij} - \overline{X})^2 \right.$$ (13.2)

The two terms on the right-hand side of equation 13.1 also have to be squared and summed across all observations. The algebraic derivation is found in Appendix D with the following result:

$$\sum_{j=1}^{k} \sum_{i=1}^{n_j} (X_{ij} - \overline{X})^2 = \sum_{j=1}^{k} \sum_{i=1}^{n_j} (X_{ij} - \overline{X}_j)^2 + \sum_{j=1}^{k} n_j(\overline{X}_j - \overline{X})^2$$ (13.3)

or

Total sum of squares = Within sum of squares + Between sum of squares

Equation 13.3 illustrates partitioning the total sum of squares (SS$_T$) into the within-group sum of squares (SS$_W$) and the between-group sum of squares (SS$_B$). Consider

$$SS_W = \sum_{j=1}^{k} \sum_{i=1}^{n_j} (X_{ij} - \overline{X}_j)^2$$ (13.4)

which is the sum of squared deviations of the original scores from the respective group mean summed over the k groups. This summing across the k groups is often referred to as "pooling" the sum of squares. Now consider

$$SS_B = \sum_{j=1}^{k} n_j(\overline{X}_j - \overline{X})^2$$ (13.5)

which is the weighted sum of squared deviations of the group means around the grand mean and is a measure of the differences among groups. If there is a substantial difference among the group means, the group means will differ considerably from the grand mean and SS$_B$ will be substantially larger than zero.

In one-way ANOVA, the total sum of squares (SS$_T$) is partitioned into two components: (1) the sum of squares within groups (SS$_W$); and (2) the sum of squares between groups (SS$_B$).

---

[2]For the student with minimal mathematical background, the use of this double summation notation is no doubt confusing. We do not use it here purposely to confuse you, but only for efficiency in notation. The two summation signs are read from the inside out.

$\sum_{i=1}^{n_j}$ says to sum the squared deviations in the first group. $\sum_{j=1}^{k}$ says to do the same for each of the k groups and then sum across all the k groups. This process will become clearer as we work through the example of this chapter.

Two Variance Estimates

In ANOVA, the null hypothesis is tested using the ratio of two variance estimates $(s_B{}^2/s_W{}^2)$. Thus far, we do not have the variance estimates; we have only sum of squares (SS_B and SS_W). Recall from Chapter 3 that the sum of squared deviations is the numerator of the estimate of the population variance. Dividing this sum of squares by the degrees of freedom associated with this estimate $(n - 1)$, we will have the unbiased estimate of the population variance. In ANOVA, the between-groups and within-groups variance estimates are found by dividing SS_B and SS_W by the degrees of freedom associated with each of these estimates. The resulting two variance estimates are commonly called *mean squares* and are denoted MS_B and MS_W, respectively.

Consider SS_B. It is calculated using the deviation of the group means from the grand mean, weighted for the number of observations in each group. Since there are k groups and since the grand mean is calculated, there are $k - 1$ degrees of freedom associated with the between-groups mean square (MS_B), or

$$s_B^2 = MS_B = \frac{SS_B}{k - 1} \tag{13.6}$$

Now consider SS_W. The *pooled estimate of the population variance* is the sum of squared deviation within each group, pooled across groups and then divided by the sum of the degrees of freedom for each group:

$$\frac{\Sigma(X_{i_1} - \overline{X}_1)^2 + \Sigma(X_{i_2} - \overline{X}_2)^2 + \ldots + (X_{ij} - \overline{X}_j)^2}{(n_1 - 1) + (n_2 - 1) + \ldots + (n_j - 1)}$$

Note that the numerator of this expression is SS_W and total degrees of freedom associated with this estimate are

$$(n_1 - 1) + (n_2 - 1) + \ldots + (n_j - 1) = N - k$$

where

$N = n_1 + n_2 + \ldots + n_j$ (the sum of the observations in each sample)
$k =$ the number of groups.

Therefore:

$$s_W{}^2 = MS_W = \frac{SS_W}{N - k} \tag{13.7}$$

A mean square (MS) is a variance estimate. The degrees of freedom associated with the two estimates of the population variance are $k - 1$ and $N - k$ for MS_B and MS_W, respectively.

An ANOVA Example

At this point, let us review briefly the chain of reasoning for testing the null hypothesis in ANOVA. The data for ANOVA consist of k (two or more) random samples that have been assigned at random to the k levels of the independent variable. In an experiment, the different experimental treatments constitute the levels of the independent variable. Following the experimental treatment, the samples exposed to the different experimental treatments reflect respective populations, and the sample means are thus the estimates of the means of the respective populations.

Step 1: Stating the Hypothesis

As mentioned above, the null hypothesis tested using ANOVA is that the population means from which the k samples were selected are equal. For the example of opinion regarding support for student research, the null hypothesis is as follows:

$$H_0: \quad \mu_1 = \mu_2 = \mu_3 = \mu_4$$

The alternative hypothesis is that *at least* one population mean differs from the rest; symbolically,

$$H_a: \quad \mu_i \neq \mu_j \quad \text{for some } i, j$$

Step 2: Computing the Test Statistic

The test statistic for ANOVA is the F ratio of the two variance estimates

$$F = \frac{MS_B}{MS_W} \tag{13.8}$$

The conventional way of reporting the results of ANOVA is in a standard tabular format; this format is outlined below:

SUMMARY ANOVA

Source	SS	df	MS	F	F_{cv}
Between					
Within					
Total					

Note that the sums of squares (SS), degrees of freedom (df), and mean squares (MS) are reported along with the F ratio and the critical value of F.

Computational Formulas

In Chapter 3, we defined the sample variance of a distribution of scores as the sum of squared deviations (SS) divided by $n - 1$. We also indicated that a raw score formula was available and directly derivable from the deviation formula. This derivation is found in Appendix D. Analogously, the computational formulas for SS_B, SS_W, and SS_T can be derived, and these formulas are presented below. First, to simplify the formulas, it is convenient to denote the sum of all the scores in the jth group as T_j:

$$T_j = \sum_{i=1}^{n_j} X_{ij}$$

(13.9)

Secondly, denote the sum of all observations in all k groups as T:

$$T = \sum_{j=1}^{k} \sum_{i=1}^{n_j} X_{ij}$$

(13.10)

Now consider SS_B, SS_W, and SS_T:

SS_B = sum of squares *between groups*

$$= \sum_{j=1}^{k} n_j(\bar{X}_j - \bar{X})^2$$

[handwritten: SSB]

[handwritten: Wheeler]
$$SS_B = \frac{(\Sigma X_1)^2}{n_1} + \frac{(\Sigma X_2)^2}{n_2} + \frac{(\Sigma X_3)^2}{n_3} + \frac{(\Sigma X_4)^2}{n_4} - \frac{(\Sigma X)^2}{N}$$

$$= \sum_{j=1}^{k} \left(\frac{T_j^2}{n_j}\right) - \frac{T^2}{N}$$

(13.11)

SS_W = sum of squares *within groups*

[handwritten: SSW]

[handwritten: Wheeler]
$$SS_W = SS_T - SS_B$$

$$= \sum_{j=1}^{k} \sum_{i=1}^{n_j} (X_{ij} - \bar{X}_j)^2$$

$$= \sum_{j=1}^{k} \sum_{i=1}^{n_j} X_{ij}^2 - \sum_{j=1}^{k} \left(\frac{T_j^2}{n_j}\right)$$

(13.12)

SS_T = sum of squares *total*

[handwritten: SST]

[handwritten: $F = \frac{MSB}{MSW}$]

[handwritten: Wheeler]
$$SS_T = \Sigma X^2 - \frac{(\Sigma X)^2}{N}$$

$$= \sum_{j=1}^{k} \sum_{i=1}^{n_j} (X_{ij} - \bar{X})^2$$

$$= \sum_{j=1}^{k} \sum_{i=1}^{n_j} X_{ij}^2 - \frac{T^2}{N}$$

(13.13)

These formulas are summarized in Table 13.3.

Applying these raw score formulas to the data for our example (see Table 13.2) yields the following:

[handwritten: P 270]

[handwritten: T = Page 214]

	Definition Formula	Computational Formula
Between Groups	$\displaystyle\sum_{j=1}^{k} n_j(\overline{X}_j - \overline{X})^2$	$\displaystyle\sum_{j=1}^{k}\left(\frac{T_j^2}{n_j}\right) - \frac{T^2}{N}$
Within Groups	$\displaystyle\sum_{j=1}^{k}\sum_{i=1}^{n_j}(X_{ij} - \overline{X}_j)^2$	$\displaystyle\sum_{j=1}^{k}\sum_{i=1}^{n_j}X_{ij}^2 - \sum_{j=1}^{k}\left(\frac{T_j^2}{n_j}\right)$
Total	$\displaystyle\sum_{j=1}^{k}\sum_{i=1}^{n_j}(X_{ij} - \overline{X})^2$	$\displaystyle\sum_{j=1}^{k}\sum_{i=1}^{n_j}X_{ij}^2 - \frac{T^2}{N}$

Table 13.3 Summary of formulas for ANOVA

$$SS_T = \sum_{j=1}^{k}\sum_{i=1}^{n_j}X_{ij}^2 - \frac{T^2}{N}$$

$$= 6676 - \frac{(436)^2}{30}$$

$$= 6676 - 6336.53$$

$$= 339.47$$

$$SS_B = \sum_{j=1}^{k}\left(\frac{T_j^2}{n_j}\right) - \frac{T^2}{N}$$

$$= \left[\frac{(114)^2}{7} + \frac{(126)^2}{10} + \frac{(108)^2}{6} + \frac{(88)^2}{7}\right] - \frac{(436)^2}{30}$$

$$= 6494.46 - 6336.53$$

$$= 157.93$$

$$SS_W = \sum_{j=1}^{k}\sum_{i=1}^{n_j}X_{ij}^2 - \sum_{j=1}^{k}\left(\frac{T_j^2}{n_j}\right)$$

$$= 6676 - 6494.46$$

$$= 181.54$$

Computing the mean squares, we find

$$MS_B = \frac{157.93}{3} = 52.64$$

$$MS_W = \frac{181.54}{26} = 6.98$$

The F ratio is then found to be

$$F = \frac{52.64}{6.98} = 7.54$$

These calculations are summarized in Table 13.4.

$$\frac{SSB}{K-1} = MSB$$

$$\frac{SSW}{N-K} = MSW$$

Source of Variation	Sum of Squares (SS)	Degrees of Freedom (df)	Mean Square (MS)	F	F_{cv}
Between	157.93	3	52.64	7.54	2.98
Within	181.54	26	6.98		
Total	339.47	29			

Table 13.4 Summary table for the example ANOVA

Step 3: Setting the Criterion for Rejecting the Null Hypothesis

The underlying distribution of the F ratio is the F distribution. There is a family of F distributions, each one a function of the degrees of freedom associated with the two variance estimates. In ANOVA, the two variance estimates are MS_B and MS_W. There are $k - 1$ degrees of freedom associated with MS_B and $N - k$ degrees of freedom associated with MS_W. Thus, the underlying distribution of the F ratio (MS_B/MS_W) is the F distribution with $k - 1$ and $N - k$ degrees of freedom, or $F_{(k-1, N-k)}$. Table 4 in the Appendix contains the critical values of F at the .05 and .01 levels for a wide range of degrees of freedom. The degrees of freedom for MS_B are read across the column headings from left to right; the degrees of freedom for MS_W are read down the rows on the left side. The intersection of the row for $N - k$ degrees of freedom and the column for $k - 1$ degrees of freedom is the critical value of F.

In our example, there are $k - 1$, or 3, degrees of freedom associated with MS_B and $N - k$, or 26, degrees of freedom associated with MS_W. From Table 13.4, we find the critical value of F to be 2.98 for $\alpha = .05$; the critical value for $\alpha = .01$ is 4.64.

Step 4: Deciding Whether to Reject or to Retain the Null Hypothesis

If the computed F value exceeds the critical value we reject H_0. The MS_B contains more variance than we can attribute to random variance, so some of this variance must be due to difference in population means. We say the F value is statistically significant and the H_0 is not tenable. If the computed F value does not exceed the critical value in the table, we fail to reject H_0.

In our example the calculated F value was 7.54 while the critical or tabled F value was 2.98. Therefore, the proper decision is to reject the null hypothesis. The probability is less than .05 that the observed F value would occur by chance if the null hypothesis is true. Another way of stating the probability is that *the probability that the observed sample means would have appeared by chance, if the population means are equal, is less than .05.*

MSB
k-1

msw
N-k
Table 4

We decide about the H_0 by comparing the computed F value with the critical F value given in the table. If the computed value exceeds the tabled value, H_0 is rejected; if not H_0 is retained.

Assumptions Underlying ANOVA

A number of assumptions are required for applying ANOVA. They are basically the parametric assumptions.

1. The samples are random samples from defined populations.
2. The samples are independent.
3. The dependent variable is measured on at least an interval scale.
4. The dependent variable is normally distributed in the population.
5. The population variances are equal. (This is the assumption of homogeneity of variance.)

These assumptions underlie the mathematical derivation of the analysis of variance. Empirical studies have shown that ANOVA is robust with regard to violations of these assumptions. A statistical procedure is said to be *robust* if the results with respect to the level of significance are little affected by a violation of one or more assumptions. ANOVA is particularly robust with respect to the assumptions of normality and interval-scale measurement and with respect to the assumption of homogeneity of variance if sample sizes are equal.

What happens when one or more of the assumptions underlying the analysis of variance are not met? Generally, failure to meet these assumptions makes the probability statement imprecise. That is, instead of operating at the designated level of significance, the actual Type I error rate may be greater or less than, say .05, depending on how the assumptions were violated. An excellent treatment of this topic is Glass, *et al.*[3] This article reports the results of a number of studies in which the assumptions were violated systematically and the effect on the Type I error rate was observed. Briefly, some of the useful findings were these:

1. When the populations sampled are not normal, the effect on the Type I error rate is minimal.
2. When measurement of the dependent variable is dichotomous or on an ordinal scale, the effect on the probability statement is not serious.
3. When the sample variances are different enough for us to conclude that the population variances are probably unequal, there may be a serious problem. With unequal sample sizes, when the larger variance is associated with the larger sample, the F ratio

[3]G. V. Glass, *et al.*, ''Consequences of Failure to Meet the Assumptions Underlying the Use of Analysis of Variance and Covariance,'' *Review of Educational Research*, 42 (1972), 237–288.

will be too conservative. When the smaller variance is associated with the larger sample, the F ratio will be too liberal. (When the α level is .05, "conservative" means that the actual rate is less than .05.) When the sample sizes are equal, the effect of heterogeneity of variance on the Type I error is minimal.

Interpreting the Result—
The Use of *Post Hoc* Multiple Comparison Tests

When a statistically significant F ratio is obtained in an ANOVA, and the null hypothesis is rejected, we conclude that at least one population mean is different from the others. However, it could be that all the population means differ or that any combination of them differs. To determine which pair or pairs of means differ, it is necessary to do some follow-up analysis. Several follow-up methods, commonly called multiple comparison procedures, can be used. In this text we discuss two such procedures:[4] the Scheffé method, which can be used with unequal sample sizes, and the Tukey method, which requires equal sample sizes.

The Scheffé Method

recommended when unequal sample sizes

The Scheffé multiple comparison procedure involves computing an F value for each combination of two means. The null hypothesis for each comparison is

$$H_0: \quad \mu_i = \mu_j \quad \text{for } i \neq j$$

The general formula for comparing any two sample means after obtaining a statistically significant result in the ANOVA is as follows:

$$F = \frac{(\overline{X}_i - \overline{X}_j)^2}{MS_W\left(\dfrac{1}{n_i} + \dfrac{1}{n_j}\right)} \tag{13.14}$$

Formula

where i and j refer to any two samples.

The computed F value is compared to the critical F value with $(k - 1)$ and $(N - k)$ degrees of freedom multiplied by $(k - 1)$. If the computed F value exceeds this product, the difference between the two sample means is statistically significant and we conclude that $\mu_i \neq \mu_j$. So, in order for two sample means to be significantly different,

$$\text{Computed } F \text{ value} > (k - 1)(\text{critical } F \text{ value})$$

---

[4]Readers who want a more comprehensive discussion of multiple comparison procedures are referred to Dennis E. Hinkle, William Wiersma, and Stephen G. Jurs, *Applied Statistics for the Behavioral Sciences* (Boston, Mass.: Houghton Mifflin, 1979), Chapter 12, pp. 268–295 or Roger E. Kirk, *Experimental Design: Procedures for the Behavioral Sciences,* (Belmont, Calif.: Brooks/Cole, 1968), pp. 69–98.

We will now illustrate the use of the Scheffé test with the example data. The critical F value is $(k - 1)$, or 3, times the tabled F value with 3 and 26 degrees of freedom. For the .05 level of significance, the critical F value is

$$F_{cv} = (3)(2.98)$$
$$= 8.94$$

The calculated F values for the pairwise comparisons follow. (Subscripts denote the means being compared.)

$$F_{1-2} = \frac{(16.286 - 12.600)^2}{6.98\left(\frac{1}{7} + \frac{1}{10}\right)} = 8.016$$

$$F_{1-3} = \frac{(16.286 - 18.000)^2}{6.98\left(\frac{1}{7} + \frac{1}{6}\right)} = 1.360$$

$$F_{1-4} = \frac{(16.286 - 12.571)^2}{6.98\left(\frac{1}{7} + \frac{1}{7}\right)} = 6.921$$

$$F_{2-3} = \frac{(12.600 - 18.000)^2}{6.98\left(\frac{1}{10} + \frac{1}{6}\right)} = 15.669$$

$$F_{2-4} = \frac{(12.600 - 12.571)^2}{6.98\left(\frac{1}{10} + \frac{1}{7}\right)} = .000$$

$$F_{3-4} = \frac{(18.000 - 12.571)^2}{6.98\left(\frac{1}{6} + \frac{1}{7}\right)} = 13.645$$

Both F_{2-3} and F_{3-4} exceed the critical F value of 8.94. Thus, in terms of the original analysis, we can conclude that the population mean for engineering students differs from the means for the sociology and history graduate student populations. The Scheffé test tends to be slightly conservative.[5] That is, it may indicate that a difference between two means is not significant when in fact it is significant. We will discuss another multiple comparison test called the Tukey method, which is not conservative. It requires equal sample sizes, however, so it will be illustrated with another example.

[5]Since the Scheffé method is statistically conservative, it has been recommended that, if the F ratio is tested at $\alpha = .05$, the post hoc tests be made at $\alpha = .10$.

Group	1	2	3	4
	10	6	16	9
	16	12	18	12
	12	9	14	14
	12	8	20	11
	18	7	17	16
ΣX	68	42	85	62
$\overline{X}$	13.60	8.40	17.00	12.40

Summary ANOVA					
Source	SS	df	MS	F	$F_{cv(.05)}$
Between	188.95	3	62.98	8.87	3.24
Within	113.60	16	7.10		
Total	302.55	19			

Table 13.5 Data to illustrate the Tukey method

The Scheffé method is the most versatile and at the same time most conservative *post hoc* multiple comparison procedure. It is recommended only when the n_j's are unequal.

The Tukey Method

The Tukey method, often called the HSD (honestly significant difference) test, is designed to make all pairwise comparisons of means while maintaining the error rate at the pre-established α level. The null hypothesis tested is

$$H_0: \quad \mu_i = \mu_j \quad \text{for } i \neq j$$

That is, each pair of population means is equal. The test statistic is Q, and it is defined as follows:

$$Q = \frac{\overline{X}_i - \overline{X}_j}{\sqrt{MS_W/n_j}} \tag{13.15}$$

This test statistic is analogous to the t statistic. The only difference is in its use of the studentized range distributions rather than the t distributions as the sampling distribution. These distributions are found in Table 5 in the Appendix.

Suppose a psychologist randomly selects 20 subjects from a population and then randomly assigns them to 4 different sets of motivating instructions (the levels of the independent variable). The dependent variable is the number of trials necessary to successfully complete a complex motor task. The resulting data are given in Table

Group	2	4	1	3
$\overline{X}$	8.40	12.40	13.60	17.00
$\overline{X}_i - \overline{X}_j$		4.00	5.20	8.60
			1.20	4.60
				3.40
$Q = \dfrac{\overline{X}_i - \overline{X}_j}{\sqrt{MS_w/n_j}}$		3.36	4.37*	7.23**
			1.01	3.87
				2.86

*$p < .05$, $Q_{cv(.05)} = 4.05$ for df $= 16$
**$p < .01$, $Q_{cv(.01)} = 5.19$ for df $= 16$

Table 13.6   Calculation of the Q statistic for the Tukey method

13.5. Note that the F ratio (8.87) exceeds the critical value of F at the .05 level, so we reject the null hypothesis ($\mu_1 = \mu_2 = \mu_3 = \mu_4$).

Calculation of the Q statistic for each of the possible pairwise comparsions is illustrated in Table 13.6. For convenience, the means of the groups are ranked from low to high. The second step involves finding the differences between the respective group means—that is, determining the numerator of Q. For example, $12.40 - 8.40 = 4.00$, $13.60 - 8.40 = 5.20$, and so on. The final step is to divide these differences by $\sqrt{MS_w/n_j} = \sqrt{7.10/5} = 1.19$. For example, $4.00/1.19 = 3.36$, $5.20/1.19 = 4.37$, and so on. We find the critical value of Q for the Tukey method by considering the maximum range in the means when comparing the smallest mean with the largest mean. In general, the maximum range is r, where r is the number of means. In our example, the range would be 4. Thus, from Table 8, the critical values of Q for ranges equal to 4 and 16 degrees of freedom are 4.05 and 5.19 for $\alpha = .05$ and .01, respectively. Assuming $\alpha = .05$, the null hypotheses $\mu_1 = \mu_2$ and $\mu_2 = \mu_3$ are rejected; the four remaining null hypotheses are retained. The psychologist can then conclude that the population means for motivating instructions 1 and 3 are different from the population mean for motivating instruction 2 in terms of the number of trials necessary to perform a complex motor task.

The Tukey method uses the Q distribution for r ranges. The Type I error rate is maintained at α for all possible pairwise comparisons.

ANOVA for Repeated Measures

There are numerous occasions in behavioral science research when the data from several samples are not independent. For example, a researcher may test the same sample of individuals at different times. Learning experiments often involve measuring the same individuals on their performances as they solve a sequence of four or five problems. The data for the various problems comprise dependent samples. Such experiments fall in the general category of *repeated-measures analysis*. The situation is analogous to the *t* test for correlated samples discussed in Chapter 11. However, the ANOVA for repeated measures is slightly different.

Repeated-measures analysis has the characteristic of scores for the same individual being dependent and scores for different individuals being independent. Accordingly, the partitioning of variation in the ANOVA needs to be adjusted somewhat so that appropriate *F* ratios are obtained. The total sum of squares is partitioned into three parts: one due to *variation among individuals* (SS_I), another due to *variation among test occasions* (SS_O), and a third part that is not due to either persons or test occasions (SS_{Res}). This remaining variation is often called the *residual variation*. The mean squares for these three components of variation are computed, and the mean square for residual variation (MS_{Res}) is used as the denominator for the *F* ratio in testing the effect of test occasion (represented by the different, dependent samples), which is the effect of primary interest.

We determine sums of squares for a repeated-measures ANOVA by using the following formulas. (In this situation, the samples have the same number of individuals, n_j, and there are *k* samples.)

$$SS_T = \sum_{j=1}^{k} \sum_{i=1}^{n_j} X_{ij}^2 - \frac{T^2}{N} \qquad (13.16)$$

where

$$\sum_{j=1}^{k} \sum_{i=1}^{n_j} X_{ij}^2 = \text{sum of the squared scores}$$

$T = $ grand total of the scores

$N = $ total number of scores or observations

$$SS_I = \sum_{i=1}^{n_j} \left(\frac{T_i^2}{k} \right) - \frac{T^2}{N} \qquad (13.17)$$

where

$T_i = $ total for the *i*th individual

$k = $ number of test occasions; that is, the number of repeated measures on each individual

$$SS_O = \sum_{j=1}^{k} \left(\frac{T_j^2}{n_j} \right) - \frac{T^2}{N} \qquad (13.18)$$

where

$T_j = $ total for the *j*th test occasion

$n_j = $ number of observations (individuals) in each sample

MS formula

Source	SS	df	MS	F
Occasions	SS_O	$k - 1$	$SS_O/k - 1$	MS_O/MS_{Res}
Individuals	SS_I	$n_j - 1$	$SS_I/n_j - 1$	
Residual	SS_{Res}	$(k - 1)(n_j - 1)$	$SS_{Res}/(k - 1)(n_j - 1)$	
Total	SS_T	$N - 1$		

Table 13.7   Summary ANOVA table for a repeated-measures analysis

Then, the sum of squares for residual variability is determined by subtraction.

$$SS_{Res} = SS_T - SS_I - SS_O$$

In order to compute mean squares, we must determine the number of degrees of freedom for each sum of squares. These degrees-of-freedom values are:

$$df_I = n_j - 1$$

$$df_O = k - 1$$

$$df_{Res} = (k - 1)(n_j - 1)$$

$$df_T = N - 1$$

As we have indicated, the F ratio of interest is the one that tests the repeated measure, the test occasion. This F ratio is determined by MS_O/MS_{Res}. The summary ANOVA table for the general case in a repeated-measures ANOVA is given in Table 13.7.

In a repeated-measures ANOVA, the total variability is partitioned into three components: (1) variation among individuals, (2) variation among test occasions, and (3) residual variation.

Repeated-Measures ANOVA Example

Suppose the data from a sample of ten people taken on three separate testing occasions are to be analyzed via a repeated-measures ANOVA. The dependent variable is performance on a learning task and, prior to each test occasion, the people are given different treatments. The primary interest of the experiment is the possible differing effects of treatments. We will use the four steps of hypothesis testing.

Step 1: Stating the Hypothesis   The null hypothesis is

$$H_0: \quad \mu_1 = \mu_2 = \mu_3$$

Person	Test 1	Test 2	Test 3		T_i	$\dfrac{T_i^2}{n_i}$
A	6	12	18		36	432.00
B	9	14	16		39	507.00
C	4	8	15		27	243.00
D	3	10	12		25	208.33
E	1	6	10		17	96.33
F	7	15	20		42	588.00
G	6	8	15		31	320.33
H	9	11	18		38	481.33
I	8	12	13		33	363.00
J	6	10	16		32	341.33

n_j: 10, 10, 10 $N = 30$ $\displaystyle\sum_{i=1}^{n_j} \frac{T_i^2}{k} = 3580.66$

T_j: 61, 106, 153 $T = 320$

$\bar{X}_j$: 6.1, 10.6, 15.3 $\dfrac{T^2}{n} = 3413.33$

$\displaystyle\sum_{i=1}^{n_j} X_{ij}^2$: 437, 1194, 2423 $\displaystyle\sum_{j=1}^{k}\sum_{i=1}^{n_j} X_{ij}^2 = 4054$

$\dfrac{T_j^2}{n_j}$: 372.10, 1123.60, 2340.90 $\displaystyle\sum_{j=1}^{k}\left(\frac{T_j^2}{n_j}\right) = 3836.60$

Table 13.8   Scores for ten persons on three test occasions

where μ_i's represent population means of testing occasions. The alternative hypothesis is

$$H_a: \quad \mu_i \neq \mu_j \quad \text{for some } i \text{ and } j$$

The level of significance is set at .05.

Step 2: Computing the Test Statistic

The data layout of Table 13.8 presents the data for this example. The column sums representing testing occasions and the row sums representing individuals are also given, along with other information provided below the data layout. The computations are as follows:

$$SS_T = 4054 - \frac{(320)^2}{30} = 640.67$$

$$SS_I = \frac{(36)^2}{3} + \frac{(39)^2}{3} + \ldots + \frac{(32)^2}{3} - \frac{(320)^2}{30} = 167.33$$

(*Note:* Individuals are persons.)

$$SS_O = \frac{(61)^2}{10} + \frac{(106)^2}{10} + \frac{(153)^2}{10} - \frac{(320)^2}{30} = 423.27$$

$$SS_{Res} = 640.67 - 167.33 - 423.27 = 50.07$$

Source	SS	df	MS	F	F_{cv}
Occasions	423.27	2	211.64	76.13	3.55
Individuals	167.33	9	18.59		
Residual	50.07	18	2.78		
Total	640.67	29			

Table 13.9   Summary table for the repeated-measures ANOVA

The corresponding degrees of freedom are df_T = 29, df_I = 9, df_O = 2, and df_{Res} = 18. The summary ANOVA table for this analysis is given in Table 13.9.

Step 3: Setting the Criterion for Rejecting H_0 To determine the critical value of the F ratio, we use Table 4 in the Appendix. The tabled F value at α = .05, with 2 and 18 degrees of freedom, is 3.55. The calculated F value of 76.13 clearly exceeds the critical value, so the test statistic is statistically significant.

Step 4: Deciding Whether to Reject or Retain the Null Hypothesis
The calculated F value exceeds the critical value, so we reject the null hypothesis. The conclusion is that the population means of the three test occasions are not all equal. In terms of probability: *The probability that the sample test occasion means occurred by chance if the population means are equal is less than .05.* A multiple comparison analysis such as the Tukey test would then be appropriate to determine which pairs of means are significantly different.

Assumptions for the Repeated-Measures ANOVA   There are a number of assumptions that underlie the use of a repeated-measures ANOVA.

1. The sample is a random sample from the population.
2. The dependent variable is normally distributed in the population.
3. The dependent variable is measured on at least an interval scale.
4. The population variances for the test occasions are equal.
5. The population correlation coefficients between pairs of test-occasion scores are equal.

When the last two assumptions are not met, the Type I error rate can be seriously affected. However, we can make an appropriate correction by merely changing the degrees of freedom from $(k - 1)$ and $(k - 1)(n_j - 1)$ to 1 and $(n_j - 1)$. This change provides a conservative statistical test of the null hypothesis, even when assumptions 4 and 5 are not met.

Summary

This chapter introduces the analysis of variance. ANOVA is a procedure for testing hypotheses about population means by partitioning variance. In one-way ANOVA, one independent variable is included in the analysis. The statistic generated is the ratio of two variance estimates, called mean squares, which has as its sampling distribution the F distribution with the appropriate number of degrees of freedom.

The basic ANOVA for independent samples is discussed. The special situation of repeated measures, often found in behavioral science research, is also considered. ANOVA tests H_0: $\mu_1 = \mu_2 = \cdots = \mu_k$, $k \geq 2$. But ANOVA is usually applied when three or more samples are involved. If the F value is statistically significant, a multiple comparison procedure can be applied to determine where the significance lies. There are numerous such procedures. Two of them, the Scheffé method and the Tukey method, are discussed here.

It should be emphasized that this chapter provides only an introduction to ANOVA. ANOVA can be extended to include two or more independent variables, and such analyses can become quite complex. The next chapter discusses two-way ANOVA, in which two independent variables are included. There are entire books written on the topic. But whatever the specific analysis, the basic idea remains the same. ANOVA is an inferential statistics procedure, and the underlying reasoning of inferential statistics and our familiar steps of hypothesis testing still apply.

Key Concepts

Type I error rate
One-way analysis of variance
Independent variable
Levels of the independent variable
Dependent variable
Within-groups variation
Between-groups variation
Variance estimates
Partitioning variation
Sum of squares

Mean squares
F-distribution
Parametric assumptions
Post hoc multiple comparison procedures
Scheffé method
Tukey method
Studentized range distributions
Repeated-measures ANOVA
Residual variation

Exercises

13.1 The following are the grade point averages for students in different class ranks. Assume that they represent random samples from the respective class ranks.

Complete the ANOVA. (Use $\alpha = .05$.) Compute both the Scheffé and the Tukey *post hoc* comparisons.

$\Sigma X = 82.3$

$\Sigma X^2 = 220.19$

$N = 32$

Freshmen		Sophomores		Juniors		Seniors	
1.9	3.61	2.7	7.29	2.4	5.76	2.8	7.84
2.4	5.76	2.4	5.76	2.1	4.41	3.1	9.61
1.6	2.56	3.3	10.89	3.4	11.56	2.3	5.29
2.0	4	2.1	4.41	3.0	9	2.3	5.29
1.8	3.24	2.7	7.29	2.1	4.41	3.8	14.44
2.1	4.41	2.6	6.76	2.7	7.29	3.1	9.61
2.5	6.25	1.9	3.61	2.6	6.76	3.2	10.24
2.4	5.76	3.2	10.24	2.8	7.84	3.0	9
16.7	35.59	20.9	56.25	21.1	57.03	23.6	71.32

a. State the hypothesis.
b. Compute the test statistic and determine the critical value.
c. Decide whether to reject or to retain the null hypothesis.
d. Interpret the results of the *post hoc* tests.

13.2 Eighteen students were randomly assigned to three groups and trained on a concept-attainment task using three different methods. The data follow. Test the null hypothesis that the population means for method are equal. (Use $\alpha = .05$.)

Method 1	Method 2	Method 3
14	18	12
10	16	15
18	15	11
12	17	17
16	13	14
15	16	16

a. State the hypothesis.
b. Compute the test statistic and determine the critical value.
c. Decide whether to reject or to retain the null hypothesis.

13.3 A researcher draws five random samples, each of size 8 from different populations and measures them on a dependent variable. Complete the following Summary ANOVA table, testing the H_0: $\mu_1 = \ldots \mu_5$ using the .01 level of significance. Would your conclusion change if you had used $\alpha = .05$?

Summary ANOVA

Source	SS	df	MS	F	F_{cv}	
Between	58	4	14.5	3.05	4.02	2.69
Within	166	35	4.74		.01	.05
Total	224					

retain   *reject*

13.4 A researcher measured the performance of 6 individuals over 5 trials on a learning task; the data follow. Determine whether performance in the population is equivalent across the trials. (Use $\alpha = .01$.) Complete the Tukey *post hoc* comparisons to determine which of the trial means differ.

Individual	Trial 1	Trial 2	Trial 3	Trial 4	Trial 5
Jones	36 6	25 5	64 8	100 10	121 11
Smith	25 5	25 5	25 5	64 8	49 7
Grey	36 6	64 8	100 10	100 10	144 12
White	16 4	16 4	16 4	25 5	25 5
Doc	25 5	25 5	36 6	100 10	144 12
Roe	36 6	49 7	64 8	64 8	100 10

(handwritten sums: 174/32   204/34   305/41   453/51   593/55)

(handwritten right margin: Ti Tj — 40 160, 30 9., 46 21, 25 6, 38 6, 39 14, 15.   820)

a. State the hypothesis.
b. Compute the test statistic and determine the critical value.
c. Decide whether to reject or to retain the null hypothesis.
d. Interpret the results of the Tukey tests.

13.5 A school principal traced a random sample of dyslexic students across three years in terms of reading comprehension scores on a standardized test. Determine whether the average achievement in the population of dyslexic students is the same across the three years. (Use the .10 level of significance.) Complete the Scheffé *post hoc* comparisons.

Student	Third grade	Fourth grade	Fifth grade
A	2.8	3.2	4.5
B	2.6	4.0	5.1
C	3.1	4.3	5.0
D	3.8	4.9	5.7
E	2.5	3.1	4.4
F	2.4	3.1	3.9
G	3.2	3.8	4.3
H	3.0	3.6	4.4

a. State the hypothesis.
b. Compute the test statistic and determine the critical value.
c. Decide whether to reject or to retain the null hypothesis.
d. Interpret the results of the Scheffé tests.

13.6 A repeated-measures ANOVA is computed to determine the effect of test occasion on a performance measure. Complete the following Summary ANOVA table. (Use $\alpha = .05$.)

Source	SS	df	MS	F	F_{cv}
Test Occasion		4	38		
Persons	260	5			
Residual	510				
Total					

13.7 A researcher randomly assigns 6 laboratory rats to each of 8 diets. After a designated period of time, the rats are given an exercise stress test. The sums and means for the group are as follows:

Group	1	2	3	4	5	6	7	8
Sums	432	468	492	510	390	456	366	372
Means	72	78	82	85	65	76	61	62

$\overline{X}$ = grand mean = 72.63

$$\sum_{j=1}^{8} \sum_{i=1}^{6} X_{ij}^2 = 262{,}364$$

Complete the ANOVA (use $\alpha = .05$). Complete the Scheffé and Tukey tests for the following comparisons.

Group 4 versus Group 6
Group 2 versus Group 7
Group 4 versus Group 7

a. State the hypothesis.
b. Compute the test statistic and determine the critical value.
c. Decide whether to reject or to retain the null hypothesis.
d. Interpret the results of the Scheffé and Tukey tests.

▦ Key Strokes for Selected Exercises

13.1 Use the Key Strokes described in the Appendix to determine ΣX and ΣX^2 for each group. The data needed are as follows:

$T_1 = 16.7$ $T = 82.3$
$T_2 = 20.9$ $\Sigma \Sigma X_{ij}^2 = 220.19$
$T_3 = 21.1$
$T_4 = 23.6$

Step 1: Determine $\Sigma\ (T_j^2/n_j)$.

Value	Key Stroke	Display
16.7 (T_1)	X^2	278.89
	÷	278.89
8 (n_1)	=	34.86125
	STO	34.86125
20.9 (T_2)	X^2	436.81
	÷	436.81
8 (n_2)	=	54.60125
	M+	54.60125
21.1 (T_3)	X^2	445.21
	÷	445.21

Value	Key Stroke	Display
$8\ (n_3)$	$=$	55.65125
	M+	55.65125
23.6 (T_4)	X^2	556.96
	$\div$	556.96
8 (n_4)	$=$	69.62
	M+	69.62
	RCL	214.73375

(Keep
in storage)

Step 2: Determine SS_B, SS_W, SS_T.

Value	Key Stroke	Display
	RCL	214.73375
	$-$	214.73375
211.67 (T^2/n)	$=$	3.06375 = SS_B
220.19 $(\Sigma\Sigma X^2)$	$-$	220.19
	RCL	214.73375
	$=$	5.45625 = SS_W
220.19 $(\Sigma\Sigma X^2)$	$-$	220.19
211.67 (T^2/n)	$=$	8.52 = SS_T

Step 3: Determine MS_B, MS_W, F.

Value	Key Stroke	Display
3.06 (SS_B)	$\div$	3.06
3 (df_B)	$=$	1.02 MS_B
5.46 (SS_W)	$\div$	5.46
28 (df_W)	$=$	0.195 MS_W
	STO	0.195
1.02 (MS_B)	$\div$	1.02
	RCL	0.195
	$=$	5.23077 = F

Step 4: Determine denominator for Scheffé tests.

Value	Key Stroke	Display
8 (n_1)	1/X	0.125
	$+$	0.125
8 (n_2)	1/X	0.125
	$=$	0.250
	$\times$	0.250
0.195 (MS_W)	$=$	0.04875
	STO	0.04875

Step 5: Scheffé tests.

Value	Key Stroke	Display
$2.0875\ (\overline{X}_1)$	−	2.0875
$2.6125\ (\overline{X}_2)$	=	−0.5250
	X^2	0.275625
	÷	0.275625
	RCL	0.04875
	=	5.6538 $= F_{1-2}$
$2.0875\ (\overline{X}_1)$	−	2.0875
$2.6375\ (\overline{X}_3)$	=	−0.5500
	X^2	0.3025
	÷	0.3025
	RCL	0.04875
	=	6.2051 $= F_{1-3}$

Repeat Key Strokes for the remaining Scheffé tests.

Step 6: Determine denominator for Tukey tests.

Value	Key Stroke	Display
$0.195\ (MS_W)$	÷	0.195
$8\ (n_j)$	=	0.024375
	$\sqrt{}$	0.1561
	STO	0.1561

Step 7: Tukey tests.

Value	Key Stroke	Display
$2.0875\ (\overline{X}_1)$	−	2.0875
$2.6125\ (\overline{X}_2)$	=	−0.525
	÷	−0.525
	RCL	0.1561
	=	−3.3627 $= Q_{1-2}$
$2.0875\ (\overline{X}_1)$	−	2.0875
$2.6375\ (\overline{X}_2)$	=	−0.5500
	÷	−0.5500
	RCL	0.1561
	=	−3.5228 $= Q_{1-3}$

Repeat Key Strokes for the remaining Tukey tests.

14

Two-Way Analysis of Variance

In Chapter 13 we discussed one-way analysis of variance as the procedure for testing the null hypothesis that the population means for more than two levels of a single independent variable are equal. In this chapter, we will extend that discussion to include the procedures for testing the null hypothesis when two independent variables, or *factors,* are considered simultaneously in a research study. Such an arrangement is referred to as a *factorial design*. We will restrict our discussion to procedures for analyzing data from factorial designs with only two factors. These procedures constitute two-way analysis of variance, or two-way ANOVA. It is possible, however, to generalize this discussion to factorial designs with more than two factors.

Factorial Design

Consider the following example of a two-factor design. A social psychologist is interested in where children would place themselves in relation to an adult stranger in various settings. He therefore has children in three age groups (6, 8, and 10 years old) place figures that represent themselves on a picture that contains an adult figure who is described as a stranger. Three pictures are used: an amusement park, a classroom, and the backyard of a home. Note that there are three levels for each independent variable, three age groups and three settings; thus there are nine possible treatment combinations (six-year-olds/amusement park, six-year-olds/classroom, and so on). It is not necessary to have the same number of levels in the two independent variables.

The design of the study is illustrated in Table 14.1. The dependent variable in this example is the distance in centimeters between the figure of the stranger and the figure the child places on the picture. The social psychologist randomly selects 18 children from each age population and assigns 6 of each age group to each of the settings; each child deals with only one setting. Thus a total of 54 children are included. We will return to this example later as a computational example.

A two-way ANOVA, often called a factorial ANOVA, includes two independent variables. The number of cells or combinations is equal to the product of the numbers of levels in the two independent variables. Thus a 3-by-3 factorial contains 9 cells.

Advantages of a Factorial Design

There are several advantages to using a factorial design. The first advantage is that of *efficiency*. Consider the foregoing example. With simultaneous analysis of two independent variables, we are in essence conducting two separate research studies at the same time. In addition to investigating how different levels of two independent variables affect the dependent variable, we can test whether levels of one independent variable affect the dependent variable in the same way across levels of the second independent variable. If the effect is not the same, we say that there is *interaction* between the two independent variables in the factorial design.

A two-way ANOVA enables the researcher to check on the possible interaction of the independent variables. Interaction is the effect of one independent variable on another.

		Setting		
		Amusement Park	Classroom	Backyard
Age	6			
	8			
	10			

Table 14.1   Two-factor design with age and setting as the independent variables

A second advantage is *control* over more than one independent variable or factor. In our example, both independent variables are considered equally important in the study. In other research settings, we may use a factorial design to control for an additional source of variation. For example, an educational researcher may be interested in the mathematics achievement of students taught with three different teaching methods. If the researcher also has the IQ scores for the students, a second independent variable can be included in the design. Suppose the researcher divides the IQ scores into four categories: (1) less than 85, (2) 85 to 99, (3) 100 to 114, and (4) 115 or greater. This design is illustrated in Table 14.2. Using this design, the researcher can examine the effects of the various teaching methods *per se* and also the effects of the various teaching methods across the levels of student IQ.

The third advantage of factorial designs was mentioned in connection with the other two advantages, but it deserves special attention. Factorial designs make it possible to investigate the interaction of two or more independent variables. In behavioral science research, the effect of a single independent variable is rarely unaffected by one or more other independent variables, so a study of the interaction among the independent variables may be the most important objective of an investigation. If only single-factor studies were conducted, the study of interaction among the independent variables would be impossible.

The advantages of factorial research designs are (1) efficiency, (2) control over additional variables, and (3) the study of the interaction among the independent variables.

The Variables in Two-Way ANOVA

As indicated in Chapter 13, there are two kinds of variables in any ANOVA: independent and dependent. In two-way ANOVA, there are two independent variables

	IQ			
	$b_1(<85)$	$b_2(85\text{–}99)$	$b_3(100\text{–}114)$	$b_4(\geq 115)$
a_1 (Method 1)				
Teaching Method a_2 (Method 2)				
a_3 (Method 3)				

Table 14.2   Two-factor design with teaching method and IQ as independent variables

and a single dependent variable. Changes in the dependent variable are, or are presumed to be, the result of changes in the independent variable(s). As in one-way ANOVA, it is assumed that the scores on the dependent variable are normally distributed in the population and, because calculating means and variances is required, that the dependent variable is measured on at least an interval or equal-unit scale. For both independent variables in a two-way ANOVA, there are multiple categories or treatment *levels*. The scores of the subjects assigned to the various combinations of these treatment levels are illustrated in Table 14.3. The symbol X_{111} refers to the score on the dependent variable of the *first* subject assigned to the *first* level of the first independent variable and to the *first* level of the second independent variable. (We have arbitrarily put the first independent variable in the rows and the second independent variable in the columns.) In general cases, X_{irc} would be the score of the *i*th subject in the *r*th row and the *c*th column. For example, X_{423} would be the score of the fourth subject in the second row and the third column.

We will assume for now that there are the same number of observations in each cell of the data matrix. Thus we will say that there are n observations in each cell. The total number of observations is the sum of all observations in all cells and is denoted N. We will also use a new symbol called the "dot subscript" to denote the means for the rows and columns. For example, $\overline{X}_1$. is the mean of all observations in the first row of the data matrix averaged across all columns; $\overline{X}_{.1}$ is the mean of all observations in the first column of the data matrix averaged across all rows. The mean of all observations in the first cell of the matrix corresponding to the first row and the first column is denoted $\overline{X}_{11}$. In general terms, $\overline{X}_{rc}$ is the cell mean for the *r*th row and the *c*th column; $\overline{X}_r$. is the row mean for the *r*th row; and $\overline{X}_{.c}$ is the column mean for the *c*th column. This notation provides a convenient way to identify specific data in the cells of the ANOVA.

Sources of Variation

In our discussion of the one-way ANOVA, the total sum of squares (SS_T) was partitioned into two parts, the variation between groups (SS_B) and the variation within

		b_1	b_2		b_c	
		X_{111}	X_{112}	$\cdots$	X_{11c}	
		X_{211}	X_{212}	$\cdots$	X_{21c}	
		$\cdot$	$\cdot$	$\cdots$	$\cdot$	
	a_1	$\cdot$	$\cdot$	$\cdots$	$\cdot$	$\overline{X}_{1\cdot}$
		$\cdot$	$\cdot$	$\cdots$	$\cdot$	
		X_{n11}	X_{n12}	$\cdots$	X_{n1c}	
		X_{121}	X_{122}	$\cdots$	X_{12c}	
		X_{221}	X_{222}	$\cdots$	X_{22c}	
		$\cdot$	$\cdot$	$\cdots$	$\cdot$	
	a_2	$\cdot$	$\cdot$	$\cdots$	$\cdot$	$\overline{X}_{2\cdot}$
		$\cdot$	$\cdot$	$\cdots$	$\cdot$	
		X_{n21}	X_{n22}	$\cdots$	X_{n2c}	
		$\cdot$	$\cdot$	$\cdots$	$\cdot$	
		$\cdot$	$\cdot$	$\cdots$	$\cdot$	
		$\cdot$	$\cdot$	$\cdots$	$\cdot$	
		X_{1r1}	X_{1r2}	$\cdots$	X_{1rc}	
		X_{2r1}	X_{2r2}	$\cdots$	X_{2rc}	
		$\cdot$	$\cdot$	$\cdots$	$\cdot$	
	a_r	$\cdot$	$\cdot$	$\cdots$	$\cdot$	$\overline{X}_{r\cdot}$
		$\cdot$	$\cdot$	$\cdots$	$\cdot$	
		X_{nr1}	X_{nr2}	$\cdots$	X_{nrc}	
		$\overline{X}_{\cdot 1}$	$\overline{X}_{\cdot 2}$	$\cdots$	$\overline{X}_{\cdot c}$	$\overline{X}$

Levels of First Independent Variables (row label for a_1 through a_r)

Table 14.3   Data layout for two-way ANOVA

groups (SS_W). These sums of squares were divided by their degrees of freedom to form mean squares. The ratio MS_B/MS_W was then used to test the hypothesis that the population means were equal. When the null hypothesis was true, this ratio (an F ratio) was approximately equal to 1.00. On the other hand, when the null hypothesis was false, the ratio was greater than 1.00. When the ratio was so much greater than 1.00 that it could not be attributed to random sampling fluctuation (in other words, it exceeded the tabled F with the appropriate degrees of freedom), the null hypothesis was rejected.

The procedures of the two-way ANOVA are very similar to those of the one-way ANOVA. SS_T is partitioned, mean squares and F ratios are calculated. The major difference is that the variation (SS_T) is partitioned into four components instead of two, as in the one-way ANOVA. The first component, the within-cells variation, is analogous to the SS_W in the one-way ANOVA. The other three components are analogous to the between-groups variation. They are (1) the variation among the row means, (2) the variation among the column means, and (3) the variation due to the interaction between the two independent variables. In the one-way ANOVA, the SS_B included variation among the group means. In the two-way ANOVA, the sum of squares for row (SS_R) includes variation among row means, the sum of squares for column (SS_C) includes variation among column means, and the sum of squares for interactions (SS_{RC}) includes variation among the cell means after the row and column variation is accounted for. The rows and columns represent the respective independent variables placed on them.

The variation in a two-way ANOVA is partitioned into four parts: (1) that due to variation within cells, (2) the independent variable on the rows, (3) the independent variable on the columns, and (4) the interaction of the independent variables.

Testing Hypotheses

The two-way ANOVA has three null hypotheses, one for each of the independent variables and one for the interaction. The null hypothesis for row population means is

$$H_0: \quad \mu_{1.} = \mu_{2.} = \cdots = \mu_{r.}$$

That is, there is no difference among the r row population means. If the null hypothesis for rows is true, the MS_R/MS_W will approximately equal 1.00. If the null hypothesis is false, this ratio will exceed 1.00 by more than can be attributed to random sampling fluctuation. We use the F distribution with the appropriate degrees of freedom, along with the level of significance, to determine whether to reject the null hypothesis.

The null hypothesis for column population means is

$$H_0: \quad \mu_{.1} = \mu_{.2} = \cdots = \mu_{.c}$$

That is, there is no difference among the c column population means. Similarly, if the ratio MS_C/MS_W exceeds the tabled F value with the appropriate degrees of freedom and level of significance, we reject this null hypothesis.

Finally, the null hypothesis for interaction is

$$H_0: \quad \text{all } (\mu_{rc} - \mu_{r.} - \mu_{.c} + \mu) = 0$$

That is, there is no difference among the rc cell means that cannot be explained by differences among either the row means or the column means or both. For example, if we find a significant difference among adjacent row means, under the null hypothesis for interaction, we would find the same difference to exist among all pairs of adjacent row means for each column. On the other hand, if the null hypothesis for interaction is false, we would find differences among the cell means that could not be explained by differences among either the row means or the column means or both. (We will continue our discussion of interaction in a later section.)

If the null hypothesis for interaction is true, the ratio MS_{RC}/MS_W will be approximately equal to 1.00. If the ratio exceeds the tabled F value, we reject the null hypothesis for interaction.

Three null hypotheses are tested in the two-way ANOVA, one for each of the independent variables and one for the interaction. The procedure for testing the null hypotheses—computing an F ratio made up of two mean squares—is the same as for a one-way ANOVA.

The Summary Table for Two-Way ANOVA

Just as in the one-way ANOVA, the sums of squares, degrees of freedom, mean squares and F ratios for a two-way ANOVA are presented in a standard tabular format. The only difference is that there are more sources of variation in the two-way ANOVA. Table 14.4 presents a two-way summary ANOVA. This table also includes a guide to the appropriate degrees of freedom, mean squares, and F ratios.

The only part of the summary table that has not yet been developed is the calculation of sums of squares for the various sources of variation. Those formulas are presented next, followed by a specific example of how the formulas are used.

Computational Formulas for Sums of Squares

The computational formulas are quite similar to those that we used in the one-way ANOVA. However, we now must compute row totals, column totals, and cell totals. Consider the following notation for the sum of the observations in each of the cells, T_{rc}:

$$T_{rc} = \sum_{i=1}^{n} X_{irc} \tag{14.1}$$

Denote the sum of the observations in each of the rows as $T_{r\cdot}$ and the sum of the observations in each of the columns as $T_{\cdot c}$:

$$T_{r\cdot} = \sum_{c=1}^{C} \sum_{i=1}^{n} X_{i\cdot c} \tag{14.2}$$

$$T_{\cdot c} = \sum_{r=1}^{R} \sum_{i=1}^{n} X_{ir\cdot} \tag{14.3}$$

Source of Variation	Sum of Squares (SS)	Degrees of Freedom (df)	Mean Square (MS)	F
Row		$r - 1$	$SS_R/(r - 1)$	MS_R/MS_W
Column		$c - 1$	$SS_C/(c - 1)$	MS_C/MS_W
R × C Interaction		$(r - 1)(c - 1)$	$SS_{RC}/[(r - 1)(c - 1)]$	MS_{RC}/MS_W
Within		$rc(n-1)$	$SS_W/rc(n - 1)$	
Total		$N - 1$		

Table 14.4   Two-way summary ANOVA table

Finally, denote the sum of all observations in the RC cells as T:

$$T = \sum_{c=1}^{C} \sum_{r=1}^{R} \sum_{i=1}^{n} X_{irc}$$

The computational formulas for the sums of squares are then as follows:

Sum of squares for rows

$$= SS_R = \frac{1}{nC} \sum_{r=1}^{R} T_{r\cdot}^2 - \frac{T^2}{N} \tag{14.4}$$

Sum of squares for columns

$$= SS_C = \frac{1}{nR} \sum_{c=1}^{C} T_{\cdot c}^2 - \frac{T^2}{N} \tag{14.5}$$

Sum of squares for interaction

$$= SS_{RC} = \frac{1}{n} \sum_{c=1}^{C} \sum_{r=1}^{R} T_{rc}^2 - \frac{1}{nC} \sum_{r=1}^{R} T_{r\cdot}^2$$

$$- \frac{1}{nR} \sum_{c=1}^{C} T_{\cdot c}^2 + \frac{T^2}{N} \tag{14.6}$$

Sum of squares within cells

$$= SS_W = \sum_{c=1}^{C} \sum_{r=1}^{R} \sum_{i=1}^{n} X_{irc}^2 - \frac{1}{n} \sum_{c=1}^{C} \sum_{r=1}^{R} T_{rc}^2 \tag{14.7}$$

Sum of squares total

$$= SS_T = \sum_{c=1}^{C} \sum_{r=1}^{R} \sum_{i=1}^{n} X_{irc}^2 - \frac{T^2}{N} \tag{14.8}$$

Completing the Example

Let us now return to the example involving the social psychologist, which was described at the beginning of this chapter. Recall that the first independent variable was the age of the child; the levels were 6, 8, and 10 years. The second independent variable was the type of setting presented to the child; the levels were an amusement park (AP), a classroom (C), and a backyard (B). The dependent variable was the distance (measured in centimeters) between an adult stranger in the picture and the child's placement of a figure representing herself or himself. The data for this example are given in Table 14.5, and a summary of the initial calculations appears in Table 14.6.

Step 1: Stating the Hypothesis

The null hypotheses that are to be tested are:

$$H_0: \quad \mu_6 = \mu_8 = \mu_{10} \qquad \text{for rows}$$

$$H_0: \quad \mu_{AP} = \mu_C = \mu_B \qquad \text{for columns}$$

$$H_0: \quad \text{all } (\mu_{rc} - \mu_{r.} - \mu_{.c} + \mu) = 0 \qquad \text{for interaction}$$

For the interaction, the hypothesis is that the population cell means are equal after the effects of age and setting have been removed. The .05 level of significance will be used.

Step 2: Computing the Test Statistic

Using Formulas 14.4 through 14.8, we calculate as follows:

$$SS_R = \frac{1}{(6)(3)} (867{,}721) - \frac{(1609)^2}{54}$$

$$= 48{,}206.72 - 47{,}942.24 = 264.48$$

$$SS_C = \frac{1}{(6)(3)} (862{,}995) - \frac{(1609)^2}{54}$$

$$= 47{,}944.17 - 47{,}942.24 = 1.93$$

$$SS_{RC} = \frac{1}{6} (291{,}677) - \frac{1}{(6)(3)} (867{,}721) - \frac{1}{(6)(3)} (862{,}995) + \frac{(1609)^2}{54}$$

$$= 48{,}612.83 - 48{,}206.72 - 47{,}944.17 + 47{,}942.24$$

$$= 404.18$$

$$SS_W = 48{,}953 - \frac{1}{6} (291{,}677)$$

$$= 48{,}953 - 48{,}612.83 = 340.17$$

$$SS_T = 48{,}953 - \frac{(1609)^2}{54}$$

$$= 48{,}953 - 47{,}942.24$$

$$= 1010.76$$

		Amusement Park		Classroom		Backyard	
		22	30	25	28	33	28
	6	28	25	31	22	30	34
		24	27	23	24	35	32
		36	31	29	31	24	27
Age	8	34	35	32	25	28	21
		29	32	26	28	23	25
		35	30	36	32	33	34
	10	29	34	35	34	30	28
		32	28	38	36	32	36

Table 14.5   Data for the two-way ANOVA example

The summary ANOVA table for this example is given in Table 14.7; note that the F ratios for testing the three null hypotheses are as follows:

$$F_R = MS_R/MS_W = \frac{132.24}{7.559} = 17.49$$

$$F_C = MS_C/MS_W = \frac{0.965}{7.559} = 0.13$$

$$F_{RC} = MS_{RC}/MS_W = \frac{101.045}{7.559} = 13.37$$

Step 3: Setting the Criterion for Rejecting H_0

The sampling distribution for the null hypotheses for both rows and columns is the F distribution with 2 and 45 degrees of freedom. We find from Table 4 in the Appendix that the critical value of F at $\alpha = 0.05$ is 5.11. The sampling distribution for the interaction hypothesis is the F distribution with 4 and 45 degrees of freedom; the critical value at $\alpha = 0.05$ is 3.77

Step 4: Deciding Whether to Reject or Retain the Null Hypothesis

For the null hypothesis for rows, H_0: $\mu_6 = \mu_8 = \mu_{10}$, the observed F ratio (17.49) exceeds the critical value (5.11); therefore, the null hypothesis is rejected, with the conclusion that the population means for children 6, 8, and 10 years of age differ. Note that, when such a hypothesis is rejected, a *post hoc* multiple comparison pro-

	Amusement Park	Classroom	Backyard	Row Totals
6	$T_{11} = 156$ $\overline{X}_{11} = 26.00$ $\Sigma X_i^2 = 4098$	$T_{12} = 153$ $\overline{X}_{12} = 25.50$ $\Sigma X_i^2 = 3959$	$T_{13} = 192$ $\overline{X}_{13} = 32.00$ $\Sigma X_i^2 = 6178$	$T_{1.} = 501$ $\overline{X}_{1.} = 27.83$ $\Sigma X_i^2 = 14,235$
Age **8**	$T_{21} = 197$ $\overline{X}_{21} = 32.83$ $\Sigma X_i^2 = 6503$	$T_{22} = 171$ $\overline{X}_{22} = 28.50$ $\Sigma X_i^2 = 4911$	$T_{23} = 148$ $\overline{X}_{23} = 24.67$ $\Sigma X_i^2 = 36.84$	$T_{2.} = 516$ $\overline{X}_{2.} = 28.67$ $\Sigma X_i^2 = 15,098$
10	$T_{31} = 188$ $\overline{X}_{31} = 31.33$ $\Sigma X_i^2 = 5930$	$T_{32} = 211$ $\overline{X}_{32} = 35.17$ $\Sigma X_i^2 = 7441$	$T_{33} = 193$ $\overline{X}_{33} = 32.17$ $\Sigma X_i^2 = 6249$	$T_{3.} = 592$ $\overline{X}_{3.} = 32.89$ $\Sigma X_i^2 = 19,620$
Column Totals	$T_{.1} = 541$ $\overline{X}_{.1} = 30.06$ $\Sigma X_i^2 = 16,531$	$T_{.2} = 535$ $\overline{X}_{.2} = 29.72$ $\Sigma X_i^2 = 16,311$	$T_{.3} = 533$ $\overline{X}_{.3} = 29.61$ $\Sigma X_i^2 = 16,111$	$T = 1,609$ $\overline{X} = 29.80$ $\Sigma\Sigma\Sigma X_{irc}^2 = 48,953$

$$\Sigma T_{r.}^2 = (501)^2 + (516)^2 + (592)^2 = 867,721$$

$$\Sigma T_{.c}^2 = (541)^2 + (535)^2 + (533)^2 = 862,995$$

$$\Sigma T_{rc}^2 = (156)^2 + (153)^2 + (192)^2 + (197)^2 + (171)^2$$
$$+ (148)^2 + (188)^2 + (211)^2 + (193)^2 = 291,677$$

$$T = \Sigma\Sigma\Sigma X_{irc} = 1609$$

$$\Sigma\Sigma\Sigma X_{irc}^2 = 48,953$$

Table 14.6   **Summary calculations for the two-way ANOVA example**

cedure, such as the Tukey method, is used to determine which pairs of means differ significantly.

For the null hypothesis for columns, H_0: $\mu_{AP} = \mu_C = \mu_B$, the observed F ratio (0.13) does *not* exceed the critical value (5.11); therefore, the null hypothesis is retained. However, the observed F ratio for the interaction hypothesis (13.37) exceeds the critical value (3.77); therefore, we reject the null hypothesis. We conclude that the differences among cell means, after accounting for row and column differences, are greater than could be attributed to sampling fluctuation alone. This significant interaction will be discussed further below.

The Meaning of the Main Effects

In our example there were two independent variables, age of the subjects and setting of the picture. The tests of the two hypotheses involving these independent variables

Source	SS	df	MS	F	F_{cv}
Rows	264.48	2	132.24	17.49	5.11
Columns	1.93	2	0.965	0.13	5.11
Interaction	404.18	4	101.045	13.37	3.77
Within Cells	340.17	45	7.559		
Total	1,010.76	53			

Table 14.7 Summary table for the two-way ANOVA example

are referred to as the tests of the *main effects*. The data given in Table 14.6 indicate that the test of the row effect was statistically significant and that the null hypothesis was rejected. For the main effect, age, the significant F ratio indicates that the differences among the means of the 6-, 8-, and 10-year-olds were too great to attribute to random sampling fluctuation if the null hypothesis is true. Therefore, the null hypothesis for rows was rejected. Note that, in interpreting the F ratio for the row main effect, we considered the row means averaged across the C columns. For the column main effect, we considered the column means averaged across the R rows.

The Meaning of Interaction

An interaction among the independent variables in a two-factor design is operating when the effect of the levels of the first independent variable on the dependent variable is *not* the same across the levels of the second independent variable. The best way to examine interaction is to present it graphically by plotting the cell means. The scale of the dependent variable is placed on the vertical axis, and the levels of one independent variable are placed on the horizontal axis. Consider the data for our example; the plot of the cell means is shown in Figure 14.1. If there is no interaction between the two independent variables, the lines connecting the cell means are parallel or nearly so. Note that, for these data, the lines are not parallel. For the classroom setting, the distance between the figure of the stranger and the figure of the child was increasingly greater for the older age groups. However, the trends for the backyard and the amusement park settings were quite different from this trend for the classroom setting, as well as different from each other. For the backyard setting, the distances for the 6- and 10-year-olds were nearly the same, whereas the distance for the 8-year-olds was much less. For the amusement park setting, the distances for the 6- and 10-year-olds were approximately the same, but the distance was much more for the 8-year-olds. As can be seen, the interpretation of this particular interaction is not easy; no pattern is apparent. However, in other studies, patterns of interaction are more easily discernible. In any case, when a significant interaction is found, the first step is to plot the cell means followed by a logical interpretation of the pattern of these means.

General Characteristics of Interaction Plots
When a nonsignificant interaction is found in a two-way ANOVA, the lines connecting the cell means in an

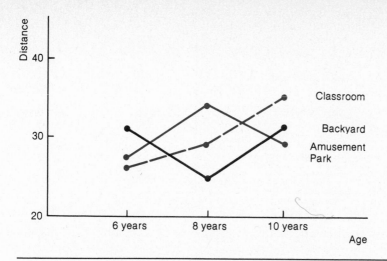

Figure 14.1   Plot of the interaction for the example data

interaction plot are nearly parallel (within sampling fluctuation). Had the interaction in our example been nonsignificant, the plot of the interaction could have looked like the plot shown in Figure 14.2(a). The fact that the lines connecting the cell means are parallel indicates that the effect of the various settings is the same for all ages. Now consider Figure 14.2(b). Here the lines are not straight across all three age levels, but the line segments connecting the cell means are parallel. Such a pattern also indicates no interaction among the two independent variables.

If the interaction is found to be nonsignificant in the ANOVA, it is not necessary to plot the cell means. However, when a significant interaction is found, the researcher should plot the cell means in order to explain the nature of the interaction. Consider the plots of the cell means shown in Figures 14.2(c) and 14.2(d). Both of these plots indicate a significant interaction, because the lines connecting the cell means are not parallel. In Figure 14.2(c), the cell means for classroom are always greater than for backyard or amusement park across all three ages, and the lines do *not* intersect. This pattern is called an *ordinal interaction*. Now consider Figure 14.2(d), wherein the lines intersect within the plot. This pattern is called a *disordinal interaction*.

The significance of the interaction is determined in the ANOVA. A nonsignificant interaction is illustrated by nearly parallel lines (within sampling fluctuation) that connect the cell means. A significant interaction is illustrated by nonparallel lines. An interaction is ordinal when the lines do not intersect within the plot; an interaction is disordinal when they do intersect within the plot.

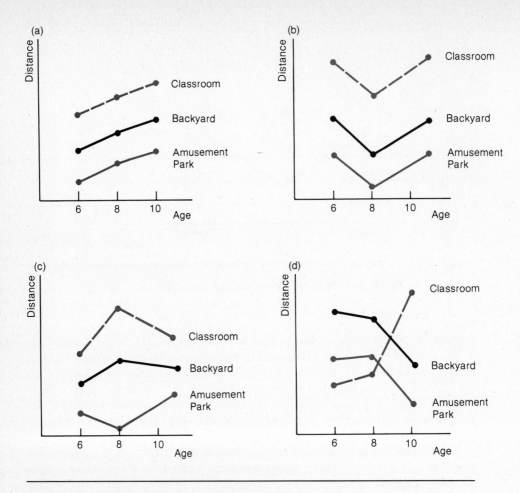

Figure 14.2 Plots of possible interactions

Assumptions of a Two-Way ANOVA

In Chapter 13 we discussed the assumptions underlying the use of ANOVA. The assumptions of two-way ANOVA are the same. The importance of the assumptions lies in the use of the F distribution as the appropriate underlying distribution for testing the null hypotheses.

1. The samples are independent, random samples from defined populations.
2. The dependent variable is measured on at least an interval scale.
3. The scores on the dependent variable are normally distributed in the population.
4. The population variances for all of the cells in the design matrix are equal (homogeneity of variance).

15	30	60		20	5	22
10	20	40		18	2	9
5	10	20		36	16	11
5	10	20		4	8	7
Proportionate				Disproportionate		

Table 14.8   Examples of proportionate and disproportionate cell frequencies

Special Topics in Two-Way ANOVA

The development of the two-way ANOVA in this chapter has assumed that the numbers of cases in each cell were equal or close to equal, or that the row or column sample sizes were proportionate. Examples of proportionate and disproportionate cell frequencies are given in Table 14.8. Procedures for computing sums of squares for disproportionate cell frequencies appear in more advanced statistics texts.

We have also presented only the fixed model ANOVA. In the fixed model, the researcher is interested only in the levels of the independent variables that appear in the analysis. There are instances (though they are rare in behavioral science research) when the levels are randomly selected from a population of possible levels. It is necessary to alter the F ratios in those situations. This topic is also left to more advanced texts.[1]

Summary

This chapter discusses two-way analysis of variance as the procedure for testing null hypotheses when two independent variables are considered simultaneously. Analyses in which two or more independent variables are considered simultaneously are called factorial designs. The advantages of factorial designs include (1) efficiency—investigating the effects of two independent variables simultaneously, (2) control—taking into consideration the variation due to a second independent variable and thus enhancing the statistical precision, and (3) making possible a study of the interaction of the two independent variables.

The concept of partitioning the total variation and the computational procedures for two-way ANOVA are a logical extension of procedures used for one-way ANOVA. The total variation is partitioned into four components: (1) the error variance, or within-cells variation, (2) variation due to the row main effect, (3) variation due to

[1]One suggested text is D. Hinkle, W. Wiersma, and S. Jurs, *Applied Statistics for the Behavioral Sciences* (Boston, Mass.: Houghton Mifflin, 1979).

the column main effect, and (4) variation due to the interaction effects after considering the main effects.

The null hypotheses tested in two-way ANOVA relate to the row effects, the column effects, and the interaction effects. The test statistic for each of the hypotheses is the F ratio. The sampling distribution of the respective F ratios is the F distribution with the appropriate number of degrees of freedom. As was the case for one-way ANOVA, when the observed F ratio exceeds the critical value of F, we reject the respective null hypothesis.

ANOVA is a widely used and powerful statistical technique. But, as we have so often emphasized, it is an inferential technique, and its application fits the chain of reasoning that underlies inferential statistics. Here, as elsewhere, we begin with a sample selected from the population and make an inference about that population on the basis of the observed sample statistics and the probability that they would occur if the null hypothesis were true. In the next chapter we will discuss inferential statistical procedures for testing hypotheses when the data do not meet all the assumptions necessary for us to use the procedures introduced in Chapters 9 through 14. Although the procedures will vary, the chain of reasoning remains the same.

Key Concepts

Factor	Main effects
Factorial design	Ordinal interaction
Interaction	Disordinal interaction

Exercises

14.1 A political scientist who was interested in attitudes toward tax reform conducted a community survey of Republicans and Democrats of various ages. One analysis of particular interest was a comparison of the attitude of young voters (age 18–20) and older voters (age 38–40). A second analysis was performed to determine whether any difference in attitude between Republicans and Democrats was consistent across these age groups. The data follow. High scores indicate a greater concern for tax reform. Complete the two-way ANOVA. (Use $\alpha = .05$.)

a. State the hypotheses.
b. Compute the test statistics and determine the critical values.
c. Decide whether to reject or to retain the null hypotheses.
d. Interpret the results.

Age Group

	18–20		38–40	
Democrats	4	4	10	10
	9	7	13	12
	7	5	9	11
	3	6	11	13
	5	2	15	8
Republicans	9	10	20	15
	9	8	22	19
	14	11	17	18
	12	10	16	20
	12	7	18	17

14.2 The following data represent the average number of trials that are required to complete a psychomotor test under the influence of 3 different drugs and at 4 different times during the day. Sixty (60) subjects were randomly assigned to the different treatment condition (5 were assigned to each cell). A partial Summary ANOVA table is also presented. Complete the Summary ANOVA table. (Use $\alpha = .01$.)

a. State the hypotheses.
b. Compute the test statistics and determine the critical values.
c. Decide whether to reject or to retain the null hypotheses.
d. Interpret the results.
e. Even though the interaction is not significant, plot the cell means.

Drug

	A	B	C	
8 A.M.	4	3	2	3
11 A.M.	5	5	2	4
4 P.M.	7	6	5	6
7 P.M.	8	6	7	7
	6	5	4	5

Summary ANOVA

Source	SS	df	MS	F	F_{cv}
Time	150				
Drug			20		
Interaction			3.33		
Within					
Total	374				

14.3 A psychologist was interested in investigating the effects of various stressors on cognitive performance of children. The independent variables in his study included three stressors (interruptions, flickering lights, and background noise) and the sex of the child (male or female). The development variable was a speeded mathematics test of 25 questions. The data follow. Complete the two-way ANOVA. (Use $\alpha = .01$.)

a. State the hypotheses.
b. Compute the test statistics and determine the critical values.
c. Decide whether to reject or to retain the null hypotheses.
d. Interpret the results.
e. Plot the cell means and discuss the interaction.

	Interruptions	Flickering Lights	Background Noise
	9	12	20
	10	11	19
Males	13	16	19
	12	14	16
	15	14	17
	8	20	15
	8	18	11
Females	10	17	13
	11	19	15
	10	20	12

▦ Key Strokes for Selected Exercises

14.1 Use the Key Strokes described in the Appendix to determine ΣX and ΣX^2 for each group. The data needed are as follows:

$$T_{11} = 52 \qquad T_{1.} = 164 \qquad T = 448$$
$$T_{12} = 112 \qquad T_{2.} = 284 \qquad \Sigma\Sigma\Sigma X^2_{irc} = 6036$$
$$T_{21} = 102 \qquad T_{.1} = 154$$
$$T_{22} = 182 \qquad T_{.2} = 294$$

Step 1: Determine $\Sigma T^2_{r.}$ and $SS_{r.}$.

Value	Key Stroke	Display
$164(T_{1.})$	X^2	26896
	STO	26896
$284(T_{2.})$	X^2	80656
	M+	80656
	RCL	107,552 $= \Sigma T^2_{r.}$

Value	Key Stroke	Display	
10(n)	$\times$	10	
2(c)	=	20	
	1/X	0.0500	
	$\times$	0.0500	
	RCL	107,552	
	=	5377.6	(save)
	$-$	5377.6	
5017.6(T^2/n)	=	360	$= SS_R$

Step 2: Determine $\Sigma T^2_{\cdot c}$ and SS_c.

154($T_{\cdot 1}$)	X^2	23716	
	STO	23716	
294($T_{\cdot 2}$)	X^2	86436	
	M+	86436	
	RCL	110,152 $= \Sigma T^2_{\cdot c}$	
10(n)	$\times$	10	
2(r)	=	20	
	1/X	0.0500	
	$\times$	0.0500	
	RCL	110,152	
	=	5507.6	(save)
	$-$	5507.6	
5017.6 (T^2/N)	=	490	$= SS_C$

Step 3: Determine ΣT^2_{rc} and SS_{rc}.

52(T_{11})	X^2	2704	
	STO	2704	
112(T_{12})	X^2	12,544	
	M+	12,544	
102(T_{21})	X^2	10,404	
	M+	10,404	
182(T_{22})	X^2	33,124	
	M+	33,124	
	RCL	58776 $= \Sigma T^2_{rc}$	
10(n)	1/X	0.100	
	$\times$	0.1000	
	RCL	58776	
	=	5877.6	(save)
	$-$	5877.6	

Value	Key Stroke	Display

$$\text{5377.6 (From Step 1)} \qquad - \qquad 500.0$$

$$\left(\frac{1}{nC}\Sigma T_{r\cdot}^2\right)$$

$$\text{5507.6 (From Step 2)} \qquad + \qquad -5007.6$$

$$\left(\frac{1}{nR}\Sigma T_{\cdot c}^2\right)$$

$$\text{5017.6 } (T^2/n) \qquad\qquad = \qquad \underline{10} \qquad = SS_{RC}$$

Step 4: Determine SS_w and SS_T.

$$\text{6036} (\Sigma\Sigma\Sigma X_{irc}^2) \qquad - \qquad 6036$$
$$\text{5877.6 (From Step 3)} \qquad = \qquad \underline{158.40} = SS_w$$

$$\left(\frac{1}{n}\Sigma T_{rc}^2\right)$$

$$\text{6036 } (\Sigma\Sigma\Sigma X_{irc}^2) \qquad - \qquad 6036$$
$$\text{5017.6 } (T^2/n) \qquad\qquad = \qquad \underline{1018.40} = SS_T$$

Step 5: Determine MS_R, MS_C, MS_{RC}, MS_W.

$$\text{360 } (SS_R) \qquad\qquad \div \qquad 360$$
$$\text{1 } (df_R) \qquad\qquad = \qquad \underline{360} = MS_R$$

$$\text{490 } (SS_C) \qquad\qquad \div \qquad 490$$
$$\text{1 } (df_C) \qquad\qquad = \qquad \underline{490} = MS_C$$

$$\text{10 } (SS_{RC}) \qquad\qquad \div \qquad 10$$
$$\text{1 } (df_{RC}) \qquad\qquad = \qquad \underline{10} = MS_{RC}$$

$$\text{158.40} \qquad\qquad \div \qquad 158.40$$
$$\text{36 } (df_W) \qquad\qquad = \qquad \underline{4.40} = MS_W$$
$$\qquad\qquad\qquad STO \qquad 4.40$$

Step 6: Determine F.

$$\text{360 } (MS_R) \qquad\qquad \div \qquad 360$$
$$\qquad\qquad\qquad RCL \qquad 4.40$$
$$\qquad\qquad\qquad = \qquad \underline{81.82} = F_R$$

$$\text{490 } (MS_C) \qquad\qquad \div \qquad 490$$
$$\qquad\qquad\qquad RCL \qquad 4.40$$
$$\qquad\qquad\qquad = \qquad \underline{111.36} = F_C$$

$$\text{10 } (MS_{RC}) \qquad\qquad \div \qquad 10$$
$$\qquad\qquad\qquad RCL \qquad 4.40$$
$$\qquad\qquad\qquad = \qquad \underline{2.27} = F_{RC}$$

15

Selected Nonparametric Tests

Suppose a political scientist is interested in studying the political party preference of registered voters who reside in different types of localities in the state. In the investigation, random samples are drawn from the following types of localities: large urban, small urban, suburban, and rural. The voters selected are then asked to respond to a questionnaire. Among the many questions in the questionnaire, the respondents are asked to indicate their age, sex, occupation, income, and political party preference. Several of these variables, such as age and income, can be considered continuous. For these two variables, we could also make the assumption that the population is normally distributed. On the other hand, the variables sex, occupation, and political party preference are discrete variables, and neither continuity nor normality can be assumed. Because the tests of hypotheses discussed in the previous chapters assume normality and thus continuity, they are not appropriate for analyzing the data for this investigation.

In the tests that we have worked with, the null hypothesis includes a specified value for the population parameter. Such tests are called *parametric tests*. As we have said, the use of these tests requires certain assumptions. They include the following: (1) the elements in the sample have been randomly selected from the population, (2) the characteristics under study in the investigation (the criterion variables) are normally distributed in the population, and (3) the null hypotheses relative to the specific values of the population parameters are true. For the $k \geq 2$ sample case, an additional assumption is made, that of equality (homogeneity) of population variances. Within this latter assumption is yet another one. We must assume that a meaningful value of the variance can be computed and thus that the measurement scale of the criterion variable must be interval (or nearly interval).

There are, however, research investigations (such as the one just outlined) for which it is impossible to meet many of the foregoing parametric assumptions. Not only would it be impossible to assume that the criterion variable (political party preference) is normally distributed in the population, but it would also be impossible to meet the assumption of homogeneity of variance for the respective populations. How can we analyze these data? The answer is to use *assumption-free tests* that have been developed over the past several decades. These tests are generally referred to as *nonparametric tests* or *distribution-free tests*.[1]

Nonparametric statistical tests are much less demanding in terms of required assumptions, so they can be used in situations wherein parametric statistics are not appropriate. In addition, one advantage of these tests is that they are relatively easy to compute and the results are relatively easy to interpret. A second advantage is that the tests are not restricted to testing hypotheses about specified values of population parameters. As we will see, nonparametric tests can be used to test hypotheses about the equality of population distributions. A third advantage is that there are no parametric alternatives for investigations dealing with small samples. This chapter presents a selected number of nonparametric tests.[2] The tests selected are analogous to the parametric tests we have discussed in previous chapters. The specific tests, along with their parametric analogs, are summarized in Table 15.1.

The χ^2 Distribution

Before we begin to discuss the various nonparametric tests, we must define the χ^2 (chi square) distributions.[3] These distributions are most frequently used in the analysis

[1]For a discussion of the differences between parametric, nonparametric, and distribution-free statistics, see L. A. Marascuilo and M. McSweeney, *Nonparametric and Distribution-Free Methods for the Social Sciences* (Monterey, Calif.: Brooks/Cole, 1977), Chapter 1.

[2]For additional nonparametric tests, see M. Hollander and D. A. Wolfe, *Nonparametric Statistical Methods* (New York: Wiley, 1973), and L. A. Marascuilo and M. McSweeney, *Nonparametric and Distribution-Free Methods for the Social Sciences*.

[3]For a more exhaustive discussion of the development of the χ^2 distribution, see L. Horowitz, *Elements of Statistics for Psychology and Education* (New York: McGraw-Hill, 1974), pp. 371–380.

Nominal Data	Ordinal Data	Parametric Analog
χ^2 One-sample case	—	z test of H_0: $P = a$
		t test of H_0: $\mu = a$
χ^2 Two-sample case (test of independence)	Median test	z test of H_0: $P_1 = P_2$
	Mann–Whitney U Test	t test of H_0: $\mu_1 = \mu_2$
χ^2 k-sample case (test of independence)	Kruskal–Wallis H Test	One-Way ANOVA
χ^2 Two-sample case, dependent samples (McNemar Test)	Wilcoxon matched-pair signed rank test	z test of H_0: $P_1 = P_2$
		t test of H_0: $\mu_1 = \mu_2$

Table 15.1 Nonparametric tests and the corresponding parametric analogs

of nominal data as the sampling distribution of the test statistic in the tests of hypotheses to be discussed in this chapter. In these analyses, we compare observed frequencies of occurrence with theoretical frequencies. The *observed frequencies* are those that are obtained empirically by the researcher though direct observation. The theoretical or *expected frequencies* are developed on the basis of some hypothesis. For example, in 200 flips of a coin, we *expect* 100 heads and 100 tails. But what if we *observed* 92 heads and 108 tails? Would we reject the hypothesis that the coin is unbiased? Or would we attribute the difference between the observed and expected frequencies of occurrence to random fluctuation?

Consider another example along the same line of reasoning. We hypothesize that we have an unbiased die; to test this hypothesis, we roll the die 300 times and observe the frequencies of occurrence of the faces. Because we hypothesize that the die is unbiased, we expect that each of the six faces will have the same frequency of occurrence; that is, each face will appear 50 times. However, suppose we observe the following frequencies of occurrence:

Face Value	Occurrence
1	42
2	55
3	38
4	57
5	64
6	44

Again, what would we conclude? Is the die biased, or can we attribute the difference to random fluctuation?

Consider a third example. Suppose a college president hypothesized that 75% of the students at the college would want to continue on the quarter system and 25% would want to change to the semester system. A survey was conducted. Of a random sample of 160 students, 99 favored the quarter system and 61 favored the semester system. The president originally hypothesized that the number of students would be divided 75% to 25%, so we would expect the frequencies to be 120 for the quarter and 40 for the semester system. Should the original hypothesis be rejected, or can the difference between the observed frequencies and the expected frequencies be attributed to chance fluctuation?

In each of these examples, the test statistic for testing the hypothesis is χ^2; that is, a χ^2 value is computed. For the purposes of this chapter, the χ^2 value is used to compare observed and expected frequencies and is defined as follows:

$$\chi^2 = \sum_{i=1}^{k} \frac{(O - E)^2}{E} \tag{15.1}$$

where

O = observed frequency

E = expected frequency

k = number of categories, groupings, or possible outcomes

The theoretical sampling distribution of χ^2 can be generated, but it is not necessary to do so. As with t distributions, there is a family of χ^2 *distributions,* each a function of the degrees of freedom associated with the sample data. As in the case of the t distributions, only a single degrees-of-freedom value is required to identify a specific χ^2 distribution. Consider the data for the coin toss example; the calculation of χ^2 is given in Table 15.2. Note that the frequencies for the two categories are not independent. To obtain the frequency of tails, we subtract the frequency of heads from the total frequency, obtaining $200 - 92$, or 108. Given one frequency, we can readily determine the other. In other words, only one frequency is free to vary in this example, so there is only 1 degree of freedom associated with the value of χ^2.

Consider the die example. The calculation of the χ^2 value is found in Table 15.3. In this example, there are 5 degrees of freedom associated with the calculation of χ^2. If any five of the frequencies are known, the sixth frequency is uniquely determined, because the total frequency must equal 300.

Unlike the t distributions, which are all symmetric, the theoretical sampling distributions of χ^2 that are associated with small degrees of freedom are positively skewed. However, as the number of degrees of freedom associated with χ^2 increases, the respective sampling distribution approaches symmetry. The χ^2 distributions for degrees of freedom 1, 3, 5, and 10 are illustrated in Figure 15.1. Note that all values of χ^2 are positive and that they range from zero to infinity.

The use of the χ^2 distributions in hypothesis testing is analogous to the use of the t or the F distributions. In the coin tossing example, the null hypothesis is that the frequencies of heads and tails are equal (100 in 200 tosses). In general, it is not required that the expected frequencies for all categories be the same.

To test a hypothesis using the χ^2 distributions, we calculate the χ^2 value and then compare it with the critical value of χ^2 for the appropriate sampling distribution (as determined by the degrees of freedom and the level of significance). If the calculated value of χ^2 exceeds the critical value, the null hypothesis is rejected, and we conclude

	O	E	$O - E$	$(O - E)^2$	$\dfrac{(O - E)^2}{E}$
Heads	92	100	-8	64	.64
Tails	108	100	$+8$	64	.64
Totals	200	200	0	—	$1.28 = \chi^2$

Table 15.2   Calculation of χ^2 for the coin toss example

that the difference between the observed and expected frequencies is too great to be explained by sampling fluctuation.

Critical Values of the χ^2 Distributions

The critical values of χ^2 for degrees of freedom 1 through 30 are found in Table 3 in the Appendix. Different percentile points in each distribution are given, though in this chapter we are primarily concerned with those associated with the more commonly used α levels, such as .05 and .01. For the coin toss example (df = 1), the critical values of χ^2 for α = .05 and α = .01 are 3.84 and 6.64, respectively. For the die example (df = 5), the critical vlaues are 11.07 and 15.09. Table 3 contains critical values for df = 1 through df = 30. This is sufficient for most research settings, although one occasionally encounters situations in which the degrees of freedom associated with a χ^2 test are greater than 30. For these situations, the following expression has a sampling distribution that is approximately normal:

$$\sqrt{2\chi^2} - \sqrt{2(\mathrm{df})} - 1 \tag{15.2}$$

Because the underlying distribution of the test statistic given by formula 15.2 is approximately normal, the critical values for the .05 and .01 levels of significance are 1.96 and 2.58, respectively.

The χ^2 distributions are a family of distributions, each of which is determined by a single degrees-of-freedom value. For degrees of freedom greater than 30, the underlying distribution of the statistic given by formula 15.2 is the normal distribution.

Now that we can use the table of χ^2 distributions, we can complete the examples. A significance level of .05 will be used. In the coin example, the computed χ^2 value of 1.28 is less than the critical value of 3.84 found in Table 3. Therefore, we retain H_0 and conclude that the coin is unbiased. The deviation of the observed heads and tails from the expected value of 100 is attributed to random fluctuation. A similar result is obtained for the die example, because the computed χ^2 value of 10.28 is less than the critical value of 11.07.

Face Value	O	E	$O - E$	$(O - E)^2$	$\dfrac{(O - E)^2}{E}$
1	42	50	-8	64	1.28
2	55	50	5	25	.50
3	38	50	-12	144	2.88
4	57	50	7	49	.98
5	64	50	14	196	3.92
6	44	50	-6	36	.72
Total	300	300	0	—	$10.28 = \chi^2$

Table 15.3 Calculation of χ^2 for the die example

For the example involving preference for the quarter system or the semester system, the χ^2 value is computed as follows:

$$\chi^2 = \frac{(99 - 120)^2}{120} + \frac{(61 - 40)^2}{40} = 14.7$$

With 1 degree of freedom, the computed χ^2 value of 14.7 is greater than the critical value of 3.84. Thus the hypothesis that 75% of the students would favor the quarter system is rejected at the .05 level of significance.

In testing a hypothesis using χ^2, if the computed χ^2 value is greater than the critical value for χ^2, we reject the hypothesis at the appropriate level of significance.

Nominal Data: One-Sample Case

The one-sample case for nominal data was illustrated in our earlier discussion of the χ^2 distributions. The one-sample case is often referred to as the "goodness-of-fit" test.[4] This terminology comes from the idea that the purpose of the test is to indicate whether the observed frequencies are a "good" fit to the expected frequencies. The fit is considered good when the observed frequencies are within sampling fluctuation of the expected frequencies and the χ^2 value is relatively small—that is, less than the critical value of χ^2 for the appropriate degrees of freedom.

Consider the public opinion poll of registered voters in which the voters were asked to indicate their political party preference. One purpose of this poll was to determine

[4]The goodness-of-fit test can be applied to other than nominal data. See Horowitz, *Elements of Statistics*, pp. 380–385.

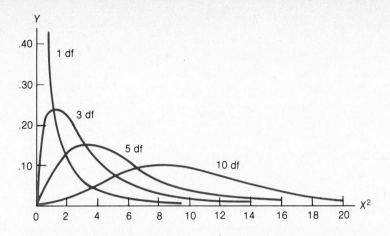

Figure 15.1 X^2 distributions for 1, 3, 5, and 10 degrees of freedom

whether the political party preference in the state had changed over the past four years. When this same poll was taken four years ago, the ratio of Republicans to Democrats to Independents was 5 : 4 : 3. The result of the more recent poll was that, of the 600 voters polled, 322 indicated that they were Republicans, 184 identified themselves as Democrats, and 94 were Independents. Let us consider this example in terms of the four steps of hypothesis testing.

Step 1: Stating the Hypothesis
In this public opinion poll, the null hypothesis tested was that the political party preference for registered voters in the state had not changed over the past four years. In other words, it was hypothesized that the ratio of Republicans to Democrats to Independents was 5 : 4 : 3. Assume that, for testing this hypotheses, the level of significance was set *a priori* at .05.

Step 2: Computing the Test Statistic
The test statistic appropriate for this example is χ^2, and the first step in the calculation is to determine the expected frequency of registered voters for each of the political parties. The ratio was hypothesized to be 5 : 4 : 3, so the expected frequencies would be 250 Republicans, 200 Democrats, and 150 Independents. The observed and expected frequencies are given in Table 15.4. Using formula 15.1, a χ^2 is found to be 42.93.

Step 3: Determining the Criterion for Rejecting H_0
In this example, there are 2 degrees of freedom associated with the test statistic. That is, once the total frequency is known and the frequencies of registered voters who are Republicans and Democrats are known, then the frequency of registered voters who are Independents is also known. Because the level of significance was established *a priori* at .05, the critical value for χ^2 for 2 degrees of freedom (found in Table 3 in the Appendix) is 5.99.

Party	O	E	$(O - E)$	$(O - E)^2$	$\dfrac{(O - E)^2}{E}$
Republicans	322	250	72	5184	20.74
Democrats	184	200	-16	256	1.28
Independents	94	150	-56	3136	20.91
Total	600	600	0	—	$42.93 = \chi^2$

Table 15.4   Calculation of χ^2 for the political party preference data

Step 4: Deciding Whether to Reject or Retain the Null Hypothesis

The calculated value of χ^2 exceeds the critical value, so the null hypothesis is rejected. In rejecting the null hypothesis, we conclude that the difference between the observed and expected frequencies in the categories of registered voters is too great to be attributable only to sampling fluctuation. Thus it would appear that political party preference has changed over the past few years. We must be aware of the fact that, by rejecting the null hypothesis, we may have made a Type I error (rejecting a true hypothesis), but the probability of our having made such an error is less than .05.

The χ^2 value is computed over all categories, and a significant χ^2 value does not indicate which of the data categories have been responsible for the statistical significance. However, inspection of the individual data categories can indicate which categories have been the major contributors to the statistical significance. We make this inspection by looking at the $(O - E)^2/E$ values for each of the categories. In this example, there were large values for both the Republican category and the Independent category. These data indicate that, over the past four years, there appears to have been a shift of Independents to the Republican Party.

A statistically significant χ^2 value indicates that the observed frequencies are not a ''good fit'' to the expected frequencies. An inspection of the difference between observed and expected frequencies in the individual data categories is needed to determine the categories contributing most to the statistical significance.

Nominal Data: Two-Sample Case

The χ^2 statistic is frequently used to compare two or more groups on a nominal variable with two or more categories. Consider the example of the college students'

	Support Quarter System	Support Semester System	
Upperclassmen	52	28	80
Lowerclassmen	47	33	80
	99	61	160

Table 15.5   Data for the university calendar example

opinion about the academic calendar (quarter system versus semester system). This example was initially used to illustrate the use of the χ^2 statistic for a one-sample case. Now let's extend the example to the two-sample case by assuming that 80 of the 160 student respondents were underclassmen (freshmen and sophomores) and that 80 were upperclassmen (juniors and seniors). The president could hypothesize that there would be no difference between the two groups in terms of the amount of support for the two academic calendars. The data used to test this hypotheses are given in a 2 × 2 contingency table in Table 15.5.

The observant reader will see that we could also test this hypothesis by using the two-sample test for independent proportions (see Chapter 12). However, this test is restricted to cases in which two groups respond only to one of two categories. What we want to consider is a statistical test that is appropriate to situations in which two samples are compared on a nominal variable that has two or more categories. For example, let's expand the political party preference opinion poll to a two-sample case by assuming that 350 of the 600 registered voters polled were males and that 250 were female. In this case, the hypothesis to be tested is one of no difference between the two groups in terms of their political party preference. The data used to test this hypothesis are given in a 2 × 3 contingency table in Table 15.6. The numbers in parentheses are the expected frequencies for each of the cells. We will outline the procedure for determining these expected frequencies.

For the latter of these two examples, the two-sample test for independent proportions is *not* appropriate. However, the χ^2 *test of independence* is appropriate for both examples. In the first example, rather than stating a hypothesis of no difference between underclassmen and upperclassmen in terms of academic calendar preference, we can use the null hypothesis that academic calendar preference is "independent" of the class status of the students. Similarly, the null hypothesis for the second example would be that political party preference is independent of the sex of the respondent. In either case, if the null hypothesis is rejected, the conclusion is that there is a difference between the respondent groups in the population. In terms of the test of independence, the conclusion would be that the responses to the various categories of the criterion variable (academic calendar preference or political party preference) is

Nominal Data: Two-Sample Case

	Republican	Democrat	Independent	
Male	170 (187.83)	112 (107.33)	68 (54.83)	350
Female	152 (134.17)	72 (76.67)	26 (39.17)	250
	322	184	94	600

Table 15.6   Data for the political party preference example

dependent on the levels of the independent variable (class status or sex of the respondent).

The χ^2 test of independence is used to test the null hypothesis that there are no differences between populations in their response to categories of a nominal variable. Specifically, the test is used to determine when the responses are independent of levels of the independent variable.

Determination of the Expected Frequencies

In the one-sample case, the expected frequencies are determined in the statement of the hypothesis. In our example, the percentage of individuals in each political party was hypothesized *a priori*. These hypothesized percentages were then multiplied by the total number of individuals in the study to determine the expected frequency in each of the political parties. For the two-sample case, the expected frequencies are computed from the marginal totals. For example, consider the data given in Table 15.6. Of the 600 respondents to the opinion poll, 322, or 53.7%, indicated that they were Republicans. If the null hypothesis is true, we would expect the same percentage for both males and females. That is, (0.537)(350) = 187.83 males and (0.537)(250) = 134.17 females would have indicated that they were Republicans. The expected frequencies for the remaining cells can be determined in a similar way.

The most convenient way to calculate the expected frequency of each cell is to multiply the total row frequency by the total column frequency corresponding to the respective cell and then to divide by the total frequency (n). That is,

$$\text{Expected frequency} = \frac{f_r f_c}{n} \tag{15.3}$$

The expected frequencies are found in parentheses in Table 15.6 for each cell of the contingency table and are calculated as follows:

	Male	Female
Republican	$\dfrac{(322)(350)}{600} = 187.83$	$\dfrac{(322)(250)}{600} = 134.17$
Democrat	$\dfrac{(184)(350)}{600} = 107.33$	$\dfrac{(184)(250)}{600} = 76.67$
Independent	$\dfrac{(94)(350)}{600} = 54.83$	$\dfrac{(94)(250)}{600} = 39.17$

Note that the sum of the expected frequencies for any row or column equals the respective row or column total. This can be a useful check of the calculation. Once the expected frequencies have been calculated, we calculate the χ^2 statistic, using formula 15.1. This calculation is illustrated in Table 15.7.

In a contingency table, the expected frequencies are determined using the marginal totals. The expected frequency of an rc cell is determined by $f_r f_c / n$, the product of the row and column frequencies divided by sample size.

Determination of the Degrees of Freedom

For the one-sample case, it is relatively simple to define the degrees of freedom associated with the χ^2 statistic. The number of degrees of freedom is equal to the number of frequency categories less one $(k - 1)$. In our earlier example, there were three political parties; thus there are $3 - 1 = 2$ degrees of freedom.

The degrees of freedom for the two-sample case are similarly defined. Consider the data given in Table 15.6. Once we compute the expected frequency for males who indicated that they are Republicans (187.83), the expected frequency for females who are Republicans is uniquely determined, because the total column frequency must equal 322. That is, $322 - 187.83 = 134.17$. Continuing across the columns, once we can determine the expected frequency of males who indicated that they are Democrats, the expected frequency for females who are Democrats can be determined by subtraction: $184 - 107.33 = 76.67$. After these expected frequencies are known, the expected frequencies for males and females who indicated that they are Independents can also be determined by subtracting the expected frequencies for Republicans and Democrats from the respective row totals. For the males, $[350 - (187.83 + 107.33)] = 54.83$ would be expected to indicate that they are Independents. For the females, $[250 - (134.17 + 76.67)] = 39.17$ would be expected to indicate that they are Independents. In short, once we compute the expected cell frequencies for the first *two* cells in the upper row, we can find the rest of the expected cell frequencies for the contingency table by subtraction from the respective row and column totals. In essence, they are uniquely determined. Thus there are *two* degrees of freedom

O	E	$O - E$	$(O - E)^2$	$\dfrac{(O - E)^2}{E}$
170	187.83	−17.83	317.91	1.69
112	107.33	4.67	21.81	0.20
68	54.83	13.17	173.45	3.16
152	134.17	17.83	317.91	2.37
72	76.67	−4.67	21.81	0.28
26	39.17	−13.17	173.45	4.43
600	600	0	—	12.13 $= \chi^2$

Table 15.7   Calculation of χ^2 for the data given in Table 15.6

associated with the χ^2 value for this example. The general formula for determining the degrees of freedom associated with the statistic for any contingency table is

$$df = (\text{rows} - 1)(\text{columns} - 1) \qquad (15.4)$$

$$= (r - 1)(c - 1)$$

For the data given in Table 15.6, df $= (2 - 1)(3 - 1) = 2$.

Example—2 × 3 Contingency Table

The data given in Table 15.6 will now be considered in the context of the four steps of hypothesis testing.

Step 1: Stating the Hypothesis
The overall purpose of the opinion poll was to determine the political party preference of the registered voters. Now that we have divided the sample into males and females, the specific hypothesis to be tested is that of no difference between male and female respondents in terms of their political party preference. In other words, the political party preference is *independent* of the sex of the respondent. This hypothesis will be tested at the .10 level of significance.

Step 2: Computing the Test Statistic
The test statistic for this example is χ^2. Using the data in Table 15.6, we find the χ^2 value to be 12.13 (see Table 15.7).

Step 3: Determining the Criterion for Rejecting H_0.
For this example, there are 2 degrees of freedom associated with the χ^2 value. Thus, at the .10 level of significance, the critical value of χ^2 is 4.61.

Step 4: Deciding Whether to Reject or Retain the Null Hypothesis
Because the calculated value of χ^2 (12.13) exceeds the critical value of χ^2 (4.61), the null hypothesis is rejected. Thus there is a difference between the political party preference of males and that of females. Inspection of the data in Table 15.6 indicates

that there were more Republican females and fewer Republican males than expected. On the other hand, there were more male Independents and fewer female Independents than expected. For both sexes, there was little difference between the observed frequencies and the expected frequencies among the Democrats.

Example—2 × 2 Contingency Table

As we have indicated, the test of the null hypothesis of no difference between two independent proportions is applicable to the data depicted in a 2×2 contingency table, and the χ^2 test of independence is equally appropriate. Consider the example of the student preference about the academic calendar at the college. The data for this example are presented again in Table 15.8. In applying the χ^2 test of independence to this example, we use formula 15.3 to calculate the expected cell frequencies and formula 15.1 to calculate the χ^2 value.

For the 2×2 contingency table, the computational formula for the χ^2 value can be simplified to

$$\chi^2 = \frac{n(AD - BC)^2}{(A + B)(C + D)(A + C)(B + D)} \tag{15.5}$$

For the data given in Table 15.8, where A, B, C, and D are the cell frequencies,

$$\chi^2 = \frac{160[(63)(23) - (57)(17)]^2}{(80)(80)(120)(40)}$$

$$= \frac{160(1449 - 969)^2}{(80)(80)(120)(40)}$$

$$= 1.20$$

Note that this χ^2 value is the same (within rounding error) as the one computed using formula 15.1. Now let's consider these data in the context of the four steps of testing hypotheses.

Step 1: Stating the Hypothesis
In this example, the null hypothesis to be tested is that preference for either the quarter system or the semester system is *independent* of the class status of the respondent. In other words, there is no difference between upperclassmen and lowerclassmen in terms of their preference for the quarter system or the semester system. The null hypothesis will be tested at the .05 level of significance.

Step 2: Computing the Test Statistic
We can use either formula 15.1 or formula 15.5 for computing the test statistic. As can be seen, $\chi^2 = 1.20$, which indicates that there are relatively small differences between the observed and the expected frequencies in the cells of the 2×2 contingency table (see Table 15.8).

Step 3: Determining the Criteria for Rejecting H_0
For a contingency table, the number of the degrees of freedom equals $(r - 1)(c - 1)$. For this example, the

	Support Quarter System			Support Semester System	
Upperclassmen	63 (60)	A	B	17 (20)	80
Lowerclassmen	57 (60)	C	D	23 (20)	80
	120			40	160

O	E	$O - E$	$(O - E)^2$	$\dfrac{(O - E)}{E}$
63	60	3	9	0.15
17	20	-3	9	0.45
57	60	-3	9	0.15
23	20	3	9	0.45
160	160	0	—	$1.20 = \chi^2$

Table 15.8   Data for calculating the χ^2 statistic for the university calendar example

number of degrees of freedom is $(2 - 1)(2 - 1) = 1$, and the critical value of χ^2 is 3.84.

Step 4: Deciding Whether to Reject or Retain the Null Hypothesis

Because the calculated value of χ^2 is less than the critical value, the null hypothesis is retained. The conclusion is that preference for either the quarter system or the semester system is independent of the class status of the college student. By retaining the null hypothesis, we conclude that there is no difference between upperclassmen and lowerclassmen in terms of their preference for the academic calendar.

Nominal Data: *k*-Sample Case

In the introduction to the two-sample case for nominal data, the χ^2 test of independence was described as one of the more frequent uses of the χ^2 statistic. The test of independence was described for the 2×2 contingency table as well as for the $2 \times c$ contingency table $(c > 2)$. The application of the χ^2 test of independence for the *k*-sample case is the logical extension of the two-sample case. Rather than two samples, there are *k* samples and hence a $k \times c$ contingency table. The general procedures for determining the expected frequencies and the χ^2 value are the same.

Consider the following example. In 1949, Hollingshead[5] reported that the members

[5]A. B. Hollingshead, *Elmtown's Youth: The Impact of Social Classes on Adolescents* (New York: Wiley, 1949).

Curriculum	Class				
	I & II	III	IV	V	
College Preparatory	23 (7.27)	40 (30.32)	16 (38.01)	2 (5.40)	81
General	11 (18.58)	75 (77.49)	107 (97.13)	14 (13.80)	207
Commercial	1 (9.15)	31 (38.19)	60 (47.86)	10 (6.80)	102
	35	146	183	26	390

O	E	$O - E$	$(O - E)^2$	$\dfrac{(O - E)^2}{E}$
23	7.27	15.73	247.43	34.03
11	18.58	−7.58	57.46	3.09
1	9.15	−8.15	66.42	7.26
40	30.32	9.68	93.70	3.09
75	77.49	−2.49	6.20	0.08
31	38.19	−7.19	51.70	1.35
16	38.01	−22.01	484.44	12.75
107	97.13	9.87	97.42	1.00
60	47.86	12.14	147.38	3.08
2	5.40	−3.40	11.56	2.14
14	13.80	.20	0.04	0.003
10	6.80	3.20	10.24	1.51
390	390	0	—	$69.383 = \chi^2$

Table 15.9 Frequency of enrollments of Elmtown youths from five social classes in their alternative high school curricula

Source: Adapted from A. B. Hollingshead, *Elmtown's Youth: The Impact of Social Classes on Adolescents* (New York: Wiley, 1949) p. 462. Table X.

of a small midwestern community could be divided into five social classes. One of his specific research questions was whether the curriculum in which high school students were enrolled was independent of the social class of the students. The social class levels and high school curricula of 390 students were determined; the data are given in Table 15.9. Note that Hollingshead combined social classes I and II; they represent the highest social classes, whereas social class V represents the lowest social class identified for the study. In this investigation, there was an inherent assumption that the 390 students were a random sample of some larger population.

Step 1: Stating the Hypothesis

One of the objectives of Hollingshead's study was to determine whether high school students from different social classes choose different high school curricula. In other words, Hollingshead wanted to test the null hypothesis that, in the population, high school curriculum chosen is independent of social class. Assume that this null hypothesis was tested at the .01 level of significance.

Step 2: Computing the Test Statistic

Using the procedures outlined in previous sections, we calculate the expected frequencies for each of the cells in the 3×4 contingency table. Applying formula 15.1, the χ^2 value is found to be 69.383.

Step 3: Determining the Criterion for Rejecting H_0

For a 3×4 contingency table, the number of degrees of freedom equals $(r - 1)(c - 1) = (3 - 1)(4 - 1) = (2)(3) = 6$. With 6 degrees of freedom at the .01 level of significance, the critical value of χ^2 (obtained from Table 3 in the Appendix) is 16.81.

Step 4: Deciding Whether to Reject or Retain the Null Hypothesis

Because the calculated value of χ^2 exceeds the critical value, the null hypothesis is rejected. The conclusion is that the high school curriculum chosen by students is dependent on their social class. The data support the contention that students from different social classes are enrolled in different high school curricula. Inspection of the data given in Table 15.9 indicates that a high percentage of those students in social classes I, II, and III are enrolled in either the college preparatory or the general curriculum, whereas a high percentage of students in social classes IV and V are enrolled in either the general and or the commercial curriculum.

Use of the $k \times c$ contingency table, for data from k independent samples, involves the same general procedures as the two-sample case. The expected frequencies are computed using the marginal totals; the χ^2 value is computed using formula 15.1.

Small Expected Frequencies in Contingency Tables

The theoretical sampling distribution of χ^2 for 1 degree of freedom was illustrated in Figure 15.1. Note that the distribution is continuous; there are no breaks in the continuity. However, when the expected frequency in any of the cells of a 2×2 contingency table is small (less than 5), the sampling distribution of χ^2 for these data may depart substantially from continuity. Thus the theoretical sampling distribution of χ^2 for 1 degree of freedom may fit the data poorly. In this situation, it has been suggested that an adjustment called the Yates correction for continuity, be applied to these data.[6] However, on the basis of a 1978 study by Camilli and Hopkins,[7] the

---

[6]W. G. Cochran, "Some Methods for Strengthening the Common χ^2 Tests," *Biometrics*, 10 (1954), pp. 417–451.

[7]Gregory Camilli and Kenneth D. Hopkins, "Applicability of Chi-square to 2×2 Contingency Tables with Small Expected Frequencies," *Psychological Bulletin*, 85 (1978), 163–67.

Yates correction is *not* recommended for the χ^2 test of independence "since its use would result in an unnecessary loss of power"—that is, a tendency to retain the null hypothesis when in fact it is false.

For contingency tables larger than 2×2, the lack of continuity in the χ^2 distribution resulting from small expected frequencies is of lesser consequence. However, it is suggested that, when more than 20% of the cells have expected frequencies less than 5 and/or any cell has an expected frequency less than 1, it may be possible to combine adjacent rows and/or columns without distorting the data.

The Contingency Coefficient

In the Hollingshead study, the null hypothesis that choice of high school curriculum is independent of social class was rejected. The conclusion was that there was a relationship between the two variables. But what is the magnitude of the relationship? The χ^2 statistic does not answer that question directly; it only provides the information we need to reject the hypothesis of independence of the two variables. In order to determine the strength of the relationship, it is necessary to compute a correlation coefficient. An appropriate correlation coefficient is the *contingency coefficient (C)*, which is computed directly from the χ^2 value. This coefficient can be computed for any size contingency table and is defined as follows:

$$C = \sqrt{\frac{\chi^2}{n + \chi^2}} \tag{15.6}$$

Applying formula 15.6 to the data given in Table 15.9, we find that

$$C = \sqrt{\frac{69.383}{390 + 69.383}} = 0.389$$

To compute the contingency coefficient, we take the square root; hence C is always positive. This is not a problem for nominal data, because directionality is not relevant. However, if the data are in some sense ordinal, the direction of the association must be observed from the contingency table.

The contingency coefficient (C) is the correlation coefficient appropriate for data in a $k \times c$ contingency table. As with any correlation coefficient, the greater the value of C, the stronger the relationship between the two variables of the contingency table.

By the usual rules of thumb for interpreting the magnitude of the correlation coefficient for the data given in Table 15.9, we interpret the value of 0.389 as indicating a low relationship between choice of high school curriculum and social class. However, those rules assume that the maximum possible value of the correlation coefficient is 1.0. The maximum value of the contingency coefficient is *not* 1.0. The procedures for determining the maximum value for a specific contingency table are complex. However, an estimate can be obtained with the following formula:

$$C_{\max} = \sqrt{\frac{k-1}{k}} \qquad (15.7)$$

where

k = number of categories in the variable that has the fewest categories.

Applying this formula to the data of the Hollingshead study, we have:

$$C_{\max} = \sqrt{\frac{3-1}{3}}$$

$$= \sqrt{\frac{2}{3}}$$

$$= 0.816$$

Rather than compare the contingency coefficient .389 to a maximum of 1.0, we compare it to .816. Thus the magnitude of the relationship between choice of high school curriculum and social class may be interpreted as moderate rather than low.

The maximum value of the contingency coefficient is not 1.0, a fact that we must keep in mind when interpreting C. We can get an estimate of the maximum value of C for a given contingency table by using a formula based on the number of categories of each of the variables.

Nominal Data: Two-Sample Case, Dependent Samples

There is a χ^2 test for dependent samples involving nominal-scale data called the *McNemar test for significance of change*.[8] The test may be used in pretest–posttest research designs wherein the same sample of individuals is categorized before and after an intervening treatment. For example, suppose the editor of a small local newspaper was interested in how a recent debate between the two candidates for the Senate affected voter preference. The editor randomly selected 38 registered voters and telephoned them before and after the debate. The results of the telephone survey are recorded in Table 15.10.

Step 1: Stating the Hypothesis   In this example, we are interested only in the cells of the contingency table that reflect a change in voter preference before and after

[8]Q. McNemar, *Psychological Statistics* (New York: Wiley, 1969), pp. 260–262.

	Before Debate		
	Republican	Democrat	
After Debate Democrat	14 *A*	*B* 6	20
Republican	16 *C*	*D* 2	18
	30	8	38

Table 15.10   Data for computing χ^2 for dependent samples in the political debate example

the debate—namely, cells A and D. The specific null hypothesis is that, in the population, there will be an equal number of changes in both directions; that is, as many registered voters will change their preference from the Republican candidate to the Democratic candidate as will change their preference from the Democratic candidate to the Republican candidate. This implies that the expected frequency in cell A will be the same as the expected frequency in cell D, or $A = D$. Thus the expected value for both cell A and cell D is $(A + D)/2$. Suppose the null hypothesis is tested at the .05 level of significance.

Step 2: Computing the Test Statistic   The test statistic for this example is again χ^2 and formula 15.1 can be applied. However, the formula will be applied only to cells A and D. Therefore, formula 15.1 reduces to

$$\chi^2 = \sum \frac{(O - E)^2}{E}$$

$$= \frac{\left(A - \dfrac{A + D}{2}\right)^2}{\dfrac{A + D}{2}} + \frac{\left(D - \dfrac{A + D}{2}\right)^2}{\dfrac{A + D}{2}}$$

$$= \frac{(A - D)^2}{A + D} \qquad\qquad (15.8)$$

Using the data given in Table 15.10, we find that

$$\chi^2 = \frac{(14 - 2)^2}{14 + 2}$$

$$= \frac{(12)^2}{16}$$

$$= 9.00$$

Step 3: Determining the Criterion for Rejecting H_0

The sampling distribution of the test statistics for the McNemar test is the χ^2 distribution with 1 degree of freedom. From Table 3 in the Appendix, we find that the critical value of χ^2 for 1 degree of freedom at the .05 level of significance to be 3.84.

Step 4: Deciding Whether to Reject or Retain the Null Hypothesis

Because the computed χ^2 value (9.00) exceeds the critical value (3.84), the null hypothesis is rejected. The conclusion is that voter preference for the two candidates changed more in one direction than in the other direction. Inspection of the data indicates that more of the voters changed their support from the Republican candidate to the Democratic candidate than changed in the other direction.

Ordinal Data: Two-Sample Case

There are several nonparametric tests of significance for the two-sample case[9] with ordinal data, but only two of them will be discussed here.[10] They are the median test and the Mann–Whitney U test. We will apply them both to the same example.

Suppose a professor of marriage and family counseling is interested in the difference in marital satisfaction between married women without children and married women with children. In a pilot study, 10 married women between the ages of 25 and 30 without children (group 1) and 10 married women in the same age group with children (group 2) were selected and administered the marital satisfaction inventory. This instrument has just recently been developed, so the professor assumes that the scale of measurement is ordinal at best. The data for this example are given in Table 15.11.

The Median Test

Step 1: Stating the Hypothesis

The null hypothesis for the *median test* is that the two samples were drawn from populations that have the same or a common median. For this example, the hypothesis is that the married women without children and the married women with children have the same median with respect to marital satisfaction. The .05 level of significance will be used to test the null hypothesis against the directional alternative—that married women without children have a higher level of marital satisfaction than married women with children. In symbols,

$$H_0: \quad Mdn_1 = Mdn_2$$

$$H_a: \quad Mdn_1 > Mdn_2$$

[9]There are several appropriate tests of significance for the one-sample case when the measurement scale of the data is ordinal. Three of these tests are (1) the one-sample sign test, (2) the Mann–Kendall test for trends, and (3) the Kolmogorov–Smirnov one-sample test. These tests are not frequently used in behavioral science research, and they will not be discussed. Descriptions of the tests are found in Hollander and Wolfe, *Nonparametric Statistical Methods,* pp. 39–47, and in James V. Bradley, *Distribution-Free Statistical Tests* (Englewood Cliffs, N.J.: Prentice-Hall, 1968), pp. 287–88, 296–303.

[10]For further discussion, see Hollander and Wolfe, *Nonparametric Statistical Methods,* pp. 83–113.

Group 1, Without Children		Group 2, With Children	
Score	Rank	Score	Rank
19	3	16	1
22	5	18	2
28	8	21	4
32	11	26	6
34	13	27	7
37	14	29	9
40	17	31	10
42	18	33	12
43	19	38	15
46	20	39	16
	$128 = R_1$		$82 = R_2$

$$\text{Median for both groups combined} = \frac{31 + 32}{2} = 31.5$$

Table 15.11 Data for the marital satisfaction example

Step 2: Computing the Test Statistic

An appropriate test statistic for this example is χ^2. In order to obtain the second dimension of the 2×2 table, we categorize each score as falling below or above the *common median* for the combined samples. The first step, then, is to determine the median for the total sample of group 1 and group 2.

$$\text{Mdn} = \frac{31 + 32}{2}$$

$$= 31.5$$

The next step is to categorize the observations in each group that fall above or below this median into a 2×2 contingency table. For the data given in Table 15.11, the contingency table is as follows:

	Group 1 Without Children	**Group 2** With Children	
Above Median	7	3	10
Below Median	3	7	10
	10	10	20

Note that seven members of group 1 and three members of group 2 have marital satisfaction scores above the median for the total sample. Applying formula 15.5 to the data given in this table, we find that

$$\chi^2 = \frac{n(AD - BC)^2}{(A + B)(C + D)(A + C)(B + D)}$$

$$= \frac{20(7 \times 7 - 3 \times 3)^2}{(7 + 3)(3 + 7)(7 + 3)(3 + 7)}$$

$$= \frac{20(49 - 9)^2}{(10)(10)(10)(10)}$$

$$= 3.20$$

Step 3: Determining the Criterion for Rejecting H_0

The sampling distribution for the test statistic for this example is χ^2 with 1 degree of freedom: $(r - 1)(c - 1) = (2 - 1)(2 - 1) = 1$. Recall that the null hypothesis (H_0: $Mdn_1 = Mdn_2$) was tested against the directional alternative (H_a: $Mdn_1 > Mdn_2$) at the .05 level of significance. The critical values of χ^2 found in Table 3 in the Appendix are for a *nondirectional* alternative hypothesis (two-tailed test). For a *directional* alternative hypothesis—that is, a one-tailed test—we must use the critical value of χ^2 at the .10 level so we have 5% of the area in the region of rejection for the right-hand tail. Therefore, the critical value of χ^2 for this example is 2.71.

Step 4: Deciding Whether to Reject or Retain the Null Hypothesis

Because the computed value of χ^2 (3.20) is greater than the critical value (2.71), the null hypothesis is rejected. The conclusion is that there is a difference between the median scores of married women without children and those with children. The null hypothesis was tested against the directional alternative, so the professor would conclude that the level of marital satisfaction was higher for married women without children than for married women with children.

The null hypothesis for the median test is that the populations from which the samples were selected have a common median.

The Mann–Whitney U Test

Step 1: Stating the Hypothesis

In the median test, the null hypothesis tested is that there is no difference between the medians of the two populations from which the samples were selected. In that case, the test statistic is sensitive only to the differences between the medians and does not take into consideration the total distributions of scores for the two groups. By contrast, the *Mann–Whitney U test* is sensitive to both the central tendency of the scores and the distribution of scores. The null hypothesis is thus stated in more general terms: There is no difference between the *scores* of the populations from which the samples were selected. For our example, the null hypothesis is that there is no difference in marital satisfaction between the population of married women without children and the population of married women with children. Again, the null hypothesis is tested against the directional alternative hypothesis at the .05 level of significance. In symbols,

$$H_0: \quad \text{attitude}_1 = \text{attitude}_2$$

$$H_a: \quad \text{attitude}_1 > \text{attitude}_2$$

Step 2: Computing the Test Statistic

The Mann–Whitney U test is an appropriate test of significance for this example. The calculation of the U statistic takes into account the central tendency of the scores and the total distribution of scores for both groups. The test statistic U is defined as the smaller of U_1 and U_2 when

$$U_1 = n_1 n_2 + \frac{n_1(n_1 + 1)}{2} - R_1 \tag{15.9}$$

$$U_2 = n_1 n_2 + \frac{n_2(n_2 + 1)}{2} - R_2$$

where

n_1 = number of observations in group 1

n_2 = number of observations in group 2

R_1 = sum of the ranks assigned to group 1

R_2 = sum of the ranks assigned to group 2

The sampling distribution of U is known and is used for testing hypotheses in the same way as the t distributions and χ^2 distributions. The critical values for one-tailed and two-tailed tests at the .05 levels are found in Table 9 in the Appendix. In using this table, remember that U equals the smaller of U_1 and U_2. In order for us to reject the null hypothesis at the .05 or the .01 level for a one-tailed test (the .10 or the .02 level for a two-tailed test), U must be *less than* the tabled value. The critical values of U for larger samples will be discussed later in this section.

To apply formula 15.9 to the data given in Table 15.11, we must assign ranks to the scores in the combined groups; these ranks are presented in Table 15.11. U_1 and U_2 are then computed as follows:

$$U_1 = (10)(10) + \frac{(10)(10 + 1)}{2} - 128$$

$$= 100 + 55 - 128$$

$$= 27$$

$$U_2 = (10)(10) + \frac{(10)(10 + 1)}{2} - 82$$

$$= 100 + 55 - 82$$

$$= 73$$

U_1 is the smaller of the two, so $U = 27$. Note that the group with the highest scores and, correspondingly, the highest sum of ranks has the smaller value of U.

Step 3: Determining the Criterion for Rejecting H_0

The sampling distribution of U for $n_1 = 10$ and $n_2 = 10$ is used to test the null hypothesis against the directional alternative. The critical values in Table 9 are for one-tailed tests at the .05 and .01 levels of significance. The critical value of U for this example is 28.

Step 4: Deciding Whether to Reject or Retain the Null Hypothesis

The calculated value of U is *less than* the critical value, so the null hypothesis is rejected in favor of the directional alternative. As before, the conclusion is that the population of married women without children exhibits a greater degree of marital satisfaction than that of married women with children. Note that this conclusion is the same as the conclusion we reached with the median test. Although both tests are appropriate for these data, the Mann–Whitney test is statistically more powerful and has been shown to be the better alternative to the two-sample t test for means. Because it is more sensitive and thus more likely to lead to rejection of the null hypothesis when it is false, the authors recommend using the Mann–Whitney test when assumptions underlying the t test cannot be adequately met.

Mann–Whitney U Test for Large Samples

When the size of the samples for both groups is greater than 20 ($n_1 > 20$ and $n_2 > 20$), the sampling distribution of U approaches the normal distribution. The mean of this sampling distribution is given by

$$\mu_U = \frac{n_1 n_2}{2} \tag{15.10}$$

where

n_1 = sample size for group 1

n_2 = sample size for group 2

and the standard deviation of the sampling distribution (the standard error of U) is given by

$$\sigma_U = \sqrt{\frac{(n_1)(n_2)(n_1 + n_2 + 1)}{12}} \tag{15.11}$$

Thus the general formula for the test statistic (discussed in Chapter 9) can be used to test the null hypothesis.

$$z = \frac{U - \mu_U}{\sigma_U} \tag{15.12}$$

$$= \frac{U - \dfrac{n_2 n_2}{2}}{\sqrt{\dfrac{(n_1)(n_2)(n_1 + n_2 + 1)}{12}}}$$

As usual, if a computed z value exceeds the critical value at a specified level of significance, the null hypothesis is rejected.

The hypothesis tested using the Mann–Whitney U test is that two population distributions are the same for a specified variable. The Mann–Whitney test is a statistically more powerful test than the median test.

Ordinal Data: *k*-Sample Case

The nonparametric analog to one-way analysis of variance for ordinal data is the *Kruskal–Wallis one-way analysis of variance*. Calculation of the test statistic for the Kruskal–Wallis test is similar to calculation of the Mann–Whitney U statistic. Consider the following example. Suppose a school psychologist in a local high school is interested in determining the effectiveness of two different counseling strategies for improving school attendance among students who have had a long history of truancy. The two methods are group counseling (group 1) and individual counseling (group 2). Group 3 serves as the control group. Seventeen students were randomly assigned to the three groups, and the attendance data for the three groups were collected at the end of the year. The data given in Table 15.12 are the numbers of days absent during the school year in which the counseling sessions took place.

Step 1: Stating the Hypothesis
The null hypothesis for the Kruskal–Wallis test is analogous to the null hypothesis for the one-way ANOVA. The null hypothesis for ANOVA is that there is no difference between the means of the k populations from which the samples were selected. The null hypothesis for the Kruskal–Wallis test, like that for the Mann–Whitney U test, is expressed in more general terms—namely, that there is no difference between the "scores" of the k populations. For this example, the null hypothesis is that, in the population, there is no difference in the number of days absent for the three groups. The alternative hypothesis of the Kruskal–Wallis test is that at least two of the k populations, or a combination of populations, differ. Assume the level of significance was set *a priori* at .05.

Step 2: Computing the Test Statistic
The test statistic for this example is the Kruskal–Wallis H; the general formula for computing H, given k samples, is

$$H = \frac{12}{N(N + 1)} \sum_{j=1}^{k} \frac{R_j^2}{n_j} - 3(N + 1) \qquad (15.13)$$

where

$$N = \sum_{j=1}^{k} n_j = \text{total number of observations}$$

n_j = number of observations in the jth sample

R_j = sum of the ranks in the jth sample

The sampling distribution of H is the χ^2 distribution with $k - 1$ degrees of freedom, where k is the number of samples. Applying formula 15.13 to the data for this example, we find that

$$H = \frac{12}{(17)(17 + 1)} \left[\frac{(29)^2}{6} + \frac{(46)^2}{5} + \frac{(78)^2}{6} \right] - 3(17 + 1)$$

$$= \frac{2}{51} [1577.37] - 54$$

$$= 7.86$$

Step 3: Determining the Criterion for Rejecting H_0 The sampling distribution of the test statistic (H) is the χ^2 distribution with $k - 1$ degrees of freedom, where k is the number of samples. In this example, the sampling distribution is χ^2 with 2 degrees of freedom, and the critical value at the .05 level of significance is 5.99.

Step 4: Deciding Whether to Reject or Retain the Null Hypothesis
Because the computed value of H is greater than the critical value, the null hypothesis is rejected and the school psychologist would conclude that, in the population, there is a difference between the effects of the different counseling strategies. Inspection of the data indicates that those students who were in group counseling were absent fewer days than those who had individual counseling. The data also indicate that both groups who were exposed to the counseling program were absent fewer days than the control students.[11]

The Kruskal–Wallis one-way analysis of variance for ranks is applicable for two or more independent samples. The null hypothesis tested is that the population distributions from which the samples were selected are the same.

Tied Ranks

For both the Mann–Whitney U test and the Kruskal–Wallis H test, *tied ranks* may occur in the data. If tied ranks do occur, we can apply a correction factor for tied ranks.[12] This correction has only a minimal effect on the calculated value of U or H, and it is not presented. However, if the number of tied ranks is excessive, the use of either the Mann–Whitney U test or the Kruskal–Wallis H test could be questionable, and another nonparametric test would be more appropriate.

[11]Multiple comparison procedures for the Kruskal–Wallis test are available and should be applied when the null hypothesis is rejected. For a discussion of these procedures, see Hollander and Wolfe, *Nonparametric Statistical Methods*, pp. 124–130.

Group Counseling		Individual Counseling		Control Group	
Score	Rank	Score	Rank	Score	Rank
52	4	66	13	63	10
46	1	49	3	65	12
62	9	64	11	58	8
48	2	53	5	70	15
57	7	68	14	71	16
54	6			73	17
	$29 = R_1$		$46 = R_2$		$78 = R_3$

Table 15.12   Data for the high school students in the different counseling groups

Ordinal Data: Two-Sample Case, Dependent Samples

The nonparametric analog of the two-sample case with dependent samples for ordinal data is the *Wilcoxon matched-pairs signed-rank test*. This test is commonly used in designs that involve either matched pairs of subjects or pretests and posttests. Consider the following example. Suppose a psychologist is interested in the aggressive behavior exhibited by young adolescents with learning disabilities before and after a series of counseling sessions. A random sample of 12 adolescents is selected for the study. The measure of aggressive behavior is the sum of subjective ratings by five trained judges. The measures are taken before and after the treatment. The data for the study are given in Table 15.13. The numerical scores are considered ordinal in nature.

Step 1: Stating the Hypothesis
As was the case for the Mann–Whitney U test and the Kruskal–Wallis H test, the null hypothesis is stated in general terms. In this example, the null hypothesis is that there is no difference in the aggressive behavior of populations of children with learning disabilities before and after a series of counseling sessions. The psychologist decides to test this null hypothesis against the directional alternative that the aggressive behavior will lessen after the counseling sessions. The level of significance is set at .01.

Step 2: Computing the Test Statistic
The statistic generated by the Wilcoxon Test is called T. For these particular data, $T = 4$. Generally, the process of determining T involves the following steps:

1. Determine the difference between the pretest and posttest scores for each individual. These are difference scores.

[12]See Sidney Siegel, *Nonparametric Statistics*, (New York: McGraw-Hill, 1956) pp. 188–189.

Child	Pretest Score	Posttest score	Difference	Rank of Difference	Ranks with Less Frequent Sign
1	36	21	15	11	
2	23	24	− 1	− 1	1
3	48	36	12	10	
4	54	30	24	12	
5	40	32	8	7	
6	32	35	− 3	− 3	3
7	50	43	7	6	
8	44	40	4	4	
9	36	30	6	5	
10	29	27	2	2	
11	33	22	11	9	
12	45	36	9	8	$T = 4$

Table 15.13   Data on the aggressive behavior of children with learning disabilities

2. Rank the absolute values of the difference scores, and then place the appropriate sign with the rank. If pretest score is larger than posttest score, the sign is positive. (For example, in the data given in Table 15.13, the difference of − 1 receives a rank of 1 but retains its minus sign, and so on.)

3. Sum the ranks with the less frequent sign. In the example there are two negative and ten positive signs, so we sum the ranks on the two negatives, obtaining $T = 4$.

Step 3: Determining the Criterion for Rejecting H_0
The sampling distribution of T was developed by Wilcoxon; critical values for one-tailed and two-tailed tests for several levels of significance are found in Table 10 in the Appendix. Referring to Table 10, we find that the critical value at the .01 level of significance for a one-tailed test for 12 matched pairs is 10. If the calculated value of the test statistic is *less than* the critical value, the null hypothesis is rejected. For most sampling distributions, the calculated value must exceed the critical value in order for us to reject the null hypothesis.

Step 4: Deciding Whether to Reject or Retain the Null Hypothesis
Because the calculated value of the test statistic (4) is less than the critical value (10), the null hypothesis is rejected. The psychologist concludes that the aggressive behavior of learning-disabled children can be reduced as a result of the counseling sessions. The conclusion relates to the population of learning-disabled children.

The Wilcoxon test is used with matched pairs of observations and tests the null hypothesis of no difference between the matched populations. This test is commonly used in a pretest–posttest design.

	Name of Test	Hypothesis Tested	Test Statistic	Underlying Distribution
	One-sample case	H_0: Goodness of fit	$\chi^2 = \sum \dfrac{(O - E)^2}{E}$	χ^2 with $k - 1$ degrees of freedom
	Two-sample case (Test of independence)	H_0: Independence	$\chi^2 = \dfrac{n(AD - BC)^2}{(A + B)(C + D)(A + C)(B + D)}$	χ^2 with 1 degree of freedom
Nominal Data	k-sample case (Test of independence)	H_0: Independence	$\chi^2 = \sum \sum \dfrac{(O - E)^2}{E}$	χ^2 with $(r - 1)(c - 1)$ degrees of freedom
	Two-sample case, dependent samples (McNemar test)	H_0: $A = D$	$\chi^2 = \dfrac{(A - D)^2}{A + D}$	χ^2 with 1 degree of freedom
	Two-sample case Median test	H_0: $Mdn_1 = Mdn_2$	$\chi^2 = \dfrac{n(AD - BC)^2}{(A + B)(C + D)(A + C)(B + D)}$	χ^2 with 1 degree of freedom
	Mann–Whitney U test (small samples)	H_0: Population distributions are the same	$U =$ smaller of U_1 and U_2 $U_1 = n_1 n_2 + \dfrac{n_1(n_1 + 1)}{2} - R_1$ $U_2 = n_1 n_2 + \dfrac{n_2(n_2 + 1)}{2} - R_2$	Sampling distribution of U
	Mann–Whitney U test (large samples)	H_0: Population distributions are the same	$z = \dfrac{U - \dfrac{(n_1)(n_2)}{2}}{\sqrt{\dfrac{(n_1)(n_2)(n_1 + n_2 + 1)}{12}}}$	Normal distribution
Ordinal Data	k-sample case Kruskal–Wallis test	H_0: Population distributions are the same	$H = \dfrac{12}{N(N + 1)} \sum \dfrac{R_j^2}{n_j} - 3(N + 1)$	χ^2 with $k - 1$ degrees of freedom
	Two-sample case, dependent samples			
	Wilcoxon test (small samples)	H_0: Population difference $= 0$	$T =$ sum or ranks with signs of lesser frequency	Sampling distribution of T
	Wilcoxon test (large samples)	H_0: Population difference $= 0$	$z = \dfrac{T - \dfrac{n(n + 1)}{4}}{\sqrt{\dfrac{n(n + 1)(2n + 1)}{24}}}$	Normal distribution

Table 15.14 Summary of nonparametric test statistics

The Wilcoxon Test for Larger Samples

When the size of the sample is large ($n > 25$), the sampling distribution of T has been shown to approximate the normal distribution. The mean of this sampling distribution is given by

$$\mu_T = \frac{n(n + 1)}{4} \tag{15.14}$$

The standard deviation of the sampling distribution (the standard error of T) is given by

$$\sigma_T = \sqrt{\frac{n(n + 1)(2n + 1)}{24}} \tag{15.15}$$

Thus we can use the general formula for the test statistic (discussed in Chapter 9) to test the null hypothesis.

$$z = \frac{T - \mu_T}{\sigma_T} \tag{15.16}$$

$$= \frac{T - \dfrac{(n)(n + 1)}{4}}{\sqrt{\dfrac{n(n + 1)(2n + 1)}{24}}}$$

This, of course, generates a score in the normal distribution. As usual, if the value of the computed z score exceeds the critical value, the null hypothesis is rejected.

Summary

This chapter presents a selected number of nonparametric statistical tests that are the analogs of the parametric tests introduced in Chapters 9, 10, 11, 12, and 13. These tests of significance are used when the assumptions underlying the use of parametric tests cannot be met. Specifically, they should be considered when the assumptions of normality and homogeneity of variance cannot reasonably be met and/or the measurement scale of the dependent variable is less than interval. The statistical tests described in this chapter are summarized in Table 15.14. This table includes the measurement scale of the dependent variable, the hypothesis tested, the computational formula of the test statistic, and its underlying distribution.

Several of the examples in this chapter involve relatively small sample sizes. These examples reflect research situations in the behavioral sciences wherein large samples are not feasible. For example, the number of subjects available may be limited because subjects are atypical or because the measurement or treatment involved in the research is expensive or otherwise demanding of time and resources. For these situations, the parametric assumptions are difficult to justify. The nonparametric tests can be used when assumptions about the population distribution are not necessary, so small samples can be used. However, statistical precision is enhanced with larger samples. For example, the assumption of normality in the population becomes less important as the sample size increases. Thus small samples should not be used when larger ones are readily available.

In the final analysis, the statistical procedures used in any research study must be appropriate to the conditions of the research and the specific hypotheses being tested. There is little point in stating hypotheses that require parametric procedures when the variables are not measured on at least an interval scale or the other parametric assumptions cannot be met. On the other hand, we would not employ nonparametric procedures if parametric procedures were applicable. When the measurement scale is at least interval and the other assumptions are tenable, the parametric procedures provide more information and can be used to test more complex hypotheses.

Key Concepts

Parametric tests
Nonparametric tests
Observed frequencies
Expected frequencies
χ^2 distributions
Test of independence
Contingency coefficient
McNemar test for significance of change

Median test
Mann–Whitney U test
Kruskal–Wallis one-way analysis of variance
Tied ranks
Wilcoxon matched-pairs signed-rank test

Exercises

15.1 A clinical psychologist is interested in the effects of 3 different methods of treating schizophrenic patients. Accordingly, 25 patients are randomly assigned to the 3 different treatments and their status is assessed after 4 weeks. The data follow. Determine whether there are any differences between the treatments. (Use $\alpha = .05$.)

	Treatment		
	1	2	3
Improved	6	5	2
Not Improved	2	4	6
	8	9	8

a. State the hypothesis.
b. Compute the test statistic.
c. Determine the criterion for rejecting the null hypothesis.
d. Decide whether to reject or to retain the null hypothesis.

15.2 A college professor contends that the distribution of grades in a freshman calculus class is 10% A, 20% B, 40% C, 20% D, and 10% F. The grades for the
436 students who were enrolled in the calculus class during the first semester
of the present academic year were as follows:

A   56
B   120
C   217
D   37
F   6

Do these grades reflect possible grade inflation? (Use $\alpha = .01$.)
a. Compute the test statistic.
b. Determine the critical value of the test statistic.
c. Interpret the results.

15.3 A local radio station was planning to revamp its program schedule. A telephone
survey was conducted and the respondents were asked to indicate what type of
music they preferred. As part of the survey, they were also asked to indicate
whether they were male or female, their age, and their marital status. The data
follow. For each of the contingency tables, determine whether there is a relationship between the type of music preferred and the demographic variable.
a. State the hypotheses.
b. Compute the test statistic.
c. Determine the criterion for rejecting the null hypothesis. (Use $\alpha = .05$.)
d. Decide whether to reject or to retain the null hypothesis.

	Classical	Jazz	Popular	
Male	41	92	29	162
Female	34	76	74	184
	75	168	103	346

	Classical	Jazz	Popular	
20–29	28	42	64	134
30–39	16	79	24	119
40–49	31	47	15	93
	75	168	103	346

	Classical	Jazz	Popular	
Single	27	47	39	113
Married	36	79	48	163
Divorced	7	36	10	53
Widowed	5	6	6	17
	75	168	103	346

Selected Nonparametric Tests

15.4 Use the data from the three contingency tables given in Exercise 15.3.
  a. Compute the contingency coefficient (C).
  b. Estimate C_{max} for each.
  c. Interpret the results.

15.5 The president of a local coal miners' union is seeking support for a wildcat strike. A vote is taken before and after a heated discussion of the major issues involved. The data are as follows:

		Before Discussion		
		No Strike	Strike	
After Discussion	Strike	16	11	27
	No Strike	17	6	23

  a. State the hypothesis.
  b. Compute the test statistic.
  c. Determine the criterion for rejecting the null hypothesis. (Use $\alpha = .05$.)
  d. Decide whether to reject or to retain the null hypothesis.

15.6 An educational psychologist was interested in whether the order in which questions are presented in a multiple-choice test has an effect on the number of correct responses. Twenty students were randomly assigned to two groups. Group I was given the test with the questions presented in the order in which the material was presented. Group II was given the same test with the items randomly arranged. The data were as follows:

Group I	Group II
10	7
13	8
15	9
16	11
19	12
21	14
22	17
23	18
25	20
26	24

Complete both the median test and Mann-Whitney U test.
  a. State the hypothesis.
  b. Compute the test statistic.
  c. Determine the criterion for rejecting the null hypothesis. (Use $\alpha = .05$.)
  d. Decide whether to reject or to retain the null hypothesis.

15.7 The director of research in a local school district conducted a study on the absentee rate of high school dropouts the year prior to their leaving school. The day-missed data from the study were as follows:

Eighth	Ninth	Tenth	Eleventh
42	37	47	31
35	40	49	44
39	32	34	38
50	33	46	
45		41	
48		43	
36			

Complete the Kruskal–Wallis Test.
a. State the hypothesis.
b. Compute the test statistic.
c. Determine the criterion for rejecting the null hypothesis. (Use $\alpha = .05$.)
d. Decide whether to reject or to retain the null hypothesis.

15.8 A computer science professor was interested in determining whether introductory computer science students are more efficient in writing simple programs using Fortran or using Basic (two widely accepted programming languages). Twelve students were asked to write a simple program in both languages, and the number of program statements required was determined. The data were as follows:

Fortran	Basic
10	17
13	11
8	18
10	9
13	16
14	10
10	22
11	20
10	23
13	24
6	14
9	23

Complete the Wilcoxon Matched-Pairs Signed-Rank test.
a. State the hypothesis.
b. Compute the test statistic.
c. Determine the criterion for rejecting the null hypothesis. (Use $\alpha = .05$.)
d. Decide whether to reject or to retain the null hypothesis.

▦ Key Strokes for Selected Exercises

15.1 Step 1: Determine the expected frequencies.

Value	Keystroke	Display
$13(f_r)$	$\times$	13
$8(f_c)$	$\div$	104
$25(n)$	$=$	4.16 (save)
13	$\times$	13
9	$\div$	117
25	$=$	4.68 (save)
13	$\times$	13
8	$\div$	104
25	$=$	4.16 (save)
12	$\times$	12
8	$\div$	96
25	$=$	3.84 (save)
12	$\times$	12
9	$\div$	108
25	$=$	4.32 (save)
12	$\times$	12
8	$\div$	96
25	$=$	3.84 (save)

Step 2: Determine χ^2.

Value	Keystroke	Display
6(O)	$-$	6
4.16(E)	$=$	1.84
	X^2	3.3856
	$\div$	3.3856
4.16(E)	$=$	0.8138
	STO	0.8138
5(O)	$-$	5
4.68(E)	$=$	0.32
	X^2	0.1024
	$\div$	0.1024
4.68(E)	$=$	0.0219
	M+	0.0219
2(O)	$-$	2
4.16(E)	$=$	-2.16
	X^2	4.6656
	$\div$	4.6656
4.16(E)	$=$	1.1215
	M+	1.1215

Value	Keystroke	Display
2(O)	−	2
3.84(E)	=	− 1.84
	X^2	3.3856
	÷	3.3856
3.84(E)	=	0.8817
	M +	0.8817
4(O)	−	4
4.32(E)	=	− 0.32
	X^2	0.1024
	÷	0.1024
4.32(E)	=	0.0237
	M +	0.0237
6(O)	−	6
3.84(E)	=	2.16
	X^2	4.6656
	÷	4.6656
3.84(E)	=	1.2150
	M +	1.2150
	RCL	4.0776 $= \chi^2$

Appendixes

Appendix A

Using a Pocket Calculator
in Statistical Analysis

    With the advent of highly reliable and reasonably inexpensive pocket calculators, it would be remiss on our part not to provide the student with efficient procedures for using these calculators in completing the exercises in both the book and workbook. Mastery of these procedures (which are called "key strokes") will remove much of the tedium of the arithmetic calculations and greatly reduce the probability of error. In both the text and the workbook we have provided a list of the key strokes that are required for completing an example of each major statistical concept. There follows here a list of the key strokes for ΣX, ΣX^2, and ΣXY; these are the most basic arithmetic calculations that will be used in many examples throughout the book and workbook.

    The many brands of calculators available on the market exhibit great variation in price. However, the major brands in the modest price range from \$10 to \$25 have all the functions necessary to complete the exercises in the book and workbook very efficiently. If you are presently considering the purchase of a calculator, we recommend one that includes at least the following functions:

$+$	Addition
$-$	Subtraction
$\times$	Multiplication
$\div$	Division
$\sqrt{}$	Square root of a number
X^2	Square of a number
$1/X$	Reciprocal of a number
CLR (or C)	Clear all information in the calculator
CE	Clear last entry without erasing previous information in calculator
$+/-$	Change sign from positive $(+)$ to negative $(-)$, or vice versa
STO	Place a quantity into memory storage for future use
RCL	Recall a quantity from memory for use
$M+$	Add a quantity to that already stored in the memory
$M-$	Subtract a quantity from that already stored in the memory

Consider the following data. They represent two scores for each of five people.

	X	Y
Person 1	12	5
Person 2	10	2
Person 3	14	-1
Person 4	9	3
Person 5	3	-3

Key strokes for computing ΣX

$$\Sigma X = 12 + 10 + 14 + 9 + 3 = 48$$

Value	Key Stroke	Display
12	+	12
10	+	22
14	+	36
9	+	45
3	=	48 $= \Sigma X$

Key strokes for computing Σ when negative numbers are included in the data

To find the sum of numbers, some of which are negative, you will have to use the $+/-$ function.

$$\Sigma Y = (5) + (2) + (-1) + (3) + (-3) = 6$$

Value	Key Stroke	Display
5	+	5
2	+	7
1	$+/-$	-1
	+	6
3	+	9
3	$+/-$	-3
	=	6 $= \Sigma Y$

Key strokes for computing ΣX^2

$$\Sigma Y^2 = 12^2 + 10^2 + 14^2 + 9^2 + 3^2$$

$$= 144 + 100 + 196 + 81 + 9$$

$$= 530$$

Value	Key Stroke	Display
12	X^2	144
	+	144
10	X^2	100
	+	244
14	X^2	196
	+	440
9	X^2	81
	+	521
3	X^2	9
	=	$530 = \Sigma X^2$

Key strokes for computing ΣX^2 when calculator does not have x^2 function

Value	Key Stroke	Display
12	×	12
12	+	144
10	×	10
10	+	244
14	×	14
14	+	440
9	×	9
9	+	521
3	×	3
3	=	$530 = \Sigma X^2$

Key strokes for computing ΣXY

For these data, you will have to use the $+/-$ function because some of the Y values are negative.

$$\Sigma XY = (12)(5) + (10)(2) + (14)(-1) + (9)(3) + (3)(-3) = 86$$

Value	Key Stroke	Display
12	×	12
5	+	60
10	×	10
2	+	80
14	×	14
1	$+/-$	-1
	+	66
9	×	9
3	+	93
3	×	3
3	$+/-$	-3
	=	$86 = \Sigma XY$

Appendix B

Glossary of Key Concepts

Absolute value   The value of a number without regard to its algebraic sign.

Alternative hypothesis   Possible outcomes not covered by the null hypothesis (H_a).

Between-groups variation   Variation between the group means and the mean of the total group; denoted s_B^2.

Binomial distribution   Distribution obtained by expanding $(X + Y)$ to the nth power $(X + Y)^n$.

Central limit theorem   Provides the mathematical basis for using the normal distribution as the sampling distribution of all sample means of a given sample size. The theorem states that this distribution of sample means (1) is normally distributed, (2) has a mean equal to μ and (3) has a variance equal to σ^2/n.

Central tendency   A central value, between the extreme scores in the distribution, around which the scores are distributed.

Chi square (χ^2) distribution   A family of distributions used as sampling distributions in both parametric and nonparametric tests of significance.

Class intervals   Intervals of scores in a frequency distribution derived by combining several scores into an interval, thus reducing the number of categories.

Cluster sampling   The random selection of clusters (groups of population members) rather than individual population members.

Coefficient of determination   The square of the correlation coefficient (r^2); a measure of the shared variance.

Conditional distribution   A distribution of error scores around the respective predicted Y score for a given value of X.

Confidence interval   A range of values that we are confident contains the population parameter.

Constant   A characteristic that assumes the same value for all members of the group under study.

Contingency coefficient   A correlation coefficient used to determine the degree of association between the two variables in a contingency table.

Continuous variable   A variable that can take on any value in the measurement scale being used.

Correlated data   Data from subjects measured under more than one condition.

Correlation   The nature, or extent, of the relationship between two variables.

Correlation coefficient   An index of the relationship between two variables.

Criterion variable   In the process of prediction, the variable for which scores are estimated, on the basis of knowledge of the scores on the predictor variable.

Critical value   The value of the sampling distribution that represents the beginning of the region of rejection.

Cumulative frequency distribution   The sum of the frequency of scores in any class interval plus the frequencies of scores in all preceding class intervals on the scale of measurement.

Data   Bits of information that are gathered on some characteristic of a group of individuals or objects under study.

Data curve   A set of data points connected by straight line segments, illustrating the relationship between two variables.

Degrees of freedom   The number of observations less the number of restrictions placed on them.

Dependent samples   A sample from a single population that is measured both under treatment conditions and under control conditions, causing the scores for each subject under one condition to be dependent on that subject's scores under the other condition.

Dependent variable   The variable that is, or is presumed to be, the result of the manipulation of the independent variable.

Descriptive statistics   Procedures used for classifying and summarizing, or describing, data.

Deviation score   The difference between a given score and the mean $x = (X - \mu)$.

Directional or one-tailed test   Test in which the region of rejection is located in one of the two tails of the sampling distribution.

Discrete variable   A variable that can take on only designated values.

Disordinal interaction   Condition that exists when the lines in an interaction plot of cell means intersect within the plot.

Equal variances (homogeneity of variance)   An assumption that the variances of the populations are equal.

Errors in prediction   The differences between the actual Y scores and the predicted scores, $\hat{Y}$.

Exact limits   Under the assumption of continuity, a range of scores that is actually represented by a single score or class interval.

Expected frequencies   The theoretical frequencies of occurrence in the categories of a contingency table that are determined from the null hypothesis.

F distribution   A family of sampling distributions that require two degrees-of-freedom values for identifying the specific distribution. The distributions are not symmetric and values range from 0 to $+ \infty$.

Factor   Another term for an independent variable.

Factorial   A mathematical operator equaling the product of all integers n through 1.

Factorial design   The design of a research study in which two or more factors are considered.

Fisher z transformation   Formula for transforming the correlation coefficient such that the sampling distribution of the transformed correlation coefficient is the normal distribution.

Frequency distribution   A tabulation of data that indicates the number of times given scores or groups of scores appear.

Frequency polygon   A graph on which the frequency of each class interval is plotted at the midpoint and the midpoints are connected with straight lines.

Graph   A pictorial representation of a set of data.

Histogram   A type of bar graph that depicts the frequencies of individual scores or scores in a class interval by the length of the bars.

Homogeneity   The extent to which the members of the group tend to be the same on the variables being investigated.

Homoscedasticity   The assumption, in predicted scores, that the standard deviations of all conditional distributions are equal.

Hypothesis testing   Determining whether some hypothesized value for un unknown population parameter is tenable, or justifiable.

Independent samples   Samples selected from two separate and distinct populations.

Independent variable   A variable that is controlled or manipulated by the researcher. A categorical variable used to form the groupings of observations.

Inferential statistics   Procedures for making generalizations about a population by studying a subset of the population, called a sample.

Interaction   Condition that exists when the effect of the levels of one independent variable is not the same across the levels of the second independent variable.

Intercept   The value on the Y axis when X equals 0.

Interval scale   A scale having distinctive and ordered categories with equivalence of interval differences.

Kurtosis   The degree of peakedness in a symmetric distribution.

Laws of probability   Laws that describe the behavior of events given certain conditions.

Level of confidence   The researcher's degree of confidence that the computed interval contains the parameter being estimated.

Level of significance   The probability of making a Type I error if H_0 is rejected.

Levels of the independent variable   Classification of groupings of observations.

Linear regression line   The mathematical equation of a straight line:

$$\hat{Y} = bX + a$$

Linear relationship   A relationship represented on a scattergram by a random scatter of points about a straight line.

Main effects The individual effects of the independent variables on the dependent variable.

Mean The arithmetic average of the scores in a distribution.

Mean deviation The average deviation from the mean for the scores in a distribution.

Mean squares The between-groups and within-groups variance estimates, found by dividing SS_B and SS_W by the degrees of freedom associated with these estimates.

Measurement The process of assigning numbers to characteristics according to a defined rule.

Median The point on the scale of measurement below which 50% of the scores fall.

Method of least squares The process of fitting a line to a scattergram of points in such a way that the sum of the squared distances from the data points to the line is a minimum.

Midpoint The point on the scale of measurement that is halfway through an interval.

Mode The most frequent score in a distribution of scores.

Multimodal distribution A distribution in which two or more scores have the same frequency, which is also the greatest frequency.

Negative correlation A pattern of points in a scattergram that tends to run from the upper left to the lower right.

Nominal scale A scale that simply classifies without ordering.

Nondirectional or two-tailed test Test in which the region of rejection is located in both tails of the underlying distribution.

Nonparametric tests Statistical tests of significance that require fewer assumptions and/or are appropriate when parametric tests are not.

Normal curve A mathematical equation that has been generated to depict the nature of the frequency distribution of a set of scores.

Normal curve equivalent (NCE) score A normalized standard score with a mean of 50 and a standard deviation of 21.

Normal distribution An underlying distribution based on the normal curve.

Normalized standard scores Scores that are transformed into percentiles and then into z scores using the properties of the standard normal curve.

Null hypothesis A statement of no difference or no relationship (H_0).

Observed frequencies The observed frequencies of occurrence in the categories of a contingency table.

Ogive The graph of a cumulative percentage distribution, which is useful for determining the various percentile points in a distribution of scores.

One-way analysis of variance Approach taken whereby only one independent variable is considered in the ANOVA.

Ordinal interaction Condition that exists when the lines in an interaction plot of cell means do not intersect within the plot.

Ordinal scale A scale that has distinctive and ordered categories.

Glossary of Key Concepts

Parameter   A descriptive measure of a population.

Parametric assumptions   Assumptions underlying the mathematical derivation of ANOVA: 1. The samples are random samples from defined populations. 2. The samples are independent. 3. The dependent variable is measured on at least an interval scale. 4. The dependent variable is normally distributed in the populations. 5. The population variances are equal (homogeneity of variance).

Parametric statistics   Procedures that require parametric assumptions.

Parametric tests   Statistical tests of hypotheses in which the null hypothesis includes a specified value for the population parameter and in which certain assumptions have been met (see *parametric assumptions*).

Partitioning variation   Partitioning the sum of squared deviations around the grand mean, $\Sigma (X - \overline{X})^2$.

Pearson product moment correlation   The index of the linear relationship between two variables; also called the Pearson r.

Percentile   The point in a distribution at or below which a given percentage of scores is located.

Percentile rank   A score that is transformed from the original measurement scale to the percentile scale.

Point estimate   The sample mean ($\overline{X}$) that represents the "best" estimate of the population value.

Pooled estimate   The estimate of the population variance that takes into consideration not only the variances of the two samples, s_1^2 and s_2^2, but also the respective sample sizes, n_1 and n_2.

Population   All members of some defined group.

Positive correlation   A pattern of points in a scattergram that tends to run from lower left to upper right and to occur when high scores on variable X are associated with high scores on variable Y.

***Post hoc* multiple comparison procedures**   Follow-up methods used to determine which pair or pairs of means differ.

Prediction   The process of estimating scores on one variable from knowledge of scores on another variable.

Predictor variable   The known variable that is used in the process of prediction.

Probability   The ratio of the number of favorable outcomes to the total number of possible outcomes for an event.

Probability sample   A sample selected in such a way that each member or element of the population has a known and nonzero probability of being included in the sample.

Proportion   The fractional part of a group that possesses some specific characteristic.

Range   The number of units on the scale of measurement necessary to include both the highest score and the lowest score.

Rank distribution   The organizing of scores to order them from highest to lowest.

Ratio scale   A scale having distinctive and ordered categories with equivalence of interval differences and a true zero point.

Real numbers   All positive and negative numbers from negative infinity ($-\infty$) to positive infinity ($+\infty$), including not only positive and negative integers but also all fractions and decimals.

Region of rejection   The area of the underlying distribution that represents those values of the sample statistic that are highly improbable if the null hypothesis is true.

Regression coefficient   The slope of the regression line.

Regression constant   The Y intercept of the regression line.

Reliability   The consistency of measurement.

Repeated-measures ANOVA   Approach taken whereby the total variation is partitioned into three components: (1) variation among individuals, (2) variation among test occasions, and (3) residual variation.

Residual variation   Variation not due to either individuals or test occasions in repeated-measures ANOVA.

Sample   A subset of a population.

Sampling distribution of the mean   The sample means ($\overline{X}$) of all possible samples of a given size.

Sampling fraction   The ratio of the size of the sample to the size of the population.

Scattergram   A graph that plots pairs of scores for each individual on the two variables.

Scheffé method   A multiple comparison procedure that involves computing an F value for each combination of two means.

Signed numbers   Real numbers with the algebraic sign added.

Simple random sample   A sample in which all population members have the same probability of being selected and the selection of each member is independent of the selection of all other members.

Skewness   The degree to which the majority of scores in a frequency distribution are located at one end of the scale of measurement with progressively fewer scores toward the opposite end of the scale.

Slope   The amount of change in Y that corresponds to a change of one unit in X.

Spearman rho coefficient   The correlation coefficient that should be used when the level of measurement for both variables is ordinal.

Standard deviation   Square root of the variance:

$$\sigma = \sqrt{\sigma^2}$$

Standard error of estimate   The standard deviation of the errors in prediction.

Standard error of the difference between means   The standard deviation of the sampling distribution of differences between two means.

Standard error of the difference between proportions   The standard deviation of the sampling distribution of differences between two proportions.

Standard error of the mean   The standard deviation of the sampling distribution.

Standard error of the proportion   The standard deviation of the sampling distribution of proportions.

Standard error of the transformed correlation coefficient   The standard deviation of the sampling distribution of z_r.

Standard normal curve   A normal curve in standard-score form with a mean of 0 and a standard deviation of 1.0.

Standard score   A score that indicates the number of standard deviations a corresponding raw score is above or below the mean.

Statistic   A descriptive measure of a sample.

Statistical estimation   The proccss of estimating a parameter from the corresponding sample statistic.

Statistical precision   The accuracy with which a confidence interval can be used to estimate a population parameter. The narrower the width of the confidence interval, the more precise the estimate.

Statistical significance   The difference between the hypothesized population parameter and the corresponding sample statistic is said to be statistically significant when the probability that the difference occurred by chance is less than the significance level (α level).

Statistics   The entire body of mathematical theory and procedures that are used to analyze data.

Stratified sampling   The selection of a sample in which the population is divided into subpopulations called strata, with all strata being represented in the sample.

Studentized range distributions   A family of distributions used in *post hoc* multiple comparison procedures.

Sum of squares   The sum of squared deviations for all scores in a distribution. $\Sigma(X_i - \mu)^2$.

Symmetric distribution   A distribution that is the same on either side of the midpoint of the scale of measurement.

Systematic sampling   A procedure for selecting a probability sample in which every nth member of the population is selected and in which $1/k$ is the sampling fraction. The first member of the sample is determined by randomly selecting an integer between 1 and k.

***t* distribution**   A family of symmetric distributions. As sample size increases, the specific t distribution increasingly approximates the normal distribution.

***T* score**   The scores in a transformed distribution with a mean of 50 and a standard deviation of 10:

$$T = (10)(z) + 50$$

Test of independence   A χ^2 test of significance used to determine whether the effects of one variable are independent of the second variable.

Test statistic   A standard score indicating the difference between the observed sample mean and the hypothesized value of the population mean.

Theoretical sampling distributions Distributions based on information from a single sample in conjunction with mathematical theory.

Tied ranks Ties among the ranks of ordinal data.

Transformed standard scores Raw scores that are transformed into a different distribution of scores with a new scale of measurement, with a predetermined mean and standard deviation.

Tukey method A multiple comparison procedure designed to make all pairwise comparisons of means while maintaining the Type I error rate at the predetermined α level.

Type I error Rejecting a hypothesis when in fact it is true.

Type II error Retaining a hypothesis when in fact it is false.

Underlying distribution The distribution of all possible outcomes of a particular event.

Validity The degree to which a test measures what it is intended to measure.

Variable A characteristic that can take on different values for different members of the group under study.

Variance A measure of the dispersion of scores in a distribution.
Deviation formula:

$$\sigma^2 = \frac{\Sigma(X - \mu)^2}{N}$$

Raw-score formula:

$$\sigma^2 = \frac{\Sigma X^2 - \frac{(\Sigma X)^2}{N}}{N}$$

Variance estimates Estimates of the variance in the population.

Variation A quantitative measure of the extent to which the scores are dispersed throughout the distribution.

Weighted average score A composite score generated by transforming individual scores into the standard scores and then applying weights.

$$\text{Weighted score} = \frac{\Sigma W_i Z_{ij}}{\Sigma W_j}$$

Weighted mean The means of the individual groups weighted by the number of observations of each group, divided by the total number in the combined group.

$$\mu = \frac{\sum_{j=1}^{k} n_j \overline{X}_j}{N}$$

Within-groups variation Variation among all subjects within a group; denoted s_w^2.

z score A standard score:

$$z = \frac{X - \mu}{\sigma}$$

Appendix C

Glossary of Symbols

Mathematical Operators

$a \neq b$	a not equal to b
$a < b$	a less than b
$a \leq b$	a less than or equal to b
$a > b$	a greater than b
$a \geq b$	a greater than or equal to b
$\sqrt{X}$	square root of X (see Table 11)
X^y	X raised to the y power
$\mid X \mid$	absolute value of X
$N!$	factorial: $(N)(N-1)(N-2) \ldots (2)(1)$
Σ	sum all scores that follow
$\sum_{i=1}^{n} X_i$	sum: $X_1 + X_2 + \ldots + X_n$

Greek Letters

α	*alpha:* Level of significance; Probability of a Type I error; Probability of rejecting H_0 when it is true.
β	*beta:* Probability of a Type II error; Probability of retaining H_0 when it is false.
χ^2	*chi square:* a family of distributions
χ^2_{cv}	critical value of χ^2
μ	*mu:* Mean of the population
ρ	*rho:* Correlation coefficient in the population
ρ_S	Spearman rho correlation coefficient
σ^2	*sigma square:* Variance of the population
σ	*sigma:* Standard deviation of the population
$\sigma_{\bar{X}}^2 = \dfrac{\sigma^2}{n}$	variance of the sampling distribution of the mean
$\sigma_{\bar{X}} = \dfrac{\sigma}{\sqrt{n}}$	standard error of the mean; standard deviation of the sampling distribution of the mean

$\sigma_{\bar{X}_1 - \bar{X}_2}$	standard error of the difference of means
σ_e^2	in ANOVA, the within-groups variation; the natural variation attributable to individual differences; error variation
σ_t^2	in ANOVA, the variation due to differential treatment effects

Latin Letters

a	in regression, the Y intercept; the regression constant
b	in regression, the slope of the line; the regression coefficient
cf	cumulative frequency
c%	cumulative percentage
CI	confidence interval
CI_{95}	95% confidence interval
E	in χ^2 tests, the expected frequency for any cell
e	in regression, the error score $(Y - \hat{Y})$
df	degrees of freedom
F	in ANOVA, the ratio of two variance estimates
F_{cv}	in ANOVA, the critical value for the appropriate F distribution
f	frequency of a score
H	the test statistic for the Kruskal-Wallis test
H_0	the null hypothesis
H_a	the alternative hypothesis
i	width of the interval in a frequency distribution
k	in one-way ANOVA, the number of levels of the independent variable; in χ^2 tests, the number of categories
M.D.	mean deviation
Mdn	median
N	number of observations in a population; in ANOVA, total number of observation in all the various groups
n	number of observations in a sample
n_j	in ANOVA, the number of observations in the jth group
NCE	normal curve equivalent score; a standardized score
O	in χ^2 tests, the observed frequency for any cell
P	proportion of the population possessing a specific characteristic
p	proportion of a sample possessing a specific characteristic
P_X	the Xth percentile of a distribution of scores
PR_X	the percentile rank of the score X in a distribution of scores
P(A)	the probability of the occurrence of the event A
P(A|B)	the probability of the occurrence of the event A, given the fact that the event B has occurred
Q	proportion of the population *not* possessing a specific characteristic
q	proportion of a sample *not* possessing a specific characteristic

Q	test statistic in the Tukey method
r	sample correlation coefficient
r^2	the coefficient of determination; the proportion of the variation in variable X that can be attributed to the variation in variable Y, or vice versa
R_1	in the Mann-Whitney U test, the sum of the ranks assigned to Group 1
R_2	in the Mann-Whitney U test, the sum of the ranks assigned to Group 2
R_j	in the Kruskal-Wallis test, the sum of the ranks assigned to the jth group
s^2	estimated variance of the population based upon the sample
s	estimated standard deviation of the population based on the sample
$s_{\bar{X}}^2$	estimated variance of the sampling distribution of the mean
$s_{\bar{X}}$	estimated standard error of the mean; estimated standard deviation of the sampling distribution of the mean
$s_{\bar{X}_1 - \bar{X}_2}$	estimated standard error of the difference of means
s_p	estimated standard error of a proportion
$s_{p_1 - p_2}$	estimated standard error of the difference of proportions
s_{z_r}	standard error of the transformed correlation coefficient (z_r)
$s_{z_{r1} - z_{r2}}$	standard error of the difference of transformed correlation coefficients
s_e	in regression, the standard error of estimate
s_B^2	in ANOVA, the between-groups variation, which contains the error variance plus the variance due to the differences among the population means
s_W^2	in ANOVA, the within-groups variation; the estimated natural variation attributable to individual differences and random sampling fluctuation
$SS = \Sigma(X - \bar{X})^2$	sum of squares; sum of squared deviations around the mean
SS_B	in one-way ANOVA, sum of squares between groups
SS_W	in ANOVA (one-way and two-way), sum of squares within groups
SS_I	in repeated-measures ANOVA, sum of squares between individuals
SS_O	in repeated-measures ANOVA, sum of squares between observations
SS_{res}	in repeated-measures ANOVA, residual sum of squares
SS_R	in two-way ANOVA, sum of squares between rows

SS_C	in two-way ANOVA, sum of squares between columns
SS_{RC}	in two-way ANOVA, interaction sum of squares
SS_T	in ANOVA, total sum of squares
T	T score; a standardized score; the test statistic in the Wilcoxon matched-pairs signed-rank test
t	t distribution; a family of distributions; the test statistic using the t distribution as the sampling distribution
t_{cv}	the critical value for the appropriate t distribution
U	the test statistic in the Mann-Whitney U test
u	the ordinate of the normal curve; the height of the curve given a value of z
X, Y	variables; scores or measures of the variables
X_i, Y_i	specific scores on variables X and Y for the ith individual
$\overline{X}, \overline{Y}$	arithmetic means for variables X and Y
$\overline{X}_j$	in one-way ANOVA, the mean for the jth group
$\overline{X}_{r\cdot}$	in two-way ANOVA, the mean for the rth row
$\overline{X}_{\cdot c}$	in two-way ANOVA, the mean for the cth column
$\overline{X}_{rc}$	in two-way ANOVA, the mean for the rcth cell
$\overline{X}_{\cdot\cdot}$	in two-way ANOVA, the grand mean
$x = (X - \overline{X})$	deviation score
$\hat{Y}$	in regression, the predicted Y score
z	standard score; the test statistic using the normal distribution as the sampling distribution
z_{cv}	the critical value for the normal distribution
z_r	the transformed correlation coefficient

Appendix D

Selected Algebraic Proofs

1. Algebraic proof that the sum of the deviations of all scores in a distribution from the mean is zero.

$$\Sigma(X_i - \mu) = \Sigma X_i - \Sigma\mu \qquad \text{Rules of summation—Rule 1}$$

$$= \Sigma X_i - N\mu \qquad \text{Rules of summation—Rule 3:}$$
Because μ is a constant for the N scores in the distribution

$$= \Sigma X_i - N\left(\frac{\Sigma X_i}{N}\right) \qquad \text{By definition } \mu = \frac{\Sigma X_i}{N}$$

$$= \Sigma X_i - \Sigma X_i \qquad \text{By dividing } N \text{ by } N$$

$$= 0$$

Thus we see that the sum of all deviations of scores from the mean in a distribution is equal to zero.

2. Algebraic proof that the sum of squared deviations from the mean is a minimum.

Consider the deviations of all scores from the value $\mu + c$, where c is any value not equal to zero; for example, $X_i - (\mu + c)$. Squaring and summing these deviations over the n observations:

$$\Sigma\{X_i - (\mu + c)\}^2 = \Sigma[(X_i - \mu) - c]^2$$

$$= \Sigma(X_i - \mu)^2 - 2c\Sigma(X_i - \mu) + \Sigma c^2 \qquad \text{Squaring the expression}$$

$$= \Sigma(X_i - \mu)^2 + \Sigma c^2 \qquad \text{Because the second term of the expression equals 0 by the first property of the mean}$$

$$= \Sigma(X_i - \mu)^2 + nc^2 \qquad \text{Rules of summation— Rule 3}$$

This expression illustrates that the sum of squared deviations around the value $\mu + c$ is made up of two parts, the sum of the squared deviations about the mean and

nc^2. Because nc^2 is always positive, the sum of squared deviations about $\mu + c$ will always be greater than the sum of squared deviations about the mean. That is, $\Sigma(X_i - \mu)^2$ is a minimum.

3. Algebraic derivation of numerator in the equation for variance of a population, using actual scores rather than deviation scores. This term (the numerator of the population variance equation) is the sum of squared deviations about the mean or just simply the sum of squares.

$$\Sigma(X_i - \mu)^2 = \Sigma(X_i^2 - 2X_i\mu + \mu^2)$$ Squaring the term $(X_i - \mu)$

$$= \Sigma X_i^2 - 2\mu\Sigma X_i + \Sigma\mu^2$$ Rules of summation—Rule 1

$$= \Sigma X_i^2 - \frac{2N}{N}\mu\Sigma X_i + N\mu^2$$ Multiplying the second term by N/N and applying Rule 3 of summation to the third term

$$= \Sigma X_i^2 - 2N\mu^2 + N\mu^2$$ Because $\Sigma X_i = N\mu$, $2N\dfrac{\Sigma X_i}{N} = 2N\mu$

$$= \Sigma X_i^2 - N\left(\frac{\Sigma X_i}{N}\right)^2$$ Because $\mu = \dfrac{\Sigma X_i}{N}$, $\mu^2 = \left(\dfrac{\Sigma X_i}{N}\right)^2$ and adding second and third terms

$$= \Sigma X_i^2 - \frac{(\Sigma X_i)^2}{N}$$ *Note:* Recall that ΣX_1^2 does *not* equal $(\Sigma X_i)^2$.

4. Algebraic proof that the mean of z scores is equal to zero.

$$\bar{z} = \frac{\Sigma z}{N}$$ By definition

$$= \frac{\Sigma\left(\dfrac{X - \mu}{\sigma}\right)}{N}$$ Because $z = \dfrac{X - \mu}{\sigma}$, by substitution

$$= \frac{1}{\sigma}\Sigma\frac{(X - \mu)}{N}$$ Rules of summation—Rule 2, because $\dfrac{1}{\sigma}$ and N are constants for a given distribution

$$= \frac{1}{\sigma}\frac{(\Sigma X - \Sigma\mu)}{N}$$ Rules of summation—Rule 1

$$= \frac{1}{\sigma}\frac{(\Sigma X - N\mu)}{N}$$ Rules of summation—Rule 3, because $\Sigma\mu = N\mu$ by substitution

$$= \frac{1}{\sigma}\left(\frac{\Sigma X}{N} - \frac{N\mu}{N}\right) = \frac{1}{\sigma}\left(\frac{\Sigma X}{N} - \mu\right)$$ Arithmetic simplification

but $\dfrac{\Sigma X}{N}$ is by definition the mean or μ. Hence,

$$\bar{z} = \frac{1}{\sigma}\,(\mu - \mu) \qquad\qquad \text{Because } \frac{\Sigma X}{N} = \mu, \text{ by substitution}$$

$$= 0$$

Thus the mean of z scores is equal to zero, regardless of what the mean of the raw-score distribution may be.

5. Algebraic proof that the variance of a distribution of z scores equals 1.0.

$\sigma_z^2 = $ variance of the distribution of z scores

$$= \frac{\Sigma(z - \bar{z})^2}{N}$$

$$= \frac{\Sigma z^2}{N} \qquad\qquad \text{Because } \bar{z} = 0 \text{ by proof 4.}$$

$$= \frac{\Sigma\left(\dfrac{X - \mu}{\sigma}\right)^2}{N} \qquad\qquad \begin{array}{l}\text{By the definition of a } z \text{ score; that is,}\\[4pt] z = \dfrac{(X - \mu)}{\sigma}\end{array}$$

$$= \frac{1}{\sigma^2} \cdot \frac{\Sigma(X - \mu)^2}{N} \qquad\qquad \text{By algebraic manipulation}$$

$$= \frac{1}{\sigma^2} \cdot \sigma^2 \qquad\qquad \text{Because } \sigma^2 = \frac{(X - \mu)^2}{N}$$

$$= 1.0$$

6. Partitioning the total sum of squares in ANOVA.

Consider the partitioning of the deviation score for the ith subject in the jth group:

$$(X_{ij} - \bar{X}) = (X_{ij} - \bar{X}_j) + (\bar{X}_j - \bar{X})$$

In order to arrive at the total sum of squares, we must square each deviation score and then sum the squared deviations across all subjects in all groups. First we square both sides of the foregoing equation.

$$(X_{ij} - \bar{X})^2 = [(X_{ij} - \bar{X}_j) + (\bar{X}_j - \bar{X})]^2$$

$$= (X_{ij} - \bar{X}_j)^2 + 2(X_{ij} - \bar{X}_j)(\bar{X}_j - \bar{X}) + (\bar{X}_j - \bar{X})^2$$

Summing these squared deviations across the n_j subjects in the jth group, we find that the total sum of squares for this group is

$$\sum_{i=1}^{n_j} (X_{ij} - \bar{X})^2 = \sum_{i=1}^{n_j} (X_{ij} - \bar{X}_j)^2 + 2(\bar{X}_j - \bar{X}) \sum_{i=1}^{n_j} (X_{ij} - \bar{X}_j) + \sum_{i=1}^{n_j} (\bar{X}_j - \bar{X})^2$$

Note that, in the second term of the right side of the equation, the 2 and the $(\overline{X}_j - \overline{X})$ are both constants and thus appear in front of the summation sign. Also, because

$$\sum_{i=1}^{n_j} (X_{ij} - \overline{X}_j) = 0$$

for any of the groups (recall from proof 1 that the sum of the deviations of all scores from the mean equals zero), the second term on the right side drops out of the equation. In addition, the third term on the right side of the equation, $(\overline{X}_j - \overline{X})^2$, is a constant. Hence

$$\sum_{i=1}^{n_j} (\overline{X}_j - \overline{X})^2 = n_j(\overline{X}_j - \overline{X})^2$$

The total sum of squares for the jth group is therefore

$$\sum_{i=1}^{n_j} (X_{ij} - \overline{X})^2 = \sum_{i=1}^{n_j} (X_{ij} - \overline{X}_j)^2 + n_j(\overline{X}_j - \overline{X})^2$$

Now, if we sum these sums of squares across the k groups, we have

$$\sum_{j=1}^{k} \sum_{i=1}^{n_j} (X_{ij} - \overline{X})^2 = \sum_{j=1}^{k} \sum_{i=1}^{n_j} (X_{ij} - \overline{X}_j)^2 + \sum_{j=1}^{k} n_j(\overline{X}_j - \overline{X})^2$$

or

$$\begin{array}{ccc} \text{Total Sum} \\ \text{of Squares} \end{array} = \begin{array}{ccc} \text{Within Sum} \\ \text{of Squares} \end{array} + \begin{array}{ccc} \text{Between Sum} \\ \text{of Squares} \end{array}$$

Appendix E

Tables

Table 1 Table of the normal curve

$\frac{x}{\sigma}$	Area	Ordinate	$\frac{x}{\sigma}$	Area	Ordinate	$\frac{x}{\sigma}$	Area	Ordinate	$\frac{x}{\sigma}$	Area	Ordinate
.00	.0000	.3989	.35	.1368	.3752	.70	.2580	.3123	1.05	.3531	.2299
.01	.0040	.3989	.36	.1406	.3739	.71	.2611	.3101	1.06	.3554	.2275
.02	.0080	.3989	.37	.1443	.3725	.72	.2642	.3079	1.07	.3577	.2251
.03	.0120	.3988	.38	.1480	.3712	.73	.2673	.3056	1.08	.3599	.2227
.04	.0160	.3986	.39	.1517	.3697	.74	.2703	.3034	1.09	.3621	.2203
.05	.0199	.3984	.40	.1554	.3683	.75	.2734	.3011	1.10	.3643	.2179
.06	.0239	.3982	.41	.1591	.3668	.76	.2764	.2989	1.11	.3665	.2155
.07	.0279	.3980	.42	.1628	.3653	.77	.2794	.2966	1.12	.3686	.2131
.08	.0319	.3977	.43	.1664	.3637	.78	.2823	.2943	1.13	.3708	.2107
.09	.0359	.3973	.44	.1700	.3621	.79	.2852	.2920	1.14	.3729	.2083
.10	.0398	.3970	.45	.1736	.3605	.80	.2881	.2897	1.15	.3749	.2059
.11	.0438	.3965	.46	.1772	.3589	.81	.2910	.2874	1.16	.3770	.2036
.12	.0478	.3961	.47	.1808	.3572	.82	.2939	.2850	1.17	.3790	.2012
.13	.0517	.3956	.48	.1844	.3555	.83	.2967	.2827	1.18	.3810	.1989
.14	.0557	.3951	.49	.1879	.3538	.84	.2995	.2803	1.19	.3830	.1965
.15	.0596	.3945	.50	.1915	.3521	.85	.3023	.2780	1.20	.3849	.1942
.16	.0636	.3939	.51	.1950	.3503	.86	.3051	.2756	1.21	.3869	.1919
.17	.0675	.3932	.52	.1985	.3485	.87	.3078	.2732	1.22	.3888	.1895
.18	.0714	.3925	.53	.2019	.3467	.88	.3106	.2709	1.23	.3907	.1872
.19	.0753	.3918	.54	.2054	.3448	.89	.3133	.2685	1.24	.3925	.1849
.20	.0793	.3910	.55	.2088	.3429	.90	.3159	.2661	1.25	.3944	.1826
.21	.0832	.3902	.56	.2123	.3410	.91	.3186	.2637	1.26	.3962	.1804
.22	.0871	.3894	.57	.2157	.3391	.92	.3212	.2613	1.27	.3980	.1781
.23	.0910	.3885	.58	.2190	.3372	.93	.3238	.2589	1.28	.3997	.1758
.24	.0948	.3876	.59	.2224	.3352	.94	.3264	.2565	1.29	.4015	.1736
.25	.0987	.3867	.60	.2257	.3332	.95	.3289	.2541	1.30	.4032	.1714
.26	.1026	.3857	.61	.2291	.3312	.96	.3315	.2516	1.31	.4049	.1691
.27	.1064	.3847	.62	.2324	.3292	.97	.3340	.2492	1.32	.4066	.1669
.28	.1103	.3836	.63	.2357	.3271	.98	.3365	.2468	1.33	.4082	.1647
.29	.1141	.3825	.64	.2389	.3251	.99	.3389	.2444	1.34	.4099	.1626
.30	.1179	.3814	.65	.2422	.3230	1.00	.3413	.2420	1.35	.4115	.1604
.31	.1217	.3802	.66	.2454	.3209	1.01	.3438	.2396	1.36	.4131	.1582
.32	.1255	.3790	.67	.2486	.3187	1.02	.3461	.2371	1.37	.4147	.1561
.33	.1293	.3778	.68	.2517	.3166	1.03	.3485	.2347	1.38	.4162	.1539
.34	.1331	.3765	.69	.2549	.3144	1.04	.3508	.2323	1.39	.4177	.1518

Source: From *Educational Statistics* by J.E. Wert. Copyright © 1938 by McGraw-Hill Book Company. Used with permission of McGraw-Hill Book Company.

Table 1 *(Continued)*

$\frac{x}{\sigma}$	Area	Ordinate	$\frac{x}{\sigma}$	Area	Ordinate	$\frac{x}{\sigma}$	Area	Ordinate	$\frac{x}{\sigma}$	Area	Ordinate
1.40	.4192	.1497	1.80	.4641	.0790	2.20	.4861	.0355	2.60	.4953	.0136
1.41	.4207	.1476	1.81	.4649	.0775	2.21	.4864	.0347	2.61	.4955	.0132
1.42	.4222	.1456	1.82	.4656	.0761	2.22	.4868	.0339	2.62	.4956	.0129
1.43	.4236	.1435	1.83	.4664	.0748	2.23	.4871	.0332	2.63	.4957	.0126
1.44	.4251	.1415	1.84	.4671	.0734	2.24	.4875	.0325	2.64	.4959	.0122
1.45	.4265	.1394	1.85	.4678	.0721	2.25	.4878	.0317	2.65	.4960	.0119
1.46	.4279	.1374	1.86	.4686	.0707	2.26	.4881	.0310	2.66	.4961	.0116
1.47	.4292	.1354	1.87	.4693	.0694	2.27	.4884	.0303	2.67	.4962	.0113
1.48	.4306	.1334	1.88	.4699	.0681	2.28	.4887	.0297	2.68	.4963	.0110
1.49	.4319	.1315	1.89	.4706	.0669	2.29	.4890	.0290	2.69	.4964	.0107
1.50	.4332	.1295	1.90	.4713	.0656	2.30	.4893	.0283	2.70	.4965	.0104
1.51	.4345	.1276	1.91	.4719	.0644	2.31	.4896	.0277	2.71	.4966	.0101
1.52	.4357	.1257	1.92	.4726	.0632	2.32	.4898	.0270	2.72	.4967	.0099
1.53	.4370	.1238	1.93	.4732	.0620	2.33	.4901	.0264	2.73	.4968	.0096
1.54	.4382	.1219	1.94	.4738	.0608	2.34	.4904	.0258	2.74	.4969	.0093
1.55	.4394	.1200	1.95	.4744	.0596	2.35	.4906	.0252	2.75	.4970	.0091
1.56	.4406	.1182	1.96	.4750	.0584	2.36	.4909	.0246	2.76	.4971	.0088
1.57	.4418	.1163	1.97	.4756	.0573	2.37	.4911	.0241	2.77	.4972	.0086
1.58	.4429	.1145	1.98	.4761	.0562	2.38	.4913	.0235	2.78	.4973	.0084
1.59	.4441	.1127	1.99	.4767	.0551	2.39	.4916	.0229	2.79	.4974	.0081
1.60	.4452	.1109	2.00	.4772	.0540	2.40	.4918	.0224	2.80	.4974	.0079
1.61	.4463	.1092	2.01	.4778	.0529	2.41	.4920	.0219	2.81	.4975	.0077
1.62	.4474	.1074	2.02	.4783	.0519	2.42	.4922	.0213	2.82	.4976	.0075
1.63	.4484	.1057	2.03	.4788	.0508	2.43	.4925	.0208	2.83	.4977	.0073
1.64	.4495	.1040	2.04	.4793	.0498	2.44	.4927	.0203	2.84	.4977	.0071
1.65	.4505	.1023	2.05	.4798	.0488	2.45	.4929	.0198	2.85	.4978	.0069
1.66	.4515	.1006	2.06	.4803	.0478	2.46	.4931	.0194	2.86	.4979	.0067
1.67	.4525	.0989	2.07	.4808	.0468	2.47	.4932	.0189	2.87	.4979	.0065
1.68	.4535	.0973	2.08	.4812	.0459	2.48	.4934	.0184	2.88	.4980	.0063
1.69	.4545	.0957	2.09	.4817	.0449	2.49	.4936	.0180	2.89	.4981	.0061
1.70	.4554	.0940	2.10	.4821	.0440	2.50	.4938	.0175	2.90	.4981	.0060
1.71	.4564	.0925	2.11	.4826	.0431	2.51	.4940	.0171	2.91	.4982	.0058
1.72	.4573	.0909	2.12	.4830	.0422	2.52	.4941	.0167	2.92	.4982	.0056
1.73	.4582	.0893	2.13	.4834	.0413	2.53	.4943	.0163	2.93	.4983	.0055
1.74	.4591	.0878	2.14	.4838	.0404	2.54	.4945	.0158	2.94	.4984	.0053
1.75	.4599	.0863	2.15	.4842	.0395	2.55	.4946	.0154	2.95	.4984	.0051
1.76	.4608	.0848	2.16	.4846	.0387	2.56	.4948	.0151	2.96	.4985	.0050
1.77	.4616	.0833	2.17	.4850	.0379	2.57	.4949	.0147	2.97	.4985	.0048
1.78	.4625	.0818	2.18	.4854	.0371	2.58	.4951	.0143	2.98	.4986	.0047
1.79	.4633	.0804	2.19	.4857	.0363	2.59	.4952	.0139	2.99	.4986	.0046
									3.00	.4987	.0044

Table 2 Critical values of the *t*-distribution

	Level of significance for one-tailed test					
	.10	.05	.025	.01	.005	.0005
	Level of significance for two-tailed test					
df	.20	.10	.05	.02	.01	.001
1	3.078	6.314	12.706	31.821	63.657	636.619
2	1.886	2.920	4.303	6.965	9.925	31.598
3	1.638	2.353	3.182	4.541	5.841	12.941
4	1.533	2.132	2.776	3.747	4.604	8.610
5	1.476	2.015	2.571	3.365	4.032	6.859
6	1.440	1.943	2.447	3.143	3.707	5.959
7	1.415	1.895	2.365	2.998	3.499	5.405
8	1.397	1.860	2.306	2.896	3.355	5.041
9	1.383	1.833	2.262	2.821	3.250	4.781
10	1.372	1.812	2.228	2.764	3.169	4.587
11	1.363	1.796	2.201	2.718	3.106	4.437
12	1.356	1.782	2.179	2.681	3.055	4.318
13	1.350	1.771	2.160	2.650	3.012	4.221
14	1.345	1.761	2.145	2.624	2.977	4.140
15	1.341	1.753	2.131	2.602	2.947	4.073
16	1.337	1.746	2.120	2.583	2.921	4.015
17	1.333	1.740	2.110	2.567	2.898	3.965
18	1.330	1.734	2.101	2.552	2.878	3.922
19	1.328	1.729	2.093	2.539	2.861	3.883
20	1.325	1.725	2.086	2.528	2.845	3.850
21	1.323	1.721	2.080	2.518	2.831	3.819
22	1.321	1.717	2.074	2.508	2.819	3.792
23	1.319	1.714	2.069	2.500	2.807	3.767
24	1.318	1.711	2.064	2.492	2.797	3.745
25	1.316	1.708	2.060	2.485	2.787	3.725
26	1.315	1.706	2.056	2.479	2.779	3.707
27	1.314	1.703	2.052	2.473	2.771	3.690
28	1.313	1.701	2.048	2.467	2.763	3.674
29	1.311	1.699	2.045	2.462	2.756	3.659
30	1.310	1.697	2.042	2.457	2.750	3.646
40	1.303	1.684	2.021	2.423	2.704	3.551
60	1.296	1.671	2.000	2.390	2.660	3.460
120	1.289	1.658	1.980	2.358	2.617	3.373
∞	1.282	1.645	1.960	2.326	2.576	3.291

Source: Taken from Table III, p. 46 of Fisher and Yates: *Statistical Tables for Biological, Agricultural and Medical Research*, published by Longman Group Ltd., London. (previously published by Oliver and Boyd, Edinburgh), and by permission of the authors and publishers.

Table 3 — Upper percentage points of the χ^2 distribution

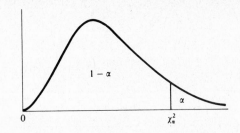

df	.99	.98	.95	.90	.80	.70	.50	.30	.20	.10	.05	.02	.01	.001
1	$.0^3157$	$.0^3628$	.00393	.0158	.0642	.148	.455	1.074	1.642	2.706	3.841	5.412	6.635	10.827
2	.0201	.0404	.103	.211	.446	.713	1.386	2.408	3.219	4.605	5.991	7.824	9.210	13.815
3	.115	.185	.352	.584	1.005	1.424	2.366	3.665	4.642	6.251	7.815	9.837	11.345	16.266
4	.297	.429	.711	1.064	1.649	2.195	3.357	4.878	5.989	7.779	9.488	11.668	13.277	18.467
5	.554	.752	1.145	1.610	2.343	3.000	4.351	6.064	7.289	9.236	11.070	13.388	15.086	20.515
6	.872	1.134	1.635	2.204	3.070	3.828	5.348	7.231	8.558	10.645	12.592	15.033	16.812	22.457
7	1.239	1.564	2.167	2.833	3.822	4.671	6.346	8.383	9.803	12.017	14.067	16.622	18.475	24.322
8	1.646	2.032	2.733	3.490	4.594	5.527	7.344	9.524	11.030	13.362	15.507	18.168	20.090	26.125
9	2.088	2.532	3.325	4.168	5.380	6.393	8.343	10.656	12.242	14.684	16.919	19.679	21.666	27.877
10	2.558	3.059	3.940	4.865	6.179	7.267	9.342	11.781	13.442	15.987	18.307	21.161	23.209	29.588
11	3.053	3.609	4.575	5.578	6.989	8.148	10.341	12.899	14.631	17.275	19.675	22.618	24.725	31.264
12	3.571	4.178	5.226	6.304	7.807	9.034	11.340	14.011	15.812	18.549	21.026	24.054	26.217	32.909
13	4.107	4.765	5.892	7.042	8.634	9.926	12.340	15.119	16.985	19.812	22.362	25.472	27.688	34.528
14	4.660	5.368	6.571	7.790	9.467	10.821	13.339	16.222	18.151	21.064	23.685	26.873	29.141	36.123
15	5.229	5.985	7.261	8.547	10.307	11.721	14.339	17.322	19.311	22.307	24.996	28.259	30.578	37.697
16	5.812	6.614	7.962	9.312	11.152	12.624	15.338	18.418	20.465	23.542	26.296	29.633	32.000	39.252
17	6.408	7.255	8.672	10.085	12.002	13.531	16.338	19.511	21.615	24.769	27.587	30.995	33.409	40.790
18	7.015	7.906	9.390	10.865	12.857	14.440	17.338	20.601	22.760	25.989	28.869	32.346	34.805	42.312
19	7.633	8.567	10.117	11.651	13.716	15.352	18.338	21.689	23.900	27.204	30.144	33.687	36.191	43.820
20	8.260	9.237	10.851	12.443	14.578	16.266	19.337	22.775	25.038	28.412	31.410	35.020	37.566	45.315
21	8.897	9.915	11.591	13.240	15.445	17.182	20.337	23.858	26.171	29.615	32.671	36.343	38.932	46.797
22	9.542	10.600	12.338	14.041	16.314	18.101	21.337	24.939	27.301	30.813	33.924	37.659	40.289	48.268
23	10.196	11.293	13.091	14.848	17.187	19.021	22.337	26.018	28.429	32.007	35.172	38.968	41.638	49.728
24	10.856	11.992	13.848	15.659	18.062	19.943	23.337	27.096	29.553	33.196	36.415	40.270	42.980	51.179
25	11.524	12.697	14.611	16.473	18.940	20.867	24.337	28.172	30.675	34.382	37.652	41.566	44.314	52.620
26	12.198	13.409	15.379	17.292	19.820	21.792	25.336	29.246	31.795	35.563	38.885	42.856	45.642	54.052
27	12.879	14.125	16.151	18.114	20.703	22.719	26.336	30.319	32.912	36.741	40.113	44.140	46.963	55.476
28	13.565	14.847	16.928	18.939	21.588	23.647	27.336	31.391	34.027	37.916	41.337	45.419	43.278	56.893
29	14.256	15.574	17.708	19.768	22.475	24.577	28.336	32.461	35.139	39.087	42.557	46.693	49.588	58.302
30	14.953	16.306	18.493	20.599	23.364	25.508	29.336	33.530	36.250	40.256	43.773	47.962	50.892	59.703

Source: Taken from Table IV, p. 47 of Fisher and Yates: *Statistical Tables for Biological, Agricultural and Medical Research,* published by Longman Group Ltd., London, (previously published by Oliver and Boyd, Edinburgh), and by permission of the authors and publishers.

For df > 30, the expression $\sqrt{2\chi^2} - \sqrt{2\,df - 1}$ may be used as a normal deviate with unit variance.

Table 4
Upper percentage points of the *F*-distribution

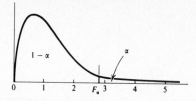

df *for de-nomi-nator*	α	\multicolumn{12}{c}{df *for numerator*}
		1	2	3	4	5	6	7	8	9	10	11	12
1	.25	5.83	7.50	8.20	8.58	8.82	8.98	9.10	9.19	9.26	9.32	9.36	9.41
	.10	39.9	49.5	53.6	55.8	57.2	58.2	58.9	59.4	59.9	60.2	60.5	60.7
	.05	161	200	216	225	230	234	237	239	241	242	243	244
2	.25	2.57	3.00	3.15	3.23	3.28	3.31	3.34	3.35	3.37	3.38	3.39	3.39
	.10	8.53	9.00	9.16	9.24	9.29	9.33	9.35	9.37	9.38	9.39	9.40	9.41
	.05	18.5	19.0	19.2	19.2	19.3	19.3	19.4	19.4	19.4	19.4	19.4	19.4
	.01	98.5	99.0	99.2	99.2	99.3	99.3	99.4	99.4	99.4	99.4	99.4	99.4
3	.25	2.02	2.28	2.36	2.39	2.41	2.42	2.43	2.44	2.44	2.44	2.45	2.45
	.10	5.54	5.46	5.39	5.34	5.31	5.28	5.27	5.25	5.24	5.23	5.22	5.22
	.05	10.1	9.55	9.28	9.12	9.01	8.94	8.89	8.85	8.81	8.79	8.76	8.74
	.01	34.1	30.8	29.5	28.7	28.2	27.9	27.7	27.5	27.3	27.2	27.1	27.1
4	.25	1.81	2.00	2.05	2.06	2.07	2.08	2.08	2.08	2.08	2.08	2.08	2.08
	.10	4.54	4.32	4.19	4.11	4.05	4.01	3.98	3.95	3.94	3.92	3.91	3.90
	.05	7.71	6.94	6.59	6.39	6.26	6.16	6.09	6.04	6.00	5.96	5.94	5.91
	.01	21.2	18.0	16.7	16.0	15.5	15.2	15.0	14.8	14.7	14.5	14.4	14.4
5	.25	1.69	1.85	1.88	1.89	1.89	1.89	1.89	1.89	1.89	1.89	1.89	1.89
	.10	4.06	3.78	3.62	3.52	3.45	3.40	3.37	3.34	3.32	3.30	3.28	3.27
	.05	6.61	5.79	5.41	5.19	5.05	4.95	4.88	4.82	4.77	4.74	4.71	4.68
	.01	16.3	13.3	12.1	11.4	11.0	10.7	10.5	10.3	10.2	10.1	9.96	9.89
6	.25	1.62	1.76	1.78	1.79	1.79	1.78	1.78	1.78	1.77	1.77	1.77	1.77
	.10	3.78	3.46	3.29	3.18	3.11	3.05	3.01	2.98	2.96	2.94	2.92	2.90
	.05	5.99	5.14	4.76	4.53	4.39	4.28	4.21	4.15	4.10	4.06	4.03	4.00
	.01	13.7	10.9	9.78	9.15	8.75	8.47	8.26	8.10	7.98	7.87	7.79	7.72
7	.25	1.57	1.70	1.72	1.72	1.71	1.71	1.70	1.70	1.69	1.69	1.69	1.68
	.10	3.59	3.26	3.07	2.96	2.88	2.83	2.78	2.75	2.72	2.70	2.68	2.67
	.05	5.59	4.74	4.35	4.12	3.97	3.87	3.79	3.73	3.68	3.64	3.60	3.57
	.01	12.2	9.55	8.45	7.85	7.46	7.19	6.99	6.84	6.72	6.62	6.54	6.47
8	.25	1.54	1.66	1.67	1.66	1.66	1.65	1.64	1.64	1.63	1.63	1.63	1.62
	.10	3.46	3.11	2.92	2.81	2.73	2.67	2.62	2.59	2.56	2.54	2.52	2.50
	.05	5.32	4.46	4.07	3.84	3.69	3.58	3.50	3.44	3.39	3.35	3.31	3.28
	.01	11.3	8.65	7.59	7.01	6.63	6.37	6.18	6.03	5.91	5.81	5.73	5.67
9	.25	1.51	1.62	1.63	1.63	1.62	1.61	1.60	1.60	1.59	1.59	1.58	1.58
	.10	3.36	3.01	2.81	2.69	2.61	2.55	2.51	2.47	2.44	2.42	2.40	2.38
	.05	5.12	4.26	3.86	3.63	3.48	3.37	3.29	3.23	3.18	3.14	3.10	3.07
	.01	10.6	8.02	6.99	6.42	6.06	5.80	5.61	5.47	5.35	5.26	5.18	5.11

				df *for numerator*										df *for de-nomi-nator*
15	20	24	30	40	50	60	100	120	200	500	∞	α	
9.49	9.58	9.63	9.67	9.71	9.74	9.76	9.78	9.80	9.82	9.84	9.85	.25	
61.2	61.7	62.0	62.3	62.5	62.7	62.8	63.0	63.1	63.2	63.3	63.3	.10	1
246	248	249	250	251	252	252	253	253	254	254	254	.05	
3.41	3.43	3.43	3.44	3.45	3.45	3.46	3.47	3.47	3.48	3.48	3.48	.25	
9.42	9.44	9.45	9.46	9.47	9.47	9.47	9.48	9.48	9.49	9.49	9.49	.10	2
19.4	19.4	19.5	19.5	19.5	19.5	19.5	19.5	19.5	19.5	19.5	19.5	.05	
99.4	99.4	99.5	99.5	99.5	99.5	99.5	99.5	99.5	99.5	99.5	99.5	.01	
2.46	2.46	2.46	2.47	2.47	2.47	2.47	2.47	2.47	2.47	2.47	2.47	.25	
5.20	5.18	5.18	5.17	5.16	5.15	5.15	5.14	5.14	5.14	5.14	5.13	.10	3
8.70	8.66	8.64	8.62	8.59	8.58	8.57	8.55	8.55	8.54	8.53	8.53	.05	
26.9	26.7	26.6	26.5	26.4	26.4	26.3	26.2	26.2	26.2	26.1	26.1	.01	
2.08	2.08	2.08	2.08	2.08	2.08	2.08	2.08	2.08	2.08	2.08	2.08	.25	
3.87	3.84	3.83	3.82	3.80	3.80	3.79	3.78	3.78	3.77	3.76	3.76	.10	4
5.86	5.80	5.77	5.75	5.72	5.70	5.69	5.66	5.66	5.65	5.64	5.63	.05	
14.2	14.0	13.9	13.8	13.7	13.7	13.7	13.6	13.6	13.5	13.5	13.5	.01	
1.89	1.88	1.88	1.88	1.88	1.88	1.87	1.87	1.87	1.87	1.87	1.87	.25	
3.24	3.21	3.19	3.17	3.16	3.15	3.14	3.13	3.12	3.12	3.11	3.10	.10	5
4.62	4.56	4.53	4.50	4.46	4.44	4.43	4.41	4.40	4.39	4.37	4.36	.05	
9.72	9.55	9.47	9.38	9.29	9.24	9.20	9.13	9.11	9.08	9.04	9.02	.01	
1.76	1.76	1.75	1.75	1.75	1.75	1.74	1.74	1.74	1.74	1.74	1.74	.25	
2.87	2.84	2.82	2.80	2.78	2.77	2.76	2.75	2.74	2.73	2.73	2.72	.10	6
3.94	3.87	3.84	3.81	3.77	3.75	3.74	3.71	3.70	3.69	3.68	3.67	.05	
7.56	7.40	7.31	7.23	7.14	7.09	7.06	6.99	6.97	6.93	6.90	6.88	.01	
1.68	1.67	1.67	1.66	1.66	1.66	1.65	1.65	1.65	1.65	1.65	1.65	.25	
2.63	2.59	2.58	2.56	2.54	2.52	2.51	2.50	2.49	2.48	2.48	2.47	.10	7
3.51	3.44	3.41	3.38	3.34	3.32	3.30	3.27	3.27	3.25	3.24	3.23	.05	
6.31	6.16	6.07	5.99	5.91	5.86	5.82	5.75	5.74	5.70	5.67	5.65	.01	
1.62	1.61	1.60	1.60	1.59	1.59	1.59	1.58	1.58	1.58	1.58	1.58	.25	
2.46	2.42	2.40	2.38	2.36	2.35	2.34	2.32	2.32	2.31	2.30	2.29	.10	8
3.22	3.15	3.12	3.08	3.04	3.02	3.01	2.97	2.97	2.95	2.94	2.93	.05	
5.52	5.36	5.28	5.20	5.12	5.07	5.03	4.96	4.95	4.91	4.88	4.86	.01	
1.57	1.56	1.56	1.55	1.55	1.54	1.54	1.53	1.53	1.53	1.53	1.53	.25	
2.34	2.30	2.28	2.25	2.23	2.22	2.21	2.19	2.18	2.17	2.17	2.16	.10	9
3.01	2.94	2.90	2.86	2.83	2.80	2.79	2.76	2.75	2.73	2.72	2.71	.05	
4.96	4.81	4.73	4.65	4.57	4.52	4.48	4.42	4.40	4.36	4.33	4.31	.01	

Table 4   *(Continued)*

df for denominator	α	df for numerator											
		1	2	3	4	5	6	7	8	9	10	11	12
10	.25	1.49	1.60	1.60	1.59	1.59	1.58	1.57	1.56	1.56	1.55	1.55	1.54
	.10	3.29	2.92	2.73	2.61	2.52	2.46	2.41	2.38	2.35	2.32	2.30	2.28
	.05	4.96	4.10	3.71	3.48	3.33	3.22	3.14	3.07	3.02	2.98	2.94	2.91
	.01	10.0	7.56	6.55	5.99	5.64	5.39	5.20	5.06	4.94	4.85	4.77	4.71
11	.25	1.47	1.58	1.58	1.57	1.56	1.55	1.54	1.53	1.53	1.52	1.52	1.51
	.10	3.23	2.86	2.66	2.54	2.45	2.39	2.34	2.30	2.27	2.25	2.23	2.21
	.05	4.84	3.98	3.59	3.36	3.20	3.09	3.01	2.95	2.90	2.85	2.82	2.79
	.01	9.65	7.21	6.22	5.67	5.32	5.07	4.89	4.74	4.63	4.54	4.46	4.40
12	.25	1.46	1.56	1.56	1.55	1.54	1.53	1.52	1.51	1.51	1.50	1.50	1.49
	.10	3.18	2.81	2.61	2.48	2.39	2.33	2.28	2.24	2.21	2.19	2.17	2.15
	.05	4.75	3.89	3.49	3.26	3.11	3.00	2.91	2.85	2.80	2.75	2.72	2.69
	.01	9.33	6.93	5.95	5.41	5.06	4.82	4.64	4.50	4.39	4.30	4.22	4.16
13	.25	1.45	1.55	1.55	1.53	1.52	1.51	1.50	1.49	1.49	1.48	1.47	1.47
	.10	3.14	2.76	2.56	2.43	2.35	2.28	2.23	2.20	2.16	2.14	2.12	2.10
	.05	4.67	3.81	3.41	3.18	3.03	2.92	2.83	2.77	2.71	2.67	2.63	2.60
	.01	9.07	6.70	5.74	5.21	4.86	4.62	4.44	4.30	4.19	4.10	4.02	3.96
14	.25	1.44	1.53	1.53	1.52	1.51	1.50	1.49	1.48	1.47	1.46	1.46	1.45
	.10	3.10	2.73	2.52	2.39	2.31	2.24	2.19	2.15	2.12	2.10	2.08	2.05
	.05	4.60	3.74	3.34	3.11	2.96	2.85	2.76	2.70	2.65	2.60	2.57	2.53
	.01	8.86	6.51	5.56	5.04	4.69	4.46	4.28	4.14	4.03	3.94	3.86	3.80
15	.25	1.43	1.52	1.52	1.51	1.49	1.48	1.47	1.46	1.46	1.45	1.44	1.44
	.10	3.07	2.70	2.49	2.36	2.27	2.21	2.16	2.12	2.09	2.06	2.04	2.02
	.05	4.54	3.68	3.29	3.06	2.90	2.79	2.71	2.64	2.59	2.54	2.51	2.48
	.01	8.68	6.36	5.42	4.89	4.56	4.32	4.14	4.00	3.89	3.80	3.73	3.67
16	.25	1.42	1.51	1.51	1.50	1.48	1.47	1.46	1.45	1.44	1.44	1.44	1.43
	.10	3.05	2.67	2.46	2.33	2.24	2.18	2.13	2.09	2.06	2.03	2.01	1.99
	.05	4.49	3.63	3.24	3.01	2.85	2.74	2.66	2.59	2.54	2.49	2.46	2.42
	.01	8.53	6.23	5.29	4.77	4.44	4.20	4.03	3.89	3.78	3.69	3.62	3.55
17	.25	1.42	1.51	1.50	1.49	1.47	1.46	1.45	1.44	1.43	1.43	1.42	1.41
	.10	3.03	2.64	2.44	2.31	2.22	2.15	2.10	2.06	2.03	2.00	1.98	1.96
	.05	4.45	3.59	3.20	2.96	2.81	2.70	2.61	2.55	2.49	2.45	2.41	2.38
	.01	8.40	6.11	5.18	4.67	4.34	4.10	3.93	3.79	3.68	3.59	3.52	3.46
18	.25	1.41	1.50	1.49	1.48	1.46	1.45	1.44	1.43	1.42	1.42	1.41	1.40
	.10	3.01	2.62	2.42	2.29	2.20	2.13	2.08	2.04	2.00	1.98	1.96	1.93
	.05	4.41	3.55	3.16	2.93	2.77	2.66	2.58	2.51	2.46	2.41	2.37	2.34
	.01	8.29	6.01	5.09	4.58	4.25	4.01	3.84	3.71	3.60	3.51	3.43	3.37
19	.25	1.41	1.49	1.49	1.47	1.46	1.44	1.43	1.42	1.41	1.41	1.40	1.40
	.10	2.99	2.61	2.40	2.27	2.18	2.11	2.06	2.02	1.98	1.96	1.94	1.91
	.05	4.38	3.52	3.13	2.90	2.74	2.63	2.54	2.48	2.42	2.38	2.34	2.31
	.01	8.18	5.93	5.01	4.50	4.17	3.94	3.77	3.63	3.52	3.43	3.36	3.30
20	.25	1.40	1.49	1.48	1.46	1.45	1.44	1.43	1.42	1.41	1.40	1.39	1.39
	.10	2.97	2.59	2.38	2.25	2.16	2.09	2.04	2.00	1.96	1.94	1.92	1.89
	.05	4.35	3.49	3.10	2.87	2.71	2.60	2.51	2.45	2.39	2.35	2.31	2.28
	.01	8.10	5.85	4.94	4.43	4.10	3.87	3.70	3.56	3.46	3.37	3.29	3.23

			df for numerator										
15	20	24	30	40	50	60	100	120	200	500	∞	α	nator
1.53	1.52	1.52	1.51	1.51	1.50	1.50	1.49	1.49	1.49	1.48	1.48	.25	
2.24	2.20	2.18	2.16	2.13	2.12	2.11	2.09	2.08	2.07	2.06	2.06	.10	10
2.85	2.77	2.74	2.70	2.66	2.64	2.62	2.59	2.58	2.56	2.55	2.54	.05	
4.56	4.41	4.33	4.25	4.17	4.12	4.08	4.01	4.00	3.96	3.93	3.91	.01	
1.50	1.49	1.49	1.48	1.47	1.47	1.47	1.46	1.46	1.46	1.45	1.45	.25	
2.17	2.12	2.10	2.08	2.05	2.04	2.03	2.00	2.00	1.99	1.98	1.97	.10	11
2.72	2.65	2.61	2.57	2.53	2.51	2.49	2.46	2.45	2.43	2.42	2.40	.05	
4.25	4.10	4.02	3.94	3.86	3.81	3.78	3.71	3.69	3.66	3.62	3.60	.01	
1.48	1.47	1.46	1.45	1.45	1.44	1.44	1.43	1.43	1.43	1.42	1.42	.25	
2.10	2.06	2.04	2.01	1.99	1.97	1.96	1.94	1.93	1.92	1.91	1.90	.10	12
2.62	2.54	2.51	2.47	2.43	2.40	2.38	2.35	2.34	2.32	2.31	2.30	.05	
4.01	3.86	3.78	3.70	3.62	3.57	3.54	3.47	3.45	3.41	3.38	3.36	.01	
1.46	1.45	1.44	1.43	1.42	1.42	1.42	1.41	1.41	1.40	1.40	1.40	.25	
2.05	2.01	1.98	1.96	1.93	1.92	1.90	1.88	1.88	1.86	1.85	1.85	.10	13
2.53	2.46	2.42	2.38	2.34	2.31	2.30	2.26	2.25	2.23	2.22	2.21	.05	
3.82	3.66	3.59	3.51	3.43	3.38	3.34	3.27	3.25	3.22	3.19	3.17	.01	
1.44	1.43	1.42	1.41	1.41	1.40	1.40	1.39	1.39	1.39	1.38	1.38	.25	
2.01	1.96	1.94	1.91	1.89	1.87	1.86	1.83	1.83	1.82	1.80	1.80	.10	14
2.46	2.39	2.35	2.31	2.27	2.24	2.22	2.19	2.18	2.16	2.14	2.13	.05	
3.66	3.51	3.43	3.35	3.27	3.22	3.18	3.11	3.09	3.06	3.03	3.00	.01	
1.43	1.41	1.41	1.40	1.39	1.39	1.38	1.38	1.37	1.37	1.36	1.36	.25	
1.97	1.92	1.90	1.87	1.85	1.83	1.82	1.79	1.79	1.77	1.76	1.76	.10	15
2.40	2.33	2.29	2.25	2.20	2.18	2.16	2.12	2.11	2.10	2.08	2.07	.05	
3.52	3.37	3.29	3.21	3.13	3.08	3.05	2.98	2.96	2.92	2.89	2.87	.01	
1.41	1.40	1.39	1.38	1.37	1.37	1.36	1.36	1.35	1.35	1.34	1.34	.25	
1.94	1.89	1.87	1.84	1.81	1.79	1.78	1.76	1.75	1.74	1.73	1.72	.10	16
2.35	2.28	2.24	2.19	2.15	2.12	2.11	2.07	2.06	2.04	2.02	2.01	.05	
3.41	3.26	3.18	3.10	3.02	2.97	2.93	2.86	2.84	2.81	2.78	2.75	.01	
1.40	1.39	1.38	1.37	1.36	1.35	1.35	1.34	1.34	1.34	1.33	1.33	.25	
1.91	1.86	1.84	1.81	1.78	1.76	1.75	1.73	1.72	1.71	1.69	1.69	.10	17
2.31	2.23	2.19	2.15	2.10	2.08	2.06	2.02	2.01	1.99	1.97	1.96	.05	
3.31	3.16	3.08	3.00	2.92	2.87	2.83	2.76	2.75	2.71	2.68	2.65	.01	
1.39	1.38	1.37	1.36	1.35	1.34	1.34	1.33	1.33	1.32	1.32	1.32	.25	
1.89	1.84	1.81	1.78	1.75	1.74	1.72	1.70	1.69	1.68	1.67	1.66	.10	18
2.27	2.19	2.15	2.11	2.06	2.04	2.02	1.98	1.97	1.95	1.93	1.92	.05	
3.23	3.08	3.00	2.92	2.84	2.78	2.75	2.68	2.66	2.62	2.59	2.57	.01	
1.38	1.37	1.36	1.35	1.34	1.33	1.33	1.32	1.32	1.31	1.31	1.30	.25	
1.86	1.81	1.79	1.76	1.73	1.71	1.70	1.67	1.67	1.65	1.64	1.63	.10	19
2.23	2.16	2.11	2.07	2.03	2.00	1.98	1.94	1.93	1.91	1.89	1.88	.05	
3.15	3.00	2.92	2.84	2.76	2.71	2.67	2.60	2.58	2.55	2.51	2.49	.01	
1.37	1.36	1.35	1.34	1.33	1.33	1.32	1.31	1.31	1.30	1.30	1.29	.25	
1.84	1.79	1.77	1.74	1.71	1.69	1.68	1.65	1.64	1.63	1.62	1.61	.10	20
2.20	2.12	2.08	2.04	1.99	1.97	1.95	1.91	1.90	1.88	1.86	1.84	.05	
3.09	2.94	2.86	2.78	2.69	2.64	2.61	2.54	2.52	2.48	2.44	2.42	.01	

Table 4 *(Continued)*

df for denominator	α	df for numerator											
		1	2	3	4	5	6	7	8	9	10	11	12
22	.25	1.40	1.48	1.47	1.45	1.44	1.42	1.41	1.40	1.39	1.39	1.38	1.37
	.10	2.95	2.56	2.35	2.22	2.13	2.06	2.01	1.97	1.93	1.90	1.88	1.86
	.05	4.30	3.44	3.05	2.82	2.66	2.55	2.46	2.40	2.34	2.30	2.26	2.23
	.01	7.95	5.72	4.82	4.31	3.99	3.76	3.59	3.45	3.35	3.26	3.18	3.12
24	.25	1.39	1.47	1.46	1.44	1.43	1.41	1.40	1.39	1.38	1.38	1.37	1.36
	.10	2.93	2.54	2.33	2.19	2.10	2.04	1.98	1.94	1.91	1.88	1.85	1.83
	.05	4.26	3.40	3.01	2.78	2.62	2.51	2.42	2.36	2.30	2.25	2.21	2.18
	.01	7.82	5.61	4.72	4.22	3.90	3.67	3.50	3.36	3.26	3.17	3.09	3.03
26	.25	1.38	1.46	1.45	1.44	1.42	1.41	1.39	1.38	1.37	1.37	1.36	1.35
	.10	2.91	2.52	2.31	2.17	2.08	2.01	1.96	1.92	1.88	1.86	1.84	1.81
	.05	4.23	3.37	2.98	2.74	2.59	2.47	2.39	2.32	2.27	2.22	2.18	2.15
	.01	7.72	5.53	4.64	4.14	3.82	3.59	3.42	3.29	3.18	3.09	3.02	2.96
28	.25	1.38	1.46	1.45	1.43	1.41	1.40	1.39	1.38	1.37	1.36	1.35	1.34
	.10	2.89	2.50	2.29	2.16	2.06	2.00	1.94	1.90	1.87	1.84	1.81	1.79
	.05	4.20	3.34	2.95	2.71	2.56	2.45	2.36	2.29	2.24	2.19	2.15	2.12
	.01	7.64	5.45	4.57	4.07	3.75	3.53	3.36	3.23	3.12	3.03	2.96	2.90
30	.25	1.38	1.45	1.44	1.42	1.41	1.39	1.38	1.37	1.36	1.35	1.35	1.34
	.10	2.88	2.49	2.28	2.14	2.05	1.98	1.93	1.88	1.85	1.82	1.79	1.77
	.05	4.17	3.32	2.92	2.69	2.53	2.42	2.33	2.27	2.21	2.16	2.13	2.09
	.01	7.56	5.39	4.51	4.02	3.70	3.47	3.30	3.17	3.07	2.98	2.91	2.84
40	.25	1.36	1.44	1.42	1.40	1.39	1.37	1.36	1.35	1.34	1.33	1.32	1.31
	.10	2.84	2.44	2.23	2.09	2.00	1.93	1.87	1.83	1.79	1.76	1.73	1.71
	.05	4.08	3.23	2.84	2.61	2.45	2.34	2.25	2.18	2.12	2.08	2.04	2.00
	.01	7.31	5.18	4.31	3.83	3.51	3.29	3.12	2.99	2.89	2.80	2.73	2.66
60	.25	1.35	1.42	1.41	1.38	1.37	1.35	1.33	1.32	1.31	1.30	1.29	1.29
	.10	2.79	2.39	2.18	2.04	1.95	1.87	1.82	1.77	1.74	1.71	1.68	1.66
	.05	4.00	3.15	2.76	2.53	2.37	2.25	2.17	2.10	2.04	1.99	1.95	1.92
	.01	7.08	4.98	4.13	3.65	3.34	3.12	2.95	2.82	2.72	2.63	2.56	2.50
120	.25	1.34	1.40	1.39	1.37	1.35	1.33	1.31	1.30	1.29	1.28	1.27	1.26
	.10	2.75	2.35	2.13	1.99	1.90	1.82	1.77	1.72	1.68	1.65	1.62	1.60
	.05	3.92	3.07	2.68	2.45	2.29	2.17	2.09	2.02	1.96	1.91	1.87	1.83
	.01	6.85	4.79	3.95	3.48	3.17	2.96	2.79	2.66	2.56	2.47	2.40	2.34
200	.25	1.33	1.39	1.38	1.36	1.34	1.32	1.31	1.29	1.28	1.27	1.26	1.25
	.10	2.73	2.33	2.11	1.97	1.88	1.80	1.75	1.70	1.66	1.63	1.60	1.57
	.05	3.89	3.04	2.65	2.42	2.26	2.14	2.06	1.98	1.93	1.88	1.84	1.80
	.01	6.76	4.71	3.88	3.41	3.11	2.89	2.73	2.60	2.50	2.41	2.34	2.27
∞	.25	1.32	1.39	1.37	1.35	1.33	1.31	1.29	1.28	1.27	1.25	1.24	1.24
	.10	2.71	2.30	2.08	1.94	1.85	1.77	1.72	1.67	1.63	1.60	1.57	1.55
	.05	3.84	3.00	2.60	2.37	2.21	2.10	2.01	1.94	1.88	1.83	1.79	1.75
	.01	6.63	4.61	3.78	3.32	3.02	2.80	2.64	2.51	2.41	2.32	2.25	2.18

				df _for numerator_									α	df _for de-nomi-nator_
15	20	24	30	40	50	60	100	120	200	500	∞		
1.36	1.34	1.33	1.32	1.31	1.31	1.30	1.30	1.30	1.29	1.29	1.28	.25	
1.81	1.76	1.73	1.70	1.67	1.65	1.64	1.61	1.60	1.59	1.58	1.57	.10	22
2.15	2.07	2.03	1.98	1.94	1.91	1.89	1.85	1.84	1.82	1.80	1.78	.05	
2.98	2.83	2.75	2.67	2.58	2.53	2.50	2.42	2.40	2.36	2.33	2.31	.01	
1.35	1.33	1.32	1.31	1.30	1.29	1.29	1.28	1.28	1.27	1.27	1.26	.25	
1.78	1.73	1.70	1.67	1.64	1.62	1.61	1.58	1.57	1.56	1.54	1.53	.10	24
2.11	2.03	1.98	1.94	1.89	1.86	1.84	1.80	1.79	1.77	1.75	1.73	.05	
2.89	2.74	2.66	2.58	2.49	2.44	2.40	2.33	2.31	2.27	2.24	2.21	.01	
1.34	1.32	1.31	1.30	1.29	1.28	1.28	1.26	1.26	1.26	1.25	1.25	.25	
1.76	1.71	1.68	1.65	1.61	1.59	1.58	1.55	1.54	1.53	1.51	1.50	.10	26
2.07	1.99	1.95	1.90	1.85	1.82	1.80	1.76	1.75	1.73	1.71	1.69	.05	
2.81	2.66	2.58	2.50	2.42	2.36	2.33	2.25	2.23	2.19	2.16	2.13	.01	
1.33	1.31	1.30	1.29	1.28	1.27	1.27	1.26	1.25	1.25	1.24	1.24	.25	
1.74	1.69	1.66	1.63	1.59	1.57	1.56	1.53	1.52	1.50	1.49	1.48	.10	28
2.04	1.96	1.91	1.87	1.82	1.79	1.77	1.73	1.71	1.69	1.67	1.65	.05	
2.75	2.60	2.52	2.44	2.35	2.30	2.26	2.19	2.17	2.13	2.09	2.06	.01	
1.32	1.30	1.29	1.28	1.27	1.26	1.26	1.25	1.24	1.24	1.23	1.23	.25	
1.72	1.67	1.64	1.61	1.57	1.55	1.54	1.51	1.50	1.48	1.47	1.46	.10	30
2.01	1.93	1.89	1.84	1.79	1.76	1.74	1.70	1.68	1.66	1.64	1.62	.05	
2.70	2.55	2.47	2.39	2.30	2.25	2.21	2.13	2.11	2.07	2.03	2.01	.01	
1.30	1.28	1.26	1.25	1.24	1.23	1.22	1.21	1.21	1.20	1.19	1.19	.25	
1.66	1.61	1.57	1.54	1.51	1.48	1.47	1.43	1.42	1.41	1.39	1.38	.10	40
1.92	1.84	1.79	1.74	1.69	1.66	1.64	1.59	1.58	1.55	1.53	1.51	.05	
2.52	2.37	2.29	2.20	2.11	2.06	2.02	1.94	1.92	1.87	1.83	1.80	.01	
1.27	1.25	1.24	1.22	1.21	1.20	1.19	1.17	1.17	1.16	1.15	1.15	.25	
1.60	1.54	1.51	1.48	1.44	1.41	1.40	1.36	1.35	1.33	1.31	1.29	.10	60
1.84	1.75	1.70	1.65	1.59	1.56	1.53	1.48	1.47	1.44	1.41	1.39	.05	
2.35	2.20	2.12	2.03	1.94	1.88	1.84	1.75	1.73	1.68	1.63	1.60	.01	
1.24	1.22	1.21	1.19	1.18	1.17	1.16	1.14	1.13	1.12	1.11	1.10	.25	
1.55	1.48	1.45	1.41	1.37	1.34	1.32	1.27	1.26	1.24	1.21	1.19	.10	120
1.75	1.66	1.61	1.55	1.50	1.46	1.43	1.37	1.35	1.32	1.28	1.25	.05	
2.19	2.03	1.95	1.86	1.76	1.70	1.66	1.56	1.53	1.48	1.42	1.38	.01	
1.23	1.21	1.20	1.18	1.16	1.14	1.12	1.11	1.10	1.09	1.08	1.06	.25	
1.52	1.46	1.42	1.38	1.34	1.31	1.28	1.24	1.22	1.20	1.17	1.14	.10	200
1.72	1.62	1.57	1.52	1.46	1.41	1.39	1.32	1.29	1.26	1.22	1.19	.05	
2.13	1.97	1.89	1.79	1.69	1.63	1.58	1.48	1.44	1.39	1.33	1.28	.01	
1.22	1.19	1.18	1.16	1.14	1.13	1.12	1.09	1.08	1.07	1.04	1.00	.25	
1.49	1.42	1.38	1.34	1.30	1.26	1.24	1.18	1.17	1.13	1.08	1.00	.10	∞
1.67	1.57	1.52	1.46	1.39	1.35	1.32	1.24	1.22	1.17	1.11	1.00	.05	
2.04	1.88	1.79	1.70	1.59	1.52	1.47	1.36	1.32	1.25	1.15	1.00	.01	

Source: Abridged from Table 18 in _Biometrika Tables for Statisticians,_ Vol. 1, 3rd ed., E. S. Pearson and H. O. Hartley, eds. (New York: Cambridge, 1966). Used with permission of the editors and the Biometrika Trustees.

Table 5 Percentage points of the studentized range

Error df	x	\multicolumn{10}{c}{r = number of means or number of steps between ordered means}									
		2	3	4	5	6	7	8	9	10	11
5	.05	3.64	4.60	5.22	5.67	6.03	6.33	6.58	6.80	6.99	7.17
	.01	5.70	6.98	7.80	8.42	8.91	9.32	9.67	9.97	10.24	10.48
6	.05	3.46	4.34	4.90	5.30	5.63	5.90	6.12	6.32	6.49	6.65
	.01	5.24	6.33	7.03	7.56	7.97	8.32	8.61	8.87	9.10	9.30
7	.05	3.34	4.16	4.68	5.06	5.36	5.61	5.82	6.00	6.16	6.30
	.01	4.95	5.92	6.54	7.01	7.37	7.68	7.94	8.17	8.37	8.55
8	.05	3.26	4.04	4.53	4.89	5.17	5.40	5.60	5.77	5.92	6.05
	.01	4.75	5.64	6.20	6.62	6.96	7.24	7.47	7.68	7.86	8.03
9	.05	3.20	3.95	4.41	4.76	5.02	5.24	5.43	5.59	5.74	5.87
	.01	4.60	5.43	5.96	6.35	6.66	6.91	7.13	7.33	7.49	7.65
10	.05	3.15	3.88	4.33	4.65	4.91	5.12	5.30	5.46	5.60	5.72
	.01	4.48	5.27	5.77	6.14	6.43	6.67	6.87	7.05	7.21	7.36
11	.05	3.11	3.82	4.26	4.57	4.82	5.03	5.20	5.35	5.49	5.61
	.01	4.39	5.15	5.62	5.97	6.25	6.48	6.67	6.84	6.99	7.13
12	.05	3.08	3.77	4.20	4.51	4.75	4.95	5.12	5.27	5.39	5.51
	.01	4.32	5.05	5.50	5.84	6.10	6.32	6.51	6.67	6.81	6.94
13	.05	3.06	3.73	4.15	4.45	4.69	4.88	5.05	5.19	5.32	5.43
	.01	4.26	4.96	5.40	5.73	5.98	6.19	6.37	6.53	6.67	6.79
14	.05	3.03	3.70	4.11	4.41	4.64	4.83	4.99	5.13	5.25	5.36
	.01	4.21	4.89	5.32	5.63	5.88	6.08	6.26	6.41	6.54	6.66
15	.05	3.01	3.67	4.08	4.37	4.59	4.78	4.94	5.08	5.20	5.31
	.01	4.17	4.84	5.25	5.56	5.80	5.99	6.16	6.31	6.44	6.55
16	.05	3.00	3.65	4.05	4.33	4.56	4.74	4.90	5.03	5.15	5.26
	.01	4.13	4.79	5.19	5.49	5.72	5.92	6.08	6.22	6.35	6.46
17	.05	2.98	3.63	4.02	4.30	4.52	4.70	4.86	4.99	5.11	5.21
	.01	4.10	4.74	5.14	5.43	5.66	5.85	6.01	6.15	6.27	6.38
18	.05	2.97	3.61	4.00	4.28	4.49	4.67	4.82	4.96	5.07	5.17
	.01	4.07	4.70	5.09	5.38	5.60	5.79	5.94	6.08	6.20	6.31
19	.05	2.96	3.59	3.98	4.25	4.47	4.65	4.79	4.92	5.04	5.14
	.01	4.05	4.67	5.05	5.33	5.55	5.73	5.89	6.02	6.14	6.25
20	.05	2.95	3.58	3.96	4.23	4.45	4.62	4.77	4.90	5.01	5.11
	.01	4.02	4.64	5.02	5.29	5.51	5.69	5.84	5.97	6.09	6.19
24	.05	2.92	3.53	3.90	4.17	4.37	4.54	4.68	4.81	4.92	5.01
	.01	3.96	4.55	4.91	5.17	5.37	5.54	5.69	5.81	5.92	6.02
30	.05	2.89	3.49	3.85	4.10	4.30	4.46	4.60	4.72	4.82	4.92
	.01	3.89	4.45	4.80	5.05	5.24	5.40	5.54	5.65	5.76	5.85
40	.05	2.86	3.44	3.79	4.04	4.23	4.39	4.52	4.63	4.73	4.82
	.01	3.82	4.37	4.70	4.93	5.11	5.26	5.39	5.50	5.60	5.69
60	.05	2.83	3.40	3.74	3.98	4.16	4.31	4.44	4.55	4.65	4.73
	.01	3.76	4.28	4.59	4.82	4.99	5.13	5.25	5.36	5.45	5.53
120	.05	2.80	3.36	3.68	3.92	4.10	4.24	4.36	4.47	4.56	4.64
	.01	3.70	4.20	4.50	4.71	4.87	5.01	5.12	5.21	5.30	5.37
∞	.05	2.77	3.31	3.63	3.86	4.03	4.17	4.29	4.39	4.47	4.55
	.01	3.64	4.12	4.40	4.60	4.76	4.88	4.99	5.08	5.16	5.23

| r = number of means or number of steps between ordered means | | | | | | | | | | Error |
12	13	14	15	16	17	18	19	20	x	df
7.32	7.47	7.60	7.72	7.83	7.93	8.03	8.12	8.21	.05	5
10.70	10.89	11.08	11.24	11.40	11.55	11.68	11.81	11.93	.01	
6.79	6.92	7.03	7.14	7.24	7.34	7.43	7.51	7.59	.05	6
9.48	9.65	9.81	9.95	10.08	10.21	10.32	10.43	10.54	.01	
6.43	6.55	6.66	6.76	6.85	6.94	7.02	7.10	7.17	.05	7
8.71	8.86	9.00	9.12	9.24	9.35	9.46	9.55	9.65	.01	
6.18	6.29	6.39	6.48	6.57	6.65	6.73	6.80	6.87	.05	8
8.18	8.31	8.44	8.55	8.66	8.76	8.85	8.94	9.03	.01	
5.98	6.09	6.19	6.28	6.36	6.44	6.51	6.58	6.64	.05	9
7.78	7.91	8.03	8.13	8.23	8.33	8.41	8.49	8.57	.01	
5.83	5.93	6.03	6.11	6.19	6.27	6.34	6.40	6.47	.05	10
7.49	7.60	7.71	7.81	7.91	7.99	8.08	8.15	8.23	.01	
5.71	5.81	5.90	5.98	6.06	6.13	6.20	6.27	6.33	.05	11
7.25	7.36	7.46	7.56	7.65	7.73	7.81	7.88	7.95	.01	
5.61	5.71	5.80	5.88	5.95	6.02	6.09	6.15	6.21	.05	12
7.06	7.17	7.26	7.36	7.44	7.52	7.59	7.66	7.73	.01	
5.53	5.63	5.71	5.79	5.86	5.93	5.99	6.05	6.11	.05	13
6.90	7.01	7.10	7.19	7.27	7.35	7.42	7.48	7.55	.01	
5.46	5.55	5.64	5.71	5.79	5.85	5.91	5.97	6.03	.05	14
6.77	6.87	6.96	7.05	7.13	7.20	7.27	7.33	7.39	.01	
5.40	5.49	5.57	5.65	5.72	5.78	5.85	5.90	5.96	.05	15
6.66	6.76	6.84	6.93	7.00	7.07	7.14	7.20	7.26	.01	
5.35	5.44	5.52	5.59	5.66	5.73	5.79	5.84	5.90	.05	16
6.56	6.66	6.74	6.82	6.90	6.97	7.03	7.09	7.15	.01	
5.31	5.39	5.47	5.54	5.61	5.67	5.73	5.79	5.84	.05	17
6.48	6.57	6.66	6.73	6.81	6.87	6.94	7.00	7.05	.01	
5.27	5.35	5.43	5.50	5.57	5.63	5.69	5.74	5.79	.05	18
6.41	6.50	6.58	6.65	6.73	6.79	6.85	6.91	6.97	.01	
5.23	5.31	5.39	5.46	5.53	5.59	5.65	5.70	5.75	.05	19
6.34	6.43	6.51	6.58	6.65	6.72	6.78	6.84	6.89	.01	
5.20	5.28	5.36	5.43	5.49	5.55	5.61	5.66	5.71	.05	20
6.28	6.37	6.45	6.52	6.59	6.65	6.71	6.77	6.82	.01	
5.10	5.18	5.25	5.32	5.38	5.44	5.49	5.55	5.59	.05	24
6.11	6.19	6.26	6.33	6.39	6.45	6.51	6.56	6.61	.01	
5.00	5.08	5.15	5.21	5.27	5.33	5.38	5.43	5.47	.05	30
5.93	6.01	6.08	6.14	6.20	6.26	6.31	6.36	6.41	.01	
4.90	4.98	5.04	5.11	5.16	5.22	5.27	5.31	5.36	.05	40
5.76	5.83	5.90	5.96	6.02	6.07	6.12	6.16	6.21	.01	
4.81	4.88	4.94	5.00	5.06	5.11	5.15	5.20	5.24	.05	60
5.60	5.67	5.73	5.78	5.84	5.89	5.93	5.97	6.01	.01	
4.71	4.78	4.84	4.90	4.95	5.00	5.04	5.09	5.13	.05	120
5.44	5.50	5.56	5.61	5.66	5.71	5.75	5.79	5.83	.01	
4.62	4.68	4.74	4.80	4.85	4.89	4.93	4.97	5.01	.05	∞
5.29	5.35	5.40	5.45	5.49	5.54	5.57	5.61	5.65	.01	

Source: Abridged from Table 29 in *Biometrika Tables for Statisticians,* Vol 1, 3rd ed., E. S. Pearson and H. O. Hartley, eds. (New York: Cambridge, 1966). Used with permission of the editors and the Biometrika Trustees.

Table 6 Transformation of r to z_r

r	z_r	r	z_r	r	z_r	r	z_r	r	z_r
.000	.000	.200	.203	.400	.424	.600	.693	.800	1.099
.005	.005	.205	.208	.405	.430	.605	.701	.805	1.113
.010	.010	.210	.213	.410	.436	.610	.709	.810	1.127
.015	.015	.215	.218	.415	.442	.615	.717	.815	1.142
.020	.020	.220	.224	.420	.448	.620	.725	.820	1.157
.025	.025	.225	.229	.425	.454	.625	.733	.825	1.172
.030	.030	.230	.234	.430	.460	.630	.741	.830	1.188
.035	.035	.235	.239	.435	.466	.635	.750	.835	1.204
.040	.040	.240	.245	.440	.472	.640	.758	.840	1.221
.045	.045	.245	.250	.445	.478	.645	.767	.845	1.238
.050	.050	.250	.255	.450	.485	.650	.775	.850	1.256
.055	.055	.255	.261	.455	.491	.655	.784	.855	1.274
.060	.060	.260	.266	.460	.497	.660	.793	.860	1.293
.065	.065	.265	.271	.465	.504	.665	.802	.865	1.313
.070	.070	.270	.277	.470	.510	.670	.811	.870	1.333
.075	.075	.275	.282	.475	.517	.675	.820	.875	1.354
.080	.080	.280	.288	.480	.523	.680	.829	.880	1.376
.085	.085	.285	.293	.485	.530	.685	.838	.885	1.398
.090	.090	.290	.299	.490	.536	.690	.848	.890	1.422
.095	.095	.295	.304	.495	.543	.695	.858	.895	1.447
.100	.100	.300	.310	.500	.549	.700	.867	.900	1.472
.105	.105	.305	.315	.505	.556	.705	.877	.905	1.499
.110	.110	.310	.321	.510	.563	.710	.887	.910	1.528
.115	.116	.315	.326	.515	.570	.715	.897	.915	1.557
.120	.121	.320	.332	.520	.576	.720	.908	.920	1.589
.125	.126	.325	.337	.525	.583	.725	.918	.925	1.623
.130	.131	.330	.343	.530	.590	.730	.929	.930	1.658
.135	.136	.335	.348	.535	.597	.735	.940	.935	1.697
.140	.141	.340	.354	.540	.604	.740	.950	.940	1.738
.145	.146	.345	.360	.545	.611	.745	.962	.945	1.783
.150	.151	.350	.365	.550	.618	.750	.973	.950	1.832
.155	.156	.355	.371	.555	.626	.755	.984	.955	1.886
.160	.161	.360	.377	.560	.633	.760	.996	.960	1.946
.165	.167	.365	.383	.565	.640	.765	1.008	.965	2.014
.170	.172	.370	.388	.570	.648	.770	1.020	.970	2.092
.175	.177	.375	.394	.575	.655	.775	1.033	.975	2.185
.180	.182	.380	.400	.580	.662	.780	1.045	.980	2.298
.185	.187	.385	.406	.585	.670	.785	1.058	.985	2.443
.190	.192	.390	.412	.590	.678	.790	1.071	.990	2.647
.195	.198	.395	.418	.595	.685	.795	1.085	.995	2.994

Table 7 Critical values of the correlation coefficient

	Level of significance for one-tailed test			
	.05	.025	.01	.005
	Level of significance for two-tailed test			
df	.10	.05	.02	.01
1	.988	.997	.9995	.9999
2	.900	.950	.980	.990
3	.805	.878	.934	.959
4	.729	.811	.882	.917
5	.669	.754	.833	.874
6	.622	.707	.789	.834
7	.582	.666	.750	.798
8	.549	.632	.716	.765
9	.521	.602	.685	.735
10	.497	.576	.658	.708
11	.476	.553	.634	.684
12	.458	.532	.612	.661
13	.441	.514	.592	.641
14	.426	.497	.574	.623
15	.412	.482	.558	.606
16	.400	.468	.542	.590
17	.389	.456	.528	.575
18	.378	.444	.516	.561
19	.369	.433	.503	.549
20	.360	.423	.492	.537
21	.352	.413	.482	.526
22	.344	.404	.472	.515
23	.337	.396	.462	.505
24	.330	.388	.453	.496
25	.323	.381	.445	.487
26	.317	.374	.437	.479
27	.311	.367	.430	.471
28	.306	.361	.423	.463
29	.301	.355	.416	.456
30	.296	.349	.409	.449
35	.275	.325	.381	.418
40	.257	.304	.358	.393
45	.243	.288	.338	.372
50	.231	.273	.322	.354
60	.211	.250	.295	.325
70	.195	.232	.274	.303
80	.183	.217	.256	.283
90	.173	.205	.242	.267
100	.164	.195	.230	.254

Source: Abridged from Table VII, p. 63 of Fisher and Yates: *Statistical Tables for Biological, Agricultural and Medical Research,* published by Longman Group Ltd., London. (previously published by Oliver and Boyd, Edinburgh), and by permission of the authors and publishers.

Table 8   Binomial coefficients

N	$\binom{N}{0}$	$\binom{N}{1}$	$\binom{N}{2}$	$\binom{N}{3}$	$\binom{N}{4}$	$\binom{N}{5}$	$\binom{N}{6}$	$\binom{N}{7}$	$\binom{N}{8}$	$\binom{N}{9}$	$\binom{N}{10}$
0	1										
1	1	1									
2	1	2	1								
3	1	3	3	1							
4	1	4	6	4	1						
5	1	5	10	10	5	1					
6	1	6	15	20	15	6	1				
7	1	7	21	35	35	21	7	1			
8	1	8	28	56	70	56	28	8	1		
9	1	9	36	84	126	126	84	36	9	1	
10	1	10	45	120	210	252	210	120	45	10	1
11	1	11	55	165	330	462	462	330	165	55	11
12	1	12	66	220	495	792	924	792	495	220	66
13	1	13	78	286	715	1287	1716	1716	1287	715	286
14	1	14	91	364	1001	2002	3003	3432	3003	2002	1001
15	1	15	105	455	1365	3003	5005	6435	6435	5005	3003
16	1	16	120	560	1820	4368	8008	11440	12870	11440	8008
17	1	17	136	680	2380	6188	12376	19448	24310	24310	19448
18	1	18	153	816	3060	8568	18564	31824	43758	48620	43758
19	1	19	171	969	3876	11628	27132	50388	75582	92378	92378
20	1	20	190	1140	4845	15504	38760	77520	125970	167960	184756

Table 9 Quantiles of the Mann-Whitney test statistic

n	p	m=2	3	4	5	6	7	8	9	10	11	12	13	14	15	16	17	18	19	20
	.001	0	0	0	0	0	0	0	0	0	0	0	0	0	0	0	0	0	0	0
	.005	0	0	0	0	0	0	0	0	0	0	0	0	0	0	0	0	0	1	1
2	.01	0	0	0	0	0	0	0	0	0	0	0	1	1	1	1	1	1	2	2
	.025	0	0	0	0	0	0	1	1	1	1	2	2	2	2	2	3	3	3	3
	.05	0	0	0	1	1	1	2	2	2	2	3	3	4	4	4	4	5	5	5
	.10	0	1	1	2	2	2	3	3	4	4	5	5	5	6	6	7	7	8	8
	.001	0	0	0	0	0	0	0	0	0	0	0	0	0	0	0	1	1	1	1
	.005	0	0	0	0	0	0	0	1	1	1	2	2	2	3	3	3	3	4	4
3	.01	0	0	0	0	0	1	1	2	2	2	3	3	3	4	4	5	5	5	6
	.025	0	0	0	1	2	2	3	3	4	4	5	5	6	6	7	7	8	8	9
	.05	0	1	1	2	3	3	4	5	5	6	6	7	8	8	9	10	10	11	12
	.10	1	2	2	3	4	5	6	6	7	8	9	10	11	11	12	13	14	15	16
	.001	0	0	0	0	0	0	0	0	1	1	1	2	2	2	3	3	4	4	4
	.005	0	0	0	0	1	1	2	2	3	3	4	4	5	6	6	7	7	8	9
4	.01	0	0	0	1	2	2	3	4	4	5	6	6	7	9	8	9	10	10	11
	.025	0	0	1	2	3	4	5	5	6	7	8	9	10	11	12	12	13	14	15
	.05	0	1	2	3	4	5	6	7	8	9	10	11	12	13	15	16	17	18	19
	.10	1	2	4	5	6	7	8	10	11	12	13	14	16	17	18	19	21	22	23
	.001	0	0	0	0	0	0	1	2	2	3	3	4	4	5	6	6	7	8	8
	.005	0	0	0	1	2	2	3	4	5	6	7	8	8	9	10	11	12	13	14
5	.01	0	0	1	2	3	4	5	6	7	8	9	10	11	12	13	14	15	16	17
	.025	0	1	2	3	4	6	7	8	9	10	12	13	14	15	16	18	19	20	21
	.05	1	2	3	5	6	7	9	10	12	13	14	16	17	19	20	21	23	24	26
	.10	2	3	5	6	8	9	11	13	14	16	18	19	21	23	24	26	28	29	31
	.001	0	0	0	0	0	0	2	3	4	5	5	6	7	8	9	10	11	12	13
	.005	0	0	1	2	3	4	5	6	7	8	10	11	12	13	14	16	17	18	19
6	.01	0	0	2	3	4	5	7	8	9	10	12	13	14	16	17	19	20	21	23
	.025	0	2	3	4	6	7	9	11	12	14	15	17	18	20	22	23	25	26	28
	.05	1	3	4	6	8	9	11	13	15	17	18	20	22	24	26	27	29	31	33
	.10	2	4	6	8	10	12	14	16	18	20	22	24	26	28	30	32	35	37	39
	.001	0	0	0	0	1	2	3	4	6	7	8	9	10	11	12	14	15	16	17
	.005	0	0	1	2	4	5	7	8	10	11	13	14	16	17	19	20	22	23	25
7	.01	0	1	2	4	5	7	8	10	12	13	15	17	18	20	22	24	25	27	29
	.025	0	2	4	6	7	9	11	13	15	17	19	21	23	25	27	29	31	33	35
	.05	1	3	5	7	9	12	14	16	18	20	22	25	27	29	31	34	36	38	40
	.10	2	5	7	9	12	14	17	19	22	24	27	29	32	34	37	39	42	44	47
	.001	0	0	0	1	2	3	5	6	7	9	10	12	13	15	16	18	19	21	22
	.005	0	0	2	3	5	7	8	10	12	14	16	18	19	21	23	25	27	29	31
8	.01	0	1	3	5	7	8	10	12	14	16	18	21	23	25	27	29	31	33	35
	.025	1	3	5	7	9	11	14	16	18	20	23	25	27	30	32	35	37	39	42
	.05	2	4	6	9	11	14	16	19	21	24	27	29	32	34	37	40	42	45	48
	.10	3	6	8	11	14	17	20	23	25	28	31	34	37	40	43	46	49	52	55

Table 9 *(Continued)*

n	p	m=2	3	4	5	6	7	8	9	10	11	12	13	14	15	16	17	18	19	20
	.001	0	0	0	2	3	4	6	8	9	11	13	15	16	18	20	22	24	26	27
	.005	0	1	2	4	6	8	10	12	14	17	19	21	23	25	28	30	32	34	37
9	.01	0	2	4	6	8	10	12	15	17	19	22	24	27	29	32	34	37	39	41
	.025	1	3	5	8	11	13	16	18	21	24	27	29	32	35	38	40	43	46	49
	.05	2	5	7	10	13	16	19	22	25	28	31	34	37	40	43	46	49	52	55
	.10	3	6	10	13	16	19	23	26	29	32	36	39	42	46	49	53	56	59	63
	.001	0	0	1	2	4	6	7	9	11	13	15	18	20	22	24	26	28	30	33
	.005	0	1	3	5	7	10	12	14	17	19	22	25	27	30	32	35	38	40	43
10	.01	0	2	4	7	9	12	14	17	20	23	25	28	31	34	37	39	42	45	48
	.025	1	4	6	9	12	15	18	21	24	27	30	34	37	40	43	46	49	53	56
	.05	2	5	8	12	15	18	21	25	28	32	35	38	42	45	49	52	56	59	63
	.10	4	7	11	14	18	22	25	29	33	37	40	44	48	52	55	59	63	67	71
	.001	0	0	1	3	5	7	9	11	13	16	18	21	23	25	28	30	33	35	38
	.005	0	1	3	6	8	11	14	17	19	22	25	28	31	34	37	40	43	46	49
11	.01	0	2	5	8	10	13	16	19	23	26	29	32	35	38	42	45	48	51	54
	.025	1	4	7	10	14	17	20	24	27	31	34	38	41	45	48	52	56	59	63
	.05	2	6	9	13	17	20	24	28	32	35	39	43	47	51	55	58	62	66	70
	.10	4	8	12	16	20	24	28	32	37	41	45	49	53	58	62	66	70	74	79
	.001	0	0	1	3	5	8	10	13	15	18	21	24	26	29	32	35	38	41	43
	.005	0	2	4	7	10	13	16	19	22	25	28	32	35	38	42	45	48	52	55
12	.01	0	3	6	9	12	15	18	22	25	29	32	36	39	43	47	50	54	57	61
	.025	2	5	8	12	15	19	23	27	30	34	38	42	46	50	54	58	62	66	70
	.05	3	6	10	14	18	22	27	31	35	39	43	48	52	56	61	65	69	73	78
	.10	5	9	13	18	22	27	31	36	40	45	50	54	59	64	68	73	78	82	87
	.001	0	0	2	4	6	9	12	15	18	21	24	27	30	33	36	39	43	46	49
	.005	0	2	4	8	11	14	18	21	25	28	32	35	39	43	46	50	54	58	61
13	.01	1	3	6	10	13	17	21	24	28	32	36	40	44	48	52	56	60	64	68
	.025	2	5	9	13	17	21	25	29	34	38	42	46	51	55	60	64	68	73	77
	.05	3	7	11	16	20	25	29	34	38	43	48	52	57	62	66	71	76	81	85
	.10	5	10	14	19	24	29	34	39	44	49	54	59	64	69	75	80	85	90	95
	.001	0	0	2	4	7	10	13	16	20	23	26	30	33	37	40	44	47	51	55
	.005	0	2	5	8	12	16	19	23	27	31	35	39	43	47	51	55	59	64	68
14	.01	1	3	7	11	14	18	23	27	31	35	39	44	48	52	57	61	66	70	74
	.025	2	6	10	14	18	23	27	32	37	41	46	51	56	60	65	70	75	79	84
	.05	4	8	12	17	22	27	32	37	42	47	52	57	62	67	72	78	83	88	93
	.10	5	11	16	21	26	32	37	42	48	53	59	54	70	75	81	86	92	98	103
	.001	0	0	2	5	8	11	15	18	22	25	29	33	37	41	44	48	52	56	60
	.005	0	3	6	9	13	17	21	25	30	34	38	43	47	52	56	61	65	70	74
15	.01	1	4	8	12	16	20	25	29	34	38	43	48	52	57	62	67	71	76	81
	.025	2	6	11	15	20	25	30	35	40	45	50	55	60	65	71	76	81	86	91
	.05	4	8	13	19	24	29	34	40	45	51	56	62	67	73	78	84	89	95	101
	.10	6	11	17	23	28	34	40	46	52	58	64	69	75	81	87	93	99	105	111

Table 9   *(Continued)*

n	p	m=2	3	4	5	6	7	8	9	10	11	12	13	14	15	16	17	18	19	20
	.001	0	0	3	6	9	12	16	20	24	28	32	36	40	44	49	53	57	61	66
	.005	0	3	6	10	14	19	23	28	32	37	42	46	51	56	61	66	71	75	80
16	.01	1	4	8	13	17	22	27	32	37	42	47	52	57	62	67	72	77	83	88
	.025	2	7	12	16	22	27	32	38	43	48	54	60	65	71	76	82	87	93	99
	.05	4	9	15	20	26	31	37	43	49	55	61	66	72	78	84	90	96	102	108
	.10	6	12	18	24	30	37	43	49	55	62	68	75	81	87	94	100	107	113	120
	.001	0	1	3	6	10	14	18	22	26	30	35	39	44	48	53	58	62	67	71
	.005	0	3	7	11	16	20	25	30	35	40	45	50	55	61	66	71	76	82	87
17	.01	1	5	9	14	19	24	29	34	39	45	50	56	61	67	72	78	83	89	94
	.025	3	7	12	18	23	29	35	40	46	52	58	64	70	76	82	88	94	100	106
	.05	4	10	16	21	27	34	40	46	52	58	65	71	78	84	90	97	103	110	116
	.10	7	13	19	26	32	39	46	53	59	66	73	80	86	93	100	107	114	121	128
	.001	0	1	4	7	11	15	19	24	28	33	38	43	47	52	57	62	67	72	77
	.005	0	3	7	12	17	22	27	32	38	43	48	54	59	65	71	76	82	88	93
18	.01	1	5	10	15	20	25	31	37	42	48	54	60	66	71	77	83	89	95	101
	.025	3	8	13	19	25	31	37	43	49	56	62	68	75	81	87	94	100	107	113
	.05	5	10	17	23	29	36	42	49	56	62	69	76	83	89	96	103	110	117	124
	.10	7	14	21	28	35	42	49	56	63	70	78	85	92	99	107	114	121	129	136
	.001	0	1	4	8	12	16	21	26	30	35	41	46	51	56	61	67	72	78	83
	.005	1	4	8	13	18	23	29	34	40	46	52	58	64	70	75	82	88	94	100
19	.01	2	5	10	16	21	27	33	39	45	51	57	64	70	76	83	89	95	102	108
	.025	3	8	14	20	26	33	39	46	53	59	66	73	79	86	93	100	107	114	120
	.05	5	11	18	24	31	38	45	52	59	66	73	81	88	95	102	110	117	124	131
	.10	8	15	22	29	37	44	52	59	67	74	82	90	98	105	113	121	129	136	144
	.001	0	1	4	8	13	17	22	27	33	38	43	49	55	60	66	71	77	83	89
	.005	1	4	9	14	19	25	31	37	43	49	55	61	68	74	80	87	93	100	106
20	.01	2	6	11	17	23	29	35	41	48	54	61	68	74	81	88	94	101	108	115
	.025	3	9	15	21	28	35	42	49	56	63	70	77	84	91	99	106	113	120	128
	.05	5	12	19	26	33	40	48	55	63	70	78	85	93	101	108	116	124	131	139
	.10	8	16	23	31	39	47	55	63	71	79	87	95	103	111	120	128	136	144	152

Source: Abridged from L. R. Verdooren, "Extended Tables of Critical Values for Wilcoxon's Test Statistic," *Biometrika* 50 (1963), pp. 177–86.
Critical regions correspond to values less than (or greater than) but not including the appropriate quantile.

Table 10 **Table of critical values of T in the Wilcoxon matched-pairs signed-ranks test**

	Level of significance for one-tailed test				Level of significance for one-tailed test		
	.025	.01	.005		.025	.01	.005
N	Level of significance for two-tailed test			N	Level of significance for two-tailed test		
	.05	.02	.01		.05	.02	.01
6	0	—	—	16	30	24	20
7	2	0	—	17	35	28	23
8	4	2	0	18	40	33	28
9	6	3	2	19	46	38	32
10	8	5	3	20	52	43	38
11	11	7	5	21	59	49	43
12	14	10	7	22	66	56	49
13	17	13	10	23	73	62	55
14	21	16	13	24	81	69	61
15	25	20	16	25	89	77	68

Source: Adapted from Table 1 of F. Wilcoxon, *Some Rapid Approximate Statistical Procedures* p. 13, copyright © 1949, 1964 Lederle Laboratories, Division of American Cyanamid Company, New York, all rights reserved, and reprinted with permission.

Table 11 Squares, square roots, and reciprocals of numbers from 1 to 1000

N	N^2	$\sqrt{N}$	$1/N$	N	N^2	$\sqrt{N}$	$1/N$	N	N^2	$\sqrt{N}$	$1/N$
1	1	1.0000	1.000000	46	2116	6.7823	.021739	91	8281	9.5394	.010989
2	4	1.4142	.500000	47	2209	6.8557	.021277	92	8464	9.5917	.010870
3	9	1.7321	.333333	48	2304	6.9282	.020833	93	8649	9.6437	.010753
4	16	2.0000	.250000	49	2401	7.0000	.020408	94	8836	9.6954	.010638
5	25	2.2361	.200000	50	2500	7.0711	.020000	95	9025	9.7468	.010526
6	36	2.4495	.166667	51	2601	7.1414	.019608	96	9216	9.7980	.010417
7	49	2.6458	.142857	52	2704	7.2111	.019231	97	9409	9.8489	.010309
8	64	2.8284	.125000	53	2809	7.2801	.018868	98	9604	9.8995	.010204
9	81	3.0000	.111111	54	2916	7.3485	.018519	99	9801	9.9499	.010101
10	100	3.1623	.100000	55	3025	7.4162	.018182	100	10000	10.0000	.010000
11	121	3.3166	.090909	56	3136	7.4833	.017857	101	10201	10.0499	.00990099
12	144	3.4641	.083333	57	3249	7.5498	.017544	102	10404	10.0995	.00980392
13	169	3.6056	.076923	58	3364	7.6158	.017241	103	10609	10.1489	.00970874
14	196	3.7417	.071429	59	3481	7.6811	.016949	104	10816	10.1980	.00961538
15	225	3.8730	.066667	60	3600	7.7460	.016667	105	11025	10.2470	.00952381
16	256	4.0000	.062500	61	3721	7.8102	.016393	106	11236	10.2956	.00943396
17	289	4.1231	.058824	62	3844	7.8740	.016129	107	11449	10.3441	.00934579
18	324	4.2426	.055556	63	3969	7.9373	.015873	108	11664	10.3923	.00925926
19	361	4.3589	.052632	64	4096	8.0000	.015625	109	11881	10.4403	.00917431
20	400	4.4721	.050000	65	4225	8.0623	.015385	110	12100	10.4881	.00909091
21	441	4.5826	.047619	66	4356	8.1240	.015152	111	12321	10.5357	.00900901
22	484	4.6904	.045455	67	4489	8.1854	.014925	112	12544	10.5830	.00892857
23	529	4.7958	.043478	68	4624	8.2462	.014706	113	12769	10.6301	.00884956
24	576	4.8990	.041667	69	4761	8.3066	.014493	114	12996	10.6771	.00877193
25	625	5.0000	.040000	70	4900	8.3666	.014286	115	13225	10.7238	.00869565
26	676	5.0990	.038462	71	5041	8.4261	.014085	116	13456	10.7703	.00862069
27	729	5.1962	.037037	72	5184	8.4853	.013889	117	13689	10.8167	.00854701
28	784	5.2915	.035714	73	5329	8.5440	.013699	118	13924	10.8628	.00847458
29	841	5.3852	.034483	74	5476	8.6023	.013514	119	14161	10.9087	.00840336
30	900	5.4772	.033333	75	5625	8.6603	.013333	120	14400	10.9545	.00833333
31	961	5.5678	.032258	76	5776	8.7178	.013158	121	14641	11.0000	.00826446
32	1024	5.6569	.031250	77	5929	8.7750	.012987	122	14884	11.0454	.00819672
33	1089	5.7446	.030303	78	6084	8.8318	.012821	123	15129	11.0905	.00813008
34	1156	5.8310	.029412	79	6241	8.8882	.012658	124	15376	11.1355	.00800452
35	1225	5.9161	.028571	80	6400	8.9443	.012500	125	15625	11.1803	.00800000
36	1296	6.0000	.027778	81	6561	9.0000	.012346	126	15876	11.2250	.00793651
37	1369	6.0828	.027027	82	6724	9.0554	.012195	127	16129	11.2694	.00787402
38	1444	6.1644	.026316	83	6889	9.1104	.012048	128	16384	11.3137	.00781250
39	1521	6.2450	.025641	84	7056	9.1652	.011905	129	16641	11.3578	.00775194
40	1600	6.3246	.025000	85	7225	9.2195	.011765	130	16900	11.4018	.00769231
41	1681	6.4031	.024390	86	7396	9.2736	.011628	131	17161	11.4455	.00763359
42	1764	6.4807	.023810	87	7569	9.3274	.011494	132	17424	11.4891	.00757576
43	1849	6.5574	.023256	88	7744	9.3808	.011364	133	17689	11.5326	.00751880
44	1936	6.6332	.022727	89	7921	9.4340	.011236	134	17956	11.5758	.00746269
45	2025	6.7082	.022222	90	8100	9.4868	.011111	135	18225	11.6190	.00740741

Source: A. L. Edwards, *Statistical Analysis*, 3rd ed. New York: Holt, Rinehart and Winston, Inc., 1969. Used by permission of the author.

Table 11 *(Continued)*

N	N²	√N	1/N	N	N²	√N	1/N	N	N²	√N	1/N
136	18496	11.6619	.00735294	181	32761	13.4536	.00552486	226	51076	15.0333	.00442478
137	18769	11.7047	.00729927	182	33124	13.4907	.00549451	227	51529	15.0665	.00440529
138	19044	11.7473	.00724638	183	33489	13.5277	.00546448	228	51984	15.0997	.00438596
139	19321	11.7898	.00719424	184	33856	13.5647	.00543478	229	52441	15.1327	.00436681
140	19600	11.8322	.00714286	185	34225	13.6015	.00540541	230	52900	15.1658	.00434783
141	19881	11.8743	.00709220	186	34596	13.6382	.00537634	231	53361	15.1987	.00432900
142	20164	11.9164	.00704225	187	34969	13.6748	.00534759	232	53824	15.2315	.00431034
143	20449	11.9583	.00699301	188	35344	13.7113	.00531915	233	54289	15.2643	.00429185
144	20736	12.0000	.00694444	189	35721	13.7477	.00529101	234	54756	15.2971	.00427350
145	21025	12.0416	.00689655	190	36100	13.7840	.00526316	235	55225	15.3297	.00425532
146	21316	12.0830	.00684932	191	36481	13.8203	.00523560	236	55696	15.3623	.00423729
147	21609	12.1244	.00680272	192	36864	13.8564	.00520833	237	56169	15.3948	.00421941
148	21904	12.1655	.00675676	193	37249	13.8924	.00518135	238	56644	15.4272	.00420168
149	22201	12.2066	.00671141	194	37636	13.9284	.00515464	239	57121	15.4596	.00418410
150	22500	12.2474	.00666667	195	38025	13.9642	.00512821	240	57600	15.4919	.00416667
151	22801	12.2882	.00662252	196	38416	14.0000	.00510204	241	58081	15.5242	.00414938
152	23104	12.3288	.00657895	197	38809	14.0357	.00507614	242	58564	15.5563	.00413223
153	23409	12.3693	.00653595	198	39204	14.0712	.00505051	243	59049	15.5885	.00411523
154	23716	12.4097	.00649351	199	39601	14.1067	.00502513	244	59536	15.6205	.00409836
155	24025	12.4499	.00645161	200	40000	14.1421	.00500000	245	60025	15.6525	.00408163
156	24336	12.4900	.00641026	201	40401	14.1774	.00497512	246	60516	15.6844	.00406504
157	24649	12.5300	.00636943	202	40804	14.2127	.00495050	247	61009	15.7162	.00404858
158	24964	12.5698	.00632911	203	41209	14.2478	.00492611	248	61504	15.7480	.00403226
159	25281	12.6095	.00628931	204	41616	14.2829	.00490196	249	62001	15.7797	.00401606
160	25600	12.6491	.00625000	205	42025	14.3178	.00487805	250	62500	15.8114	.00400000
161	25921	12.6886	.00621118	206	42436	14.3527	.00485437	251	63001	15.8430	.00398406
162	26244	12.7279	.00617284	207	42849	14.3875	.00483092	252	63504	15.8745	.00396825
163	26569	12.7671	.00613497	208	43264	14.4222	.00480769	253	64009	15.9060	.00395257
164	26896	12.8062	.00609756	209	43681	14.4568	.00478469	254	64516	15.9374	.00393701
165	27225	12.8452	.00606061	210	44100	14.4914	.00476190	255	65025	15.9687	.00392157
166	27556	12.8841	.00602410	211	44521	14.5258	.00473934	256	65536	16.0000	.00390625
167	27889	12.9228	.00598802	212	44944	14.5602	.00471698	257	66049	16.0312	.00389105
168	28224	12.9615	.00595238	213	45369	14.5945	.00469484	258	66564	16.0624	.00387597
169	28561	13.0000	.00591716	214	45796	14.6287	.00467290	259	67081	16.0935	.00386100
170	28900	13.0384	.00588235	215	46225	14.6629	.00465116	260	67600	16.1245	.00384615
171	29241	13.0767	.00584795	216	46656	14.6969	.00462963	261	68121	16.1555	.00383142
172	29584	13.1149	.00581395	217	47089	14.7309	.00460829	262	68644	16.1864	.00381679
173	29929	13.1529	.00578035	218	47524	14.7648	.00458716	263	69169	16.2173	.00380228
174	30276	13.1909	.00574713	219	47961	14.7986	.00456621	264	69696	16.2481	.00378788
175	30625	13.2288	.00571429	220	48400	14.8324	.00454545	265	70225	16.2788	.00377358
176	30976	13.2665	.00568182	221	48841	14.8661	.00452489	266	70756	16.3095	.00375940
177	31329	13.3041	.00564972	222	49284	14.8997	.00450450	267	71289	16.3401	.00374532
178	31684	13.3417	.00561798	223	49729	14.9332	.00448430	268	71824	16.3707	.00373134
179	32041	13.3791	.00558659	224	50176	14.9666	.00446429	269	72361	16.4012	.00371747
180	32400	13.4164	.00555556	225	50625	15.0000	.00444444	270	72900	16.4317	.00370370

Table 11 *(Continued)*

N	N^2	$\sqrt{N}$	$1/N$	N	N^2	$\sqrt{N}$	$1/N$	N	N^2	$\sqrt{N}$	$1/N$
271	73441	16.4621	.00369004	316	99856	17.7764	.00316456	361	130321	19.0000	.00277008
272	73984	16.4924	.00367647	317	100489	17.8045	.00315457	362	131044	19.0263	.00276243
273	74529	16.5227	.00366300	318	101124	17.8326	.00314465	363	131769	19.0526	.00275482
274	75076	16.5529	.00364964	319	101761	17.8606	.00313480	364	132496	19.0788	.00274725
275	75625	16.5831	.00363636	320	102400	17.8885	.00312500	365	133225	19.1050	.00273973
276	76176	16.6132	.00362319	321	103041	17.9165	.00311526	366	133956	19.1311	.00273224
277	76729	16.6433	.00361011	322	103684	17.9444	.00310559	367	134689	19.1572	.00272480
278	77284	16.6733	.00359712	323	104329	17.9722	.00309598	368	135424	19.1833	.00271739
279	77841	16.7033	.00358423	324	104976	18.0000	.00308642	369	136161	19.2094	.00271003
280	78400	16.7332	.00357143	325	105625	18.0278	.00307692	370	136900	19.2354	.00270270
281	78961	16.7631	.00355872	326	106276	18.0555	.00306748	371	137641	19.2614	.00269542
282	79524	16.7929	.00354610	327	106929	18.0831	.00305810	372	138384	19.2873	.00268817
283	80089	16.8226	.00353357	328	107584	18.1108	.00304878	373	139129	19.3132	.00268097
284	80656	16.8523	.00352113	329	108241	18.1384	.00303951	374	139876	19.3391	.00267380
285	81225	16.8819	.00350877	330	108900	18.1659	.00303030	375	140625	19.3649	.00266667
286	81796	16.9115	.00349650	331	109561	18.1934	.00302115	376	141376	19.3907	.00265957
287	82369	16.9411	.00348432	332	110224	18.2209	.00301205	377	142129	19.4165	.00265252
288	82944	16.9706	.00347222	333	110889	18.2483	.00300300	378	142884	19.4422	.00264550
289	83521	17.0000	.00346021	334	111556	18.2757	.00299401	379	143641	19.4679	.00263852
290	84100	17.0294	.00344828	335	112225	18.3030	.00298507	380	144400	19.4936	.00263158
291	84681	17.0587	.00343643	336	112896	18.3303	.00297619	381	145161	19.5192	.00262467
292	85264	17.0880	.00342466	337	113569	18.3576	.00296736	382	145924	19.5448	.00261780
293	85849	17.1172	.00341297	338	114244	18.3848	.00295858	383	146689	19.5704	.00261097
294	86436	17.1464	.00340136	339	114921	18.4120	.00294985	384	147456	19.5959	.00260417
295	87025	17.1756	.00338983	340	115600	18.4391	.00294118	385	148225	19.6214	.00259740
296	87616	17.2047	.00337838	341	116281	18.4662	.00293255	386	148996	19.6469	.00259067
297	88209	17.2337	.00336700	342	116964	18.4932	.00292398	387	149769	19.6723	.00258398
298	88804	17.2627	.00335570	343	117649	18.5203	.00291545	388	150544	19.6977	.00257732
299	89401	17.2916	.00334448	344	118336	18.5472	.00290698	389	151321	19.7231	.00257069
300	90000	17.3205	.00333333	345	119025	18.5742	.00289855	390	152100	19.7484	.00256410
301	90601	17.3494	.00332226	346	119716	18.6011	.00289017	391	152881	19.7737	.00255754
302	91204	17.3781	.00331126	347	120409	18.6279	.00288184	392	153664	19.7990	.00255102
303	91809	17.4069	.00330033	348	121104	18.6548	.00287356	393	154449	19.8242	.00254453
304	92416	17.4356	.00328947	349	121801	18.6815	.00286533	394	155236	19.8494	.00253807
305	93025	17.4642	.00328947	350	122500	18.7083	.00285714	395	156025	19.8746	.00253165
306	93636	17.4929	.00326797	351	123201	18.7350	.00284900	396	156816	19.8997	.00252525
307	94249	17.5214	.00325733	352	123904	18.7617	.00284091	397	157609	19.9249	.00251889
308	94864	17.5499	.00321675	353	124609	18.7883	.00283286	398	158404	19.9499	.00251256
309	95481	17.5784	.00323625	354	125316	18.8149	.00282486	399	159201	19.9750	.00250627
310	96100	17.6068	.00322581	355	126025	18.8414	.00281690	400	160000	20.0000	.00250000
311	96721	17.6352	.00321543	356	126736	18.8680	.00280899	401	160801	20.0250	.00249377
312	97344	17.6635	.00320513	357	127449	18.8944	.00280112	402	161604	20.0499	.00248756
313	97969	17.6918	.00319489	358	128164	18.9209	.00279330	403	162409	20.0749	.00248139
314	98596	17.7200	.00318471	359	128881	18.9473	.00278552	404	163216	20.0998	.00247525
315	99225	17.7482	.00317460	360	129600	18.9737	.00277778	405	164025	20.1246	.00246914

Table 11 *(Continued)*

N	N^2	$\sqrt{N}$	$1/N$	N	N^2	$\sqrt{N}$	$1/N$	N	N^2	$\sqrt{N}$	$1/N$
406	164836	20.1494	.00246305	451	203401	21.2368	.00221729	496	246016	22.2711	.00201613
407	165649	20.1742	.00245700	452	204304	21.2603	.00221239	497	247009	22.2935	.00201207
408	166464	20.1990	.00245098	453	205209	21.2838	.00220751	498	248004	22.3159	.00200803
409	167281	20.2237	.00244499	454	206116	21.3073	.00220264	499	249001	22.3383	.00200401
410	168100	20.2485	.00243902	455	207025	21.3307	.00219870	500	250000	22.3607	.00200000
411	168921	20.2731	.00243309	456	207936	21.3542	.00219298	501	251001	22.3830	.00199601
412	169744	20.2978	.00242718	457	208849	21.3776	.00218818	502	252004	22.4054	.00199203
413	170569	20.3224	.00242131	458	209764	21.4009	.00218341	503	253009	22.4277	.00198807
414	171396	20.3470	.00241546	459	210681	21.4243	.00217865	504	254016	22.4499	.00198413
415	172225	20.3715	.00240964	460	211600	21.4476	.00217391	505	255025	22.4722	.00198020
416	173056	20.3961	.00240385	461	212521	21.4709	.00216920	506	256036	22.4944	.00197628
417	173889	20.4206	.00239808	462	213444	21.4942	.00216450	507	257049	22.5167	.00197239
418	174724	20.4450	.00239234	463	214369	21.5174	.00215983	508	258064	22.5389	.00196850
419	175561	20.4695	.00238663	464	215296	21.5407	.00215517	509	259081	22.5610	.00196464
420	176400	20.4939	.00238095	465	216225	21.5639	.00215054	510	260100	22.5832	.00196078
421	177241	20.5183	.00237530	466	217156	21.5870	.00214592	511	261121	22.6053	.00195695
422	178084	20.5426	.00236967	467	218089	21.6102	.00214133	512	262144	22.6274	.00195312
423	178929	20.5670	.00236407	468	219024	21.6333	.00213675	513	263169	22.6495	.00194932
424	179776	20.5913	.00235849	469	219961	21.6564	.00213220	514	264196	22.6716	.00194553
425	180625	20.6155	.00235294	470	220900	21.6795	.00212766	515	265225	22.6936	.00194175
426	181476	20.6398	.00234742	471	221841	21.7025	.00212314	516	266256	22.7156	.00193798
427	182329	20.6640	.00234192	472	222784	21.7256	.00211864	517	267289	22.7376	.00193424
428	183184	20.6882	.00233645	473	223729	21.7486	.00211416	518	268324	22.7596	.00193050
429	184041	20.7123	.00233100	474	224676	21.7715	.00210970	519	269361	22.7816	.00192678
430	184900	20.7364	.00232558	475	225625	21.7945	.00210526	520	270400	22.8035	.00192308
431	185761	20.7605	.00232019	476	226576	21.8174	.00210084	521	271441	22.8254	.00191939
432	186624	20.7846	.00231481	477	227529	21.8403	.00209644	522	272484	22.8473	.00191571
433	187489	20.8087	.00230947	478	228484	21.8632	.00209205	523	273529	22.8692	.00191205
434	188356	20.8327	.00230415	479	229441	21.8861	.00208768	524	274576	22.8910	.00190840
435	189225	20.8567	.00229885	480	230400	21.9089	.00208333	525	275625	22.9129	.00190476
436	190096	20.8806	.00229358	481	231361	21.9317	.00207900	526	276676	22.9347	.00190114
437	190969	20.9045	.00228833	482	232324	21.9545	.00207469	527	277729	22.9565	.00189753
438	191844	20.9284	.00228311	483	233289	21.9773	.00207039	528	278784	22.9783	.00189394
439	192721	20.9523	.00227790	484	234256	22.0000	.00206612	529	279841	23.0000	.00189036
440	193600	20.9762	.00227273	485	235225	22.0227	.00206186	530	280900	23.0217	.00188679
441	194481	21.0000	.00226757	486	236196	22.0454	.00205761	531	281961	23.0434	.00188324
442	195364	21.0238	.00226244	487	237169	22.0681	.00205339	532	283024	23.0651	.00187970
443	196249	21.0476	.00225734	488	238144	22.0907	.00204918	533	284089	23.0868	.00187617
444	197136	21.0713	.00225225	489	239121	22.1133	.00204499	534	285156	23.1084	.00187266
445	198025	21.0950	.00224719	490	240100	22.1359	.00204082	535	286225	23.1301	.00186916
446	198916	21.1187	.00224215	491	241081	22.1585	.00203666	536	287296	23.1517	.00186567
447	199809	21.1424	.00223714	492	242064	22.1811	.00203252	537	288369	23.1733	.00186220
448	200704	21.1660	.00223214	493	243049	22.2036	.00202840	538	289444	23.1948	.00185874
449	201601	21.1896	.00222717	494	244036	22.2261	.00202429	539	290521	23.2164	.00185529
450	202500	21.2132	.00222222	495	245025	22.2486	.00202020	540	291600	23.2379	.00185185

Table 11   (*Continued*)

N	N^2	$\sqrt{N}$	$1/N$	N	N^2	$\sqrt{N}$	$1/N$	N	N^2	$\sqrt{N}$	$1/N$
541	292681	23.2594	.00184843	586	343396	24.2074	.00170648	631	398161	25.1197	.00158479
542	293764	23.2809	.00184502	587	344569	24.2281	.00170358	632	399424	25.1396	.00158228
543	294849	23.3024	.00184162	588	345744	24.2487	.00170068	633	400689	25.1595	.00157978
544	295936	23.3238	.00183824	589	346921	24.2693	.00169779	634	401956	25.1794	.00157729
545	297025	23.3452	.00183486	590	348100	24.2899	.00169492	635	403225	25.1992	.00157480
546	298116	23.3666	.00183150	591	349281	24.3105	.00169205	636	404496	25.2190	.00157233
547	299209	23.3880	.00182815	592	350464	24.3311	.00168919	637	405769	25.2389	.00156986
548	300304	23.4094	.00182482	593	351649	24.3516	.00168634	638	407044	25.2587	.00156740
549	301401	23.4307	.00182149	594	352836	24.3721	.00168350	639	408321	25.2784	.00156495
550	302500	23.4521	.00181818	595	354025	24.3926	.00168067	640	409600	25.2982	.00156250
551	303601	23.4734	.00181488	596	355216	24.4131	.00167785	641	410881	25.3180	.00156006
552	304704	23.4947	.00181159	597	356409	24.4336	.00167504	642	412164	25.3377	.00155763
553	305809	23.5160	.00180832	598	357604	24.4540	.00167224	643	413449	25.3574	.00155521
554	306916	23.5372	.00180505	599	358801	24.4745	.00166945	644	414736	25.3772	.00155280
555	308025	23.5584	.00180180	600	360000	24.4949	.00166667	645	416025	25.3969	.00155039
556	309136	23.5797	.00179856	601	361201	24.5153	.00166389	646	417316	25.4165	.00154799
557	310249	23.6008	.00179533	602	302404	24.5357	.00166113	647	418609	25.4362	.00154560
558	311364	23.6220	.00179211	603	363609	24.5561	.00165837	648	419904	25.4558	.00154321
559	312481	23.6432	.00178891	604	364816	24.5764	.00165563	649	421201	25.4755	.00154083
560	313600	23.6643	.00178571	605	366025	24.5967	.00165289	650	422500	25.4951	.00153846
561	314721	23.6854	.00178253	606	367236	24.6171	.00165017	651	423801	25.5147	.00153610
562	315844	23.7065	.00177936	607	368449	24.6374	.00164745	652	425104	25.5343	.00153374
563	316969	23.7276	.00177620	608	369664	24.6577	.00164474	653	426409	25.5539	.00153139
564	318096	23.7487	.00177305	609	370881	24.6779	.00164204	654	427716	25.5734	.00152905
565	319225	23.7697	.00176991	610	372100	24.6982	.00163934	655	429025	25.5930	.00152672
566	320356	23.7908	.00176678	611	373321	24.7184	.00163666	656	430336	25.6125	.00152439
567	321489	23.8118	.00176367	612	374544	24.7386	.00163399	657	431649	25.6320	.00152207
568	322624	23.8328	.00176056	613	375769	24.7588	.00163132	658	432964	25.6515	.00151976
569	323761	23.8537	.00175747	614	376996	24.7790	.00162866	659	434281	25.6710	.00151745
570	324900	23.8747	.00175439	615	378225	24.7992	.00162602	660	435600	25.6905	.00151515
571	326041	23.8956	.00175131	616	379456	24.8193	.00162338	661	436921	25.7099	.00151286
572	327184	23.9165	.00164825	617	380689	24.8395	.00162075	662	438244	25.7294	.00151057
573	328329	23.9374	.00174520	618	381924	24.8596	.00161812	663	439569	25.7488	.00150830
574	329476	23.9583	.00174216	619	383161	24.8797	.00161551	664	440896	25.7682	.00150602
575	330625	23.9792	.00173913	620	384400	24.8998	.00161290	665	442225	25.7876	.00150376
576	331776	24.0000	.00173611	621	385641	24.9199	.00161031	666	443556	25.8070	.00150150
577	332929	24.0208	.00173310	622	386884	24.9399	.00160772	667	444889	25.8263	.00149925
578	334084	24.0416	.00173010	623	388129	24.9600	.00160514	668	446224	25.8457	.00149701
579	335241	24.0624	.00172712	624	389376	24.9800	.00160256	669	447561	25.8650	.00149477
580	336400	24.0832	.00172414	625	390625	25.0000	.00160000	670	448900	25.8844	.00149254
581	337561	24.1039	.00172117	626	391876	25.0200	.00159744	671	450241	25.9037	.00149031
582	338724	24.1247	.00171821	627	393129	25.0400	.00159490	672	451584	25.9230	.00148810
583	339889	24.1454	.00171527	628	394384	25.0599	.00159236	673	452929	25.9422	.00148588
584	341056	24.1661	.00171233	629	395641	25.0799	.00158983	674	454276	25.9615	.00148368
585	342225	24.1868	.00170940	630	396900	25.0998	.00158730	675	455625	25.9808	.00148148

Table 11 *(Continued)*

N	N^2	$\sqrt{N}$	$1/N$	N	N^2	$\sqrt{N}$	$1/N$	N	N^2	$\sqrt{N}$	$1/N$
676	456976	26.0000	.00147929	721	519841	26.8514	.00138696	766	586756	27.6767	.00130548
677	458329	26.0192	.00147710	722	521284	26.8701	.00138504	767	588289	27.6948	.00130378
678	459684	26.0384	.00147493	723	522729	26.8887	.00138313	768	589824	27.7128	.00130208
679	461041	26.0576	.00147275	724	524176	26.9072	.00138122	769	591361	27.7308	.00130039
680	462400	26.0768	.00147059	725	525625	26.9258	.00137931	770	592900	27.7489	.00129870
681	463761	26.0960	.00146843	726	527076	26.9444	.00137741	771	594441	27.7669	.00129702
682	465124	26.1151	.00146628	727	528529	26.9629	.00137552	772	595984	27.7849	.00129534
683	466489	26.1343	.00146413	728	529984	26.9815	.00137363	773	597529	27.8029	.00129366
684	467856	26.1534	.00146199	729	531441	27.0000	.00137174	774	599076	27.8209	.00129199
685	469225	26.1725	.00145985	730	532900	27.0185	.00136986	775	600625	27.8388	.00129032
686	470596	26.1916	.00145773	731	534361	27.0370	.00136799	776	602176	27.8568	.00128866
687	471969	26.2107	.00145560	732	535824	27.0555	.00136612	777	603729	27.8747	.00128700
688	473344	26.2298	.00145349	733	537289	27.0740	.00136426	778	605284	27.8927	.00128535
689	474721	26.2488	.00145138	734	538756	27.0924	.00136240	779	606841	27.9106	.00128370
690	476100	26.2679	.00144928	735	540225	27.1109	.00136054	780	608400	27.9285	.00128205
691	477481	26.2869	.00144718	736	541696	27.1293	.00135870	781	609961	27.9464	.00128041
692	478864	26.3059	.00144509	737	543169	27.1477	.00135685	782	611524	27.9643	.00127877
693	480249	26.3249	.00144300	738	544644	27.1662	.00135501	783	613089	27.9821	.00127714
694	481636	26.3439	.00144092	739	546121	27.1846	.00135318	784	614656	28.0000	.00127551
695	483025	26.3629	.00143885	740	547600	27.2029	.00135135	785	616225	28.0179	.00127389
696	484416	26.3818	.00143678	741	549081	27.2213	.00134953	786	617796	28.0357	.00127226
697	485809	26.4008	.00143472	742	550564	27.2397	.00134771	787	619369	28.0535	.00127065
698	487204	26.4197	.00143266	743	552049	27.2580	.00134590	788	620944	28.0713	.00126904
699	488601	26.4386	.00143062	744	553536	27.2764	.00134409	789	622521	28.0891	.00126743
700	490000	26.4575	.00142857	745	555025	27.2947	.00134228	790	624100	28.1069	.00126582
701	491401	26.4764	.00142653	746	556516	27.3130	.00134048	791	625681	28.1247	.00126422
702	492804	26.4953	.00142450	747	558009	27.3313	.00133869	792	627264	28.1425	.00126263
703	494209	26.5141	.00142248	748	559504	27.3496	.00133690	793	628849	28.1603	.00126103
704	495616	26.5330	.00142045	749	561001	27.3679	.00133511	794	630436	28.1780	.00125945
705	497025	26.5518	.00141844	750	562500	27.3861	.00133333	795	632025	28.1957	.00125786
706	498436	26.5707	.00141643	751	564001	27.4044	.00133156	796	633616	28.2135	.00125628
707	499849	26.5895	.00141443	752	565504	27.4226	.00132979	797	635209	28.2312	.00125471
708	501264	26.6083	.00141243	753	567009	27.4408	.00132802	798	636804	28.2489	.00125313
709	502681	26.6271	.00141044	754	568516	27.4591	.00132626	799	638401	28.2666	.00125156
710	504100	26.6458	.00140845	755	570025	27.4773	.00132450	800	640000	28.2843	.00125000
711	505521	26.6646	.00140647	756	571536	27.4955	.00132275	801	641601	28.3019	.00124844
712	506944	26.6833	.00140449	757	573049	27.5136	.00132100	802	643204	28.3196	.00124688
713	508369	26.7021	.00140252	758	574564	27.5318	.00131926	803	644809	28.3373	.00124533
714	509796	26.7208	.00140056	759	576081	27.5500	.00131752	804	646416	28.3549	.00124378
715	511225	26.7395	.00139860	760	577600	27.5681	.00131579	805	648025	28.3725	.00124224
716	512656	26.7582	.00139665	761	579121	27.5862	.00131406	806	649636	28.3901	.00124069
717	514089	26.7769	.00139470	762	580644	27.6043	.00131234	807	651249	28.4077	.00123916
718	515524	26.7955	.00139276	763	582169	27.6225	.00131062	808	625864	28.4253	.00123762
719	516961	26.8142	.00139082	764	583696	27.6405	.00130890	809	654481	28.4429	.00123609
720	518400	26.8328	.00138889	765	585225	27.6586	.00130719	810	656100	28.4605	.00123457

Table 11 *(Continued)*

N	N^2	$\sqrt{N}$	$1/N$	N	N^2	$\sqrt{N}$	$1/N$	N	N^2	$\sqrt{N}$	$1/N$
811	657721	28.4781	.00123305	856	732736	29.2575	.00116822	901	811801	30.0167	.00110988
812	659344	28.4956	.00123153	857	734449	29.2746	.00116686	902	813604	30.0333	.00110865
813	660969	28.5132	.00123001	858	736164	29.2916	.00116550	903	815409	30.0500	.00110742
814	662596	28.5307	.00122850	859	737881	29.3087	.00116414	904	817216	30.0666	.00110619
815	664225	28.5482	.00122699	860	739600	29.3258	.00116279	905	819025	30.0832	.00110497
816	665856	28.5657	.00122549	861	741321	29.3428	.00116144	906	820836	30.0998	.00110375
817	667489	28.5832	.00122399	862	743044	29.3598	.00116009	907	822649	30.1164	.00110254
818	669124	28.6007	.00122249	863	744769	29.3769	.00115875	908	824464	30.1330	.00110132
819	670761	28.6182	.00122100	864	746496	29.3939	.00115741	909	826281	30.1496	.00110011
820	672400	28.6356	.00121951	865	748225	29.4109	.00115607	910	828100	30.1662	.00109890
821	674041	28.6531	.00121803	866	749956	29.4279	.00115473	911	829921	30.1828	.00109769
822	675684	28.6705	.00121655	867	751689	29.4449	.00115340	912	831744	30.1993	.00109649
823	677329	28.6880	.00121507	868	753424	29.4618	.00115207	913	833569	30.2159	.00109529
824	678976	28.7054	.00121359	869	755161	29.4788	.00115075	914	835396	30.2324	.00109409
825	680625	28.7228	.00121212	870	756900	29.4958	.00114943	915	837225	30.2490	.00109290
826	682276	28.7402	.00121065	871	758641	29.5127	.00114811	916	839056	30.2655	.00109170
827	683929	28.7576	.00120919	872	760384	29.5296	.00114679	917	840889	30.2820	.00109051
828	685584	28.7750	.00120773	873	762129	29.5466	.00114548	918	842724	30.2985	.00108932
829	687241	28.7924	.00120627	874	763876	29.5635	.00114416	919	844561	30.3150	.00108814
830	688900	28.8097	.00120482	875	765625	29.5804	.00114286	920	846400	30.3315	.00108696
831	690561	28.8271	.00120337	876	767376	29.5973	.00114155	921	848241	30.3480	.00108578
832	692224	28.8444	.00120192	877	769129	29.6142	.00114025	922	850084	30.3645	.00108460
833	693889	28.8617	.00120048	878	770884	29.6311	.00113895	923	851929	30.3809	.00108342
834	695556	28.8791	.00119904	879	772641	29.6479	.00113766	924	853776	30.3974	.00108225
835	697225	28.8964	.00119760	880	774400	29.6848	.00113636	925	855625	30.4138	.00108108
836	698896	28.9137	.00119617	881	776161	29.6816	.00113507	926	857476	30.4302	.00107991
837	700569	28.9310	.00119474	882	777924	29.6985	.00113379	927	859329	30.4467	.00107875
838	702244	28.9482	.00119332	883	779689	29.7153	.00113250	928	861184	30.4631	.00107759
839	703921	28.9655	.00119190	884	781456	29.7321	.00113122	929	863041	30.4795	.00107643
840	705600	28.9828	.00119048	885	783225	29.7489	.00112994	930	864900	30.4959	.00107527
841	707281	29.0000	.00118906	886	784996	29.7658	.00112867	931	866761	30.5123	.00107411
842	708964	29.0172	.00118765	887	786769	29.7825	.00112740	932	868624	30.5287	.00107296
843	710649	29.0345	.00118624	888	788544	29.7993	.00112613	933	870489	30.5450	.00107181
844	712336	29.0517	.00118483	889	790321	29.8161	.00112486	934	872356	30.5614	.00107066
845	714025	29.0689	.00118343	890	792100	29.8329	.00112360	935	874225	30.5778	.00106952
846	715716	29.0861	.00118203	891	793881	29.8496	.00112233	936	876096	30.5941	.00106838
847	717409	29.1033	.00118064	892	795664	29.8664	.00112108	937	877969	30.6105	.00106724
848	719104	29.1204	.00117925	893	797449	29.8831	.00111982	938	879844	30.6268	.00106610
849	720801	29.1376	.00117786	894	799236	29.8998	.00111857	939	881721	30.6431	.00106496
850	722500	29.1548	.00117647	895	801025	29.9166	.00111732	940	883600	30.6594	.00106383
851	724201	29.1719	.00117509	896	802816	29.9333	.00111607	941	885481	30.6757	.00106270
852	725904	29.1890	.00117371	897	804609	29.9500	.00111483	942	887364	30.6920	.00106157
853	727609	29.2062	.00117233	898	806404	29.9666	.00111359	943	889249	30.7083	.00106045
854	729316	29.2233	.00117096	899	808201	29.9833	.00111235	944	891136	30.7246	.00105932
855	731025	29.2404	.00116959	900	810000	30.0000	.00111111	945	893025	30.7409	.00105820

Table 11 *(Continued)*

N	N^2	$\sqrt{N}$	$1/N$	N	N^2	$\sqrt{N}$	$1/N$	N	N^2	$\sqrt{N}$	$1/N$
946	894916	30.7571	.00105708	966	933156	31.0805	.00103520	986	972196	31.4006	.00101420
947	896809	30.7734	.00105597	967	935089	31.0966	.00103413	987	974169	31.4166	.00101317
948	898704	30.7896	.00105485	968	937024	31.1127	.00103306	988	976144	31.4325	.00101215
949	900601	30.8058	.00105374	969	938961	31.1288	.00103199	989	978121	31.4484	.00101112
950	902500	30.8221	.00105263	970	940900	31.1448	.00103093	990	980100	31.4643	.00101010
951	904401	30.8383	.00105152	971	942841	31.1609	.00102987	991	982081	31.4802	.00100908
952	906304	30.8545	.00105042	972	944784	31.1769	.00102881	992	984064	31.4960	.00100806
953	908209	30.8707	.00104932	973	946729	31.1929	.00102775	993	986049	31.5119	.00100705
954	910116	30.8869	.00104822	974	948676	31.2090	.00102669	994	988036	31.5278	.00100604
955	912025	30.9031	.00104712	975	950625	31.2250	.00102564	995	990025	31.5436	.00100503
956	913936	30.9192	.00104603	976	952576	31.2410	.00102459	996	992016	31.5595	.00100402
957	915849	30.9354	.00104493	977	954529	31.2570	.00102354	997	994009	31.5753	.00103842
958	917764	30.9516	.00104384	978	956484	31.2730	.00102249	998	996004	31.5911	.00100200
959	919681	30.9677	.00104275	979	958441	31.2890	.00102145	999	998001	31.6070	.00100100
960	921600	30.9839	.00104167	980	960400	31.3050	.00102041	1000	1000000	31.6228	.00100000
961	923521	31.0000	.00104058	981	962361	31.3209	.00101937				
962	925444	31.0161	.00103950	982	964324	31.3369	.00101833				
963	927369	31.0322	.00103842	983	966289	31.3528	.00101729				
964	929296	31.0483	.00103734	984	968256	31.3688	.00101626				
965	931225	31.0644	.00103627	985	970225	31.3847	.00101523				

Table 12   **Random digits**

Row number										
00000	10097	32533	76520	13586	34673	54876	80959	09117	39292	74945
00001	37542	04805	64894	74296	24805	24037	20636	10402	00822	91665
00002	08422	68953	19645	09303	23209	02560	15953	34764	35080	33606
00003	99019	02529	09376	70715	38311	31165	88676	74397	04436	27659
00004	12807	99970	80157	36147	64032	36653	98951	16877	12171	76833
00005	66065	74717	34072	76850	36697	36170	65813	39885	11199	29170
00006	31060	10805	45571	82406	35303	42614	86799	07439	23403	09732
00007	85269	77602	02051	65692	68665	74818	73053	85247	18623	88579
00008	63573	32135	05325	47048	90553	57548	28468	28709	83491	25624
00009	73769	45753	03529	64778	35808	34282	60935	20344	35273	88435
00010	98520	17767	14905	68607	22109	40558	60970	93433	50500	73998
00011	11805	05431	39808	27732	50725	68248	29405	24201	52775	67851
00012	83452	99634	06288	98033	13746	70078	18475	40610	68711	77817
00013	88685	40200	86507	58401	36766	67951	90364	76493	29609	11062
00014	99594	67348	87517	64969	91826	08928	93785	61368	23478	34113
00015	65481	17674	17468	50950	58047	76974	73039	57186	40218	16544
00016	80124	35635	17727	08015	45318	22374	21115	78253	14385	53763
00017	74350	99817	77402	77214	43236	00210	45521	64237	96286	02655
00018	69916	26803	66252	29148	36936	87203	76621	13990	94400	56418
00019	09893	20505	14225	68514	46427	56788	96297	78822	54382	14598
00020	91499	14523	68479	27686	46162	83554	94750	89923	37089	20048
00021	80336	94598	26940	36858	70297	34135	53140	33340	42050	82341
00022	44104	81949	85157	47954	32979	26575	57600	40881	22222	06413
00023	12550	73742	11100	02040	12860	74697	96644	89439	28707	25815
00024	63606	49329	16505	34484	40219	52563	43651	77082	07207	31790
00025	61196	90446	26457	47774	51924	33729	65394	59593	42582	60527
00026	15474	45266	95270	79953	59367	83848	82396	10118	33211	59466
00027	94557	28573	67897	54387	54622	44431	91190	42592	92927	45973
00028	42481	16213	97344	08721	16868	48767	03071	12059	25701	46670
00029	23523	78317	73208	89837	68935	91416	26252	29663	05522	82562
00030	04493	52494	75246	33824	45862	51025	61962	79335	65337	12472
00031	00549	97654	64051	88159	96119	63896	54692	82391	23287	29529
00032	35963	15307	26898	09354	33351	35462	77974	50024	90103	39333
00033	59808	08391	45427	26842	83609	49700	13021	24892	78565	20106
00034	46058	85236	01390	92286	77281	44077	93910	83647	70617	42941
00035	32179	00597	87379	25241	05567	07007	86743	17157	85394	11838
00036	69234	61406	20117	45204	15956	60000	18743	92423	97118	96338
00037	19565	41430	01758	75379	40419	21585	66674	36806	84962	85207
00038	45155	14938	19476	07246	43667	94543	59047	90033	20826	69541
00039	94864	31994	36168	10851	34888	81553	01540	35456	05014	51176

Source: RAND Corporation, *A Million Random Digits,* Glencoe, Ill.: Free Press of Glencoe, 1955.

Table 12   (*Continued*)

Row number										
00040	98086	24826	45240	28404	44999	08896	39094	73407	35441	31880
00041	33185	16232	41941	50949	89435	48581	88695	41994	37548	73043
00042	80951	00406	96382	70774	20151	23387	25016	25298	94624	61171
00043	79752	49140	71961	28296	69861	02591	74852	20539	00387	59579
00044	18633	32537	98145	06571	31010	24674	05455	61427	77938	91936
00045	74029	43902	77557	32270	97790	17119	52527	58021	80814	51748
00046	54178	45611	80993	37143	05335	12969	56127	19255	36040	90324
00047	11664	49883	52079	84827	59381	71539	09973	33440	88461	23356
00048	48324	77928	31249	64710	02295	36870	32307	57546	15020	09994
00049	69074	94138	87637	91976	35584	04401	10518	21615	01848	76938
00050	09188	20097	32825	39527	04220	86304	83389	87374	64278	58044
00051	90045	85497	51981	50654	94938	81997	91870	76150	68476	64659
00052	73189	50207	47677	26269	62290	64464	27124	67018	41361	82760
00053	75768	76490	20971	87749	90429	12272	95375	05871	93823	43178
00054	54016	44056	66281	31003	00682	27398	20714	53295	07706	17813
00055	08358	69910	78542	42785	13661	58873	04618	97553	31223	08420
00056	28306	03264	81333	10591	40510	07893	32604	60475	94119	01840
00057	53840	86233	81594	13628	51215	90290	28466	68795	77762	20791
00058	91757	53741	61613	62669	50263	90212	55781	76514	83483	47055
00059	89415	92694	00397	58391	12607	17646	48949	72306	94541	37408
00060	77513	03820	86864	29901	68414	82774	51908	13980	72893	55507
00061	19502	37174	69979	20288	55210	29773	74287	75251	65344	67415
00062	21818	59313	93278	81757	05686	73156	07082	85046	31853	38452
00063	51474	66499	68107	23621	94049	91345	42836	09191	08007	45449
00064	99559	68331	62535	24170	69777	12830	74819	78142	43860	72834
00065	33713	48007	93584	72869	51926	64721	58303	29822	93174	93972
00066	85274	86893	11303	22970	28834	34137	73515	90400	71148	43643
00067	84133	89640	44035	52166	73852	70091	61222	60561	62327	18423
00068	56732	16234	17395	96131	10123	91622	85496	57560	81604	18880
00069	65138	56806	87648	85261	34313	65861	45875	21069	85644	47277
00070	38001	02176	81719	11711	71602	92937	74219	64049	65584	49698
00071	37402	96397	01304	77586	56271	10086	47324	62605	40030	37438
00072	97125	40348	87083	31417	21815	39250	75237	62047	15501	29578
00073	21826	41134	47143	34072	64638	85902	49139	06441	03856	54552
00074	73135	42742	95719	09035	85794	74296	08789	88156	64691	19202
00075	07638	77929	03061	18072	96207	44156	23821	99538	04713	66994
00076	60528	83441	07954	19814	59175	20695	05533	52139	61212	06455
00077	83596	35655	06958	92983	05128	09719	77433	53783	92301	50498
00078	10850	62746	99599	10507	13499	06319	53075	71839	06410	19362
00079	39820	98952	43622	63147	64421	80814	43800	09351	31024	73167

Table 12 (*Continued*)

Row number										
00080	59580	06478	75569	78800	88835	54486	23768	06156	04111	08408
00081	38508	07341	23793	48763	90822	97022	17719	04207	95954	49953
00082	30692	70668	94688	16127	56196	80091	82067	63400	05462	69200
00083	65443	95659	18238	27437	49632	24041	08337	65676	96299	90836
00084	27267	50264	13192	72294	07477	44606	17985	48911	97341	30358
00085	91307	06991	19072	24210	36699	53728	28825	35793	28976	66252
00086	68434	94688	84473	13622	62126	98408	12843	82590	09815	93146
00087	48908	15877	54745	24591	35700	04754	83824	52692	54130	55160
00088	06913	45197	42672	78601	11883	09528	63011	98901	14974	40344
00089	10455	16019	14210	33712	91342	37821	88325	80851	43667	70883
00090	12883	97343	65027	61184	04285	01392	17974	15077	90712	26769
00091	21778	30976	38807	36961	31649	42096	63281	02023	08816	47449
00092	19523	59515	65122	59659	86283	68258	69572	13798	16435	91529
00093	67245	52670	35583	16563	79246	86686	76463	34222	26655	90802
00094	60584	47377	07500	37992	45134	26529	26760	83637	41326	44344
00095	53853	41377	36066	94850	58838	73859	49364	73331	96240	43642
00096	24637	38736	74384	89342	52623	07992	12369	18601	03742	83873
00097	83080	12451	38992	22815	07759	51777	97377	27585	51972	37867
00098	16444	24334	36151	99073	27493	70939	85130	32552	54846	54759
00099	60790	18157	57178	65762	11161	78576	45819	52979	65130	04860
00100	03991	10461	93716	16894	66083	24653	84609	58232	88618	19161
00101	38555	95554	32886	59780	08355	60860	29735	47762	71299	23853
00102	17546	73704	92052	46215	55121	29281	59076	07936	27954	58909
00103	32643	52861	95819	06831	00911	98936	76355	93779	80863	00514
00104	69572	68777	39510	35905	14060	40619	29549	69616	33564	60780
00105	24122	66591	27699	06494	14845	46672	61958	77100	90899	75754
00106	61196	30231	92962	61773	41839	55382	17267	70943	78038	70267
00107	30532	21704	10274	12202	39685	23309	10061	68829	55986	66485
00108	03788	97599	75867	20717	74416	53166	35208	33374	87539	08823
00109	48228	63379	85783	47619	53152	67433	35663	52972	16818	60311
00110	60365	94653	35075	33949	42614	29297	01918	28316	98953	73231
00111	83799	42402	56623	34442	34994	41374	70071	14736	09958	18065
00112	32960	07405	36409	83232	99385	41600	11133	07586	15917	06253
00113	19322	53845	57620	52606	66497	68646	78138	66559	19640	99413
00114	11220	94747	07399	37408	48509	23929	27482	45476	85244	35159
00115	31751	57260	68980	05339	15470	48355	88651	22596	03152	19121
00116	88492	99382	14454	04504	20094	98977	74843	93413	22109	78508
00117	30934	47744	07481	83828	73788	06533	28597	20405	94205	20380
00118	22888	48893	27499	98748	60530	45128	74022	84617	82037	10268
00119	78212	16993	35902	91386	44372	15486	65741	14014	87481	37220

Table 12 *(Continued)*

Row number										
00120	41849	84547	46850	52326	34677	58300	74910	64345	19325	81549
00121	46352	33049	69248	93460	45305	07521	61318	31855	14413	70951
00122	11087	96294	14013	31792	59747	67277	76503	34513	39663	77544
00123	52701	08337	56303	87315	16520	69676	11654	99893	02181	68161
00124	57275	36898	81304	48585	68652	27376	92852	55866	88448	03584
00125	20857	73156	70284	24326	79375	95220	01159	63267	10622	48391
00126	15633	84924	90415	93614	33521	26665	55823	47641	86225	31704
00127	92694	48297	39904	02115	59589	49067	66821	41575	49767	04037
00128	77613	19019	88152	00080	20554	91409	96277	48257	50816	97616
00129	38688	32486	45134	63545	59404	72059	43947	51680	43852	59693
00130	25163	01889	70014	15021	41290	67312	71857	15957	68971	11403
00131	65251	07629	37239	33295	05870	01119	92784	26340	18477	65622
00132	36815	43625	18637	37509	82444	99005	04921	73701	14707	93997
00133	64397	11692	05327	82162	20247	81759	45197	25332	83745	22567
00134	04515	25624	95096	67946	48460	85558	15191	18782	16930	33361
00135	83761	60873	43253	84145	60833	25983	01291	41349	20368	07126
00136	14387	06345	80854	09279	43529	06318	38384	74761	41196	37480
00137	51321	92246	80088	77074	88722	56736	66164	49431	66919	31678
00138	72472	00008	80890	18002	94813	31900	54155	83436	35352	54131
00139	05466	55306	93128	18464	74457	90561	72848	11834	79982	68416
00140	39528	72484	82474	25593	48545	35247	18619	13674	18611	19241
00141	81616	18711	53342	44276	75122	11724	74627	73707	58319	15997
00142	07586	16120	82641	22820	92904	13141	32392	19763	61199	67940
00143	90767	04235	13574	17200	69902	63742	78464	22501	18627	90872
00144	40188	28193	29593	88627	94972	11598	62095	36787	00441	58997
00145	34414	82157	86887	55087	19152	00023	12302	80783	32624	68691
00146	63439	75363	44989	16822	36024	00867	76378	41605	65961	73488
00147	67049	09070	93399	45547	94458	74284	05041	49807	20288	34060
00148	79495	04146	52162	90286	54158	34243	46978	35482	59362	95938
00149	91704	30552	04737	21031	75051	93029	47665	64382	99782	93478
00150	94015	46874	32444	48277	59820	96163	64654	25843	41145	42820
00151	74108	88222	88570	74015	25704	91035	01755	14750	48968	38603
00152	62880	87873	95160	59221	22304	90314	72877	17334	39283	04149
00153	11748	12102	80580	41867	17710	59621	06554	07850	73950	79552
00154	17944	05600	60478	03343	25852	58905	57216	39618	49856	99326
00155	66067	42792	95043	52680	46780	56487	09971	59481	37006	22186
00156	54244	91030	45547	70818	59849	96169	61459	21647	87417	17198
00157	30945	57589	31732	57260	47670	07654	46376	25366	94746	49580
00158	69170	37403	86995	90307	94304	71803	26825	05511	12459	91314
00159	08345	88975	35841	85771	08105	59987	87112	21476	14713	71181

Table 12 *(Continued)*

Row number										
00160	27767	43584	85301	88977	29490	69714	73035	41207	74699	09310
00161	13025	14338	54066	15243	47724	66733	47431	43905	31048	56699
00162	80217	36292	98525	24335	24432	24896	43277	58874	11466	16082
00163	10875	62004	90391	61105	57411	06368	53856	30743	08670	84741
00164	54127	57326	26629	19087	24472	88779	30540	27886	61732	75454
00165	60311	42824	37301	42678	45990	43242	17374	52003	70707	70214
00166	49739	71484	92003	98086	76668	73209	59202	11973	02902	33250
00167	78626	51594	16453	94614	39014	97066	83012	09832	25571	77628
00168	66692	13986	99837	00582	81232	44987	09504	96412	90193	79568
00169	44071	28091	07362	97703	76447	42537	98524	97831	65704	09514
00170	41468	85149	49554	17994	14924	39650	95294	00556	70481	06905
00171	94559	37559	49678	53119	70312	05682	66986	34099	74474	20740
00172	41615	70360	64114	58660	90850	64618	80620	51790	11436	38072
00173	50273	93113	41794	86861	24781	89683	55411	85667	77535	99892
00174	41396	80504	90670	08289	40902	05069	95083	06783	28102	57816
00175	25807	24260	71529	78920	72682	07385	90726	57166	98884	08583
00176	06170	97965	88302	98041	21443	41808	68984	83620	89747	98882
00177	60808	54444	74412	81105	01176	28838	36421	16489	18059	51061
00178	80940	44893	10408	36222	80582	71944	92638	40333	67054	16067
00179	19516	90120	46759	71643	13177	55292	21036	82808	77501	97427
00180	49386	54480	23604	23554	21785	41101	91178	10174	29420	90438
00181	06312	88940	15995	69321	47458	64809	98189	81851	29651	84215
00182	60942	00307	11897	92674	40405	68032	96717	54244	10701	41393
00183	92329	98932	78284	46347	71209	92061	39448	93136	25722	08564
00184	77936	63574	31384	51924	85561	29671	58137	17820	22751	36518
00185	38101	77756	11657	13897	95889	57067	47648	13885	70669	93406
00186	39641	69457	91339	22502	92613	89719	11947	56203	19324	20504
00187	84054	40455	99396	63680	67667	60631	69181	96845	38525	11600
00188	47468	03577	57649	63266	24700	71594	14004	23153	69249	05747
00189	43321	31370	28977	23896	76479	68562	62342	07589	08899	05985
00190	64281	61826	18555	64937	13173	33365	78851	16499	87064	13075
00191	66847	70495	32350	02985	86716	38746	26313	77463	55387	72681
00192	72461	33230	21529	53424	92581	02262	78438	66276	18396	73538
00193	21032	91050	13058	16218	12470	56500	15292	76139	59526	52113
00194	95362	67011	06651	16136	01016	00857	55018	56374	35824	71708
00195	49712	97380	10404	55452	34030	60726	75211	10271	36633	68424
00196	58275	61764	97586	54716	50259	46345	87195	46092	26787	60939
00197	89514	11788	68224	23417	73959	76145	30342	40277	11049	72049
00198	15472	50669	48139	36732	46874	37088	63465	09819	58869	35220
00199	12120	86124	51247	44302	60883	52109	21437	36786	49226	77837

Appendix F

Answers to Exercises

Chapter 1

1.1 a. 110
     b. 75
     c. 209
     d. 5
     e. 19
     f. 412
     g. 992
     h. 37

1.2 a. 17
     b. -11
     c. 147
     d. 196

1.3 a. 125
     b. 256
     c. 9
     d. 10
     e. 64
     f. X^3Y^3
     g. 125
     h. 11

1.4 a. 112
     b. 53
     c. 2
     d. $X^2 + 2XY + Y^2 - X - Y$
     e. 72

1.5 a. 47
     b. 12
     c. 36
     d. 28

1.6 54

1.7 145

1.8 a. continuous
    b. continuous
    c. discrete
    d. discrete
    e. continuous

1.9 a. nominal
    b. ordinal
    c. interval
    d. ratio
    e. ratio
    f. nominal
    g. ratio
    h. interval
    i. interval
    j. nominal
    k. ordinal
    l. ratio
    m. ordinal
    n. ratio
    o. ratio

1.10 a. -7
    b. 19
    c. 14.025
    d. -7.65
    e. 78
    f. -12
    g. -5.098
    h. 18.30
    i. 5.1
    j. 15.30
    k. 3

Chapter 3

3.1 a. $\mu = 12.48$
    Mdn $= 13$
    Mo $= 14$
    b. Range $= 14$
    $\sigma^2 = 11.53$
    $\sigma = 3.40$

3.2 a. $\overline{X} = 10$

3.4 a. $\overline{X} = 77.4$ Mdn $= 79.5$
    b. $s^2 = 95.78$
    c. $s = 9.79$

3.6 a. $\mu_T = 69.965$
  b. $\mu_M = 70.569$
    $\mu_F = 69.482$
  c. $\mu_C = 60.870$
    $\mu_O = 80.043$
3.8 a. $\mu = 46.78$
  b. $\mu = 46.91$
3.9 $\Sigma (X_i - \mu)^2 = 54.40$
    $\Sigma (X_i - 7)^2 = 56.00$
3.10 $X_{10} = 37$

Chapter 4

4.1 a. $P_{15} = 103.20$
    $P_{45} = 113.26$
    $P_{80} = 125.39$
  b. $PR_{144} = 95.98$
    $PR_{115} = 51.24$
    $PR_{96} = 5.67$
4.3 a. $\overline{X} = 20.25$
    $s^2 = 25.66$
    $s = 5.07$
  b. $z_{23} = .542$
    $z_{20} = -.049$
    $z_{25} = .937$
    $z_{18} = -.444$
    $z_{18} = -.444$
    $z_{15} = -1.036$
    $z_{12} = -1.627$
    $z_{14} = -1.233$
    $z_{28} = 1.529$
    $z_{24} = .740$
    $z_{26} = 1.134$
    $z_{20} = -.049$
  c. $\overline{z} = 0$
    $s_z^2 = .998$ (round to 1.0)
    $s_z = .999$ (round to 1.0)
4.4 $T_{23} = 55.42$
  $T_{20} = 49.51$
  $T_{25} = 59.37$
  $T_{18} = 45.56$
  $T_{18} = 45.56$
  $T_{15} = 39.64$
  $T_{12} = 33.73$
  $T_{14} = 37.67$

$T_{28} = 65.29$
$T_{24} = 57.40$
$T_{26} = 61.34$
$T_{20} = 49.51$
$\bar{T} = 50$
$s_T^2 = 99.80$ (round to 100)
$s_T = 9.99$ (round to 10)

4.5 $z_V = -.50$
$z_M = .75$ $z_c = -.192$
$z_{GPA} = -.30$
$z_V = 1.50$
$z_M = 1.25$ $z_c = 1.388$
$z_{GPA} = 1.36$

4.6 $z_E = 1.00$ $T_E = 60$
$z_M = 1.00$ $T_M = 60$
$z_S = -.50$ $T_S = 45$

4.7 $\bar{X} = 24.25$
$s^2 = 25.66$
$s = 5.07$
z scores same as in 4.3

4.8 $z = 2.20$
$X = 27.24$

Chapter 5

5.1 a. $z_{56} = .75$
$z_{38} = -1.50$
$z_{63} = 1.625$
b. 145.1 (approximately 145)
87.25 (approximately 87)
174.6 (approximately 175)
c. 79.35 (approximately 79)
386.7 (approximately 387)
20.05 (approximately 20)
d. 345.75 (approximately 346)
42.25 (approximately 42)
447.2 (approximately 447)
e. $P_{35} = 46.92$
$P_{80} = 56.736$
$PR_{55} = 73.41$
$PR_{36} = 4.01$

5.2 a. 71.4
32.7
50.0
45.8

b. 1.4
   -1.2
   0.3
   -0.5
c. 61.2
   40.4
   52.4
   46.0

5.3 *John:*
National Norms $PR_{80} = 66.16$
Large City Norms $PR_{80} = 78.81$
Mary:
National Norms $PR_{65} = 20.24$
Large City Norms $PR_{65} = 42.07$

5.4 A: $P_{90} = 87.30$ (approximately 87) 87 and above
B: $P_{70} = 80.03$ (approximately 80) 80-87
C: $P_{30} = 69.97$ (approximately 70) 70-79
D: $P_{10} = 62.70$ (approximately 63) 63-67
F: 62 and below

5.5 $PR_{60} = 96.96$ $NCE_{60} = 89.375$
$PR_{43} = 32.03$ $NCE_{43} = 40.193$
$PR_{33} = 3.99$ $NCE_{33} = 13.208$

5.6   a. $z_{30} = -1.583$; area $= .0567$; approximately 6%
     b. $z_{50} = 1.143$; area $= .1265$; approximately 13%
     c. $z_{35} = -.542$ $X' = 37.027$

Chapter 6

6.1   a. negatively correlated
     b. positively correlated
     c. positively correlated
6.2 $\rho_S = -0.429$
6.3 $r = 0.568$
6.4 $r = -0.41$
6.5 $r = 0.483$
6.6 $r_{YZ} = 1.00$
6.7 $\rho_S = 0.345$
6.8 $r = -0.429$
6.9 $r = 0.0$

Chapter 7

7.1   a. 31.3
     b. 1.635
     c. -270.2

7.2 $\hat{Y} = 0.428X + 19.022$
7.3 Maximum value of $b = 2.2$
7.4 $\hat{Y} = -2.95X + 13.65$
7.5 $s_e = 2.94$
7.6 $\hat{Y} = 0.625X + 171.25$
7.7 $s_e = 17.32$; $z = 0.938$; area $= 0.1741$
7.8 $s_e = 4.77$; $z = -0.09$; area $= 0.4641$
7.9 $b = 0$; $a = \bar{Y}$
7.10 $r = 0.87$

Chapter 8

8.1 a. $\frac{1}{10}$
 b. $\frac{2}{10} = \frac{1}{5}$
 c. $\frac{5}{10} = \frac{1}{2}$
 d. $\frac{3}{10}$
8.2 a. 1.0
 b. $\frac{6}{10} = \frac{3}{5}$
 c. $\frac{7}{10}$
8.3 a. $.25$
 b. $.15$
 c. $.15$
8.4 a. $.2778$
 b. $.1665$
 c. $.1665$
8.5 a. $\frac{1}{52}$
 b. $\frac{1}{13}$
 c. $\frac{1}{4}$
 d. $.0003698$ with replacement
  $.0003771$ without replacement
 e. $.000455$ with replacement
  $.000181$ without replacement
 f. $.000455$ with replacement
  $.000483$ without replacement
8.6 $P(0) = .03125$
 $P(1) = .15625$
 $P(2) = .31250$
 $P(3) = .31250$
 $P(4) = .15625$
 $P(5) = .03125$
8.7 a. $P(0) = .05631$
 b. $P(1) = .18771$
 c. $P(2) = .28158$
 d. $P(<5) = .921878$

8.8　a. .2119
　　b. .1151
　　c. .2347
　　d. .0907
8.9　a. .9429
　　b. .0823
　　c. .4562
　　d. .6772
8.10 a. The distribution is normal.

$\mu = 120$

$\sigma_{\bar{X}} = 2.083$

　　b. The distribution is normal.

$\mu = 120$

$\sigma_{\bar{X}} = 1.25$

8.11 a. .2509 (interpolated)
　　b. .1938 (interpolated)
　　c. .6495 (interpolated)
　　d. .6133 (interpolated)
8.12 a. .1314
　　b. .0749
　　c. .7389
　　d. .6844
8.13 a. .9250 (interpolated)
　　b. <.0001
　　c. .8194 (interpolated)
　　d. <.0001

Chapter 9

9.1 a. H_0: $\mu = 19{,}000$　　H_a: $\mu > 19{,}000$
　　b. $t = 7.10$ (rounded)
　　c. $t_{cv} = +1.672$
　　d. Reject H_0: $p < .05$.
9.2 a. H_0: $\mu = 10$　　H_a: $\mu > 10$
　　b. $t = 2.27$ (rounded)
　　c. $t_{cv} = +2.65$
　　d. Retain H_0: $p > .01$.
9.3 a. One-tailed test
　　b. $\alpha = .05$
　　c. $n = 400$
9.4 a. H_0: $\mu = 5.4$　　H_a: $\mu \neq 5.4$
　　b. $t = -2.55$ (rounded)
　　c. $t_{cv} = \pm 2.06$
　　d. Reject H_0: $p < .05$.

9.5 a. H_0: $\mu = 1.6$ H_a: $\mu < 1.6$

    b. $t = -1.10$ (rounded)

    c. $t_{cv} = -2.718$

    d.   Retain H_0: $p > .01$.

9.6 a. Nothing can be said; t_{cv} is smaller for a one-tailed test than for a two-tailed test.

    b. H_0: $\mu = a$ will be rejected in favor of H_a: $\mu > a$.

9.7 a. Type I error

    b.   Use a smaller α.

Chapter 10

10.1 $CI_{95} = 21,200 \pm (2.00)(309.84)$

         $= (20,580.32, 21,819.68)$

10.2 $CI_{99} = 11.64 \pm (3.012)(0.72)$

           $(9.47, 13.81)$

10.3 First estimate: $n = 96.04$ (using $t_{cv} = 1.96$)

     Final estimate: $n = 98.01$ (using $t_{cv} = 1.98$)

10.4 $CI_{95} = 5.2 \pm (2.06)(0.08)$

         $= (5.04, 5.36)$

10.5 $CI_{99} = 1.49 \pm (3.106)(0.10)$

         $= (1.18, 1.80)$

10.6 $CI_{90} = (23.46, 38.94)$

     $CI_{99} = (18.46, 43.94)$

10.7 For $n = 150$: $CI_{95} = (22.226, 22.974)$

     For $n = 10$: $CI_{95} = (20.926, 24.274)$

Chapter 11

11.1 a. H_0: $\mu_1 = \mu_2$ H_a: $\mu_1 > \mu_2$

    b. $t = 2.80$ (rounded)

    c. $t_{cv} = +1.68$

    d. Reject H_0: $p < .05$.

    e. $CI_{95} = (0.113, 0.687)$

11.2 a. H_0: $\mu_1 = \mu_2$ H_a: $\mu_1 \neq \mu_2$

    b. $t = -0.656$ (rounded)

    c. $t_{cv} = \pm 2.878$

    d. Retain H_0: $p > .01$.

    e. $CI_{99} = (-4.31, 2.71)$

11.3 a. H_0: $\mu_1 = \mu_2$ H_a: $\mu_1 > \mu_2$

    b. $t = 2.37$ (rounded)

    c. $t_{cv} = +1.895$

    d. Reject H_0: $p < .05$.

    e. $CI_{95} = (0.007, 3.743)$

11.4 a. H_0: $\mu_1 = \mu_2$ H_a: $\mu_1 < \mu_2$
b. $t = -1.01$ (rounded)
c. $t_{cv} = -1.676$
d. Retain H_0: $p > .05$.
e. $CI_{95} = (-11.96, 3.96)$
11.5 a. H_0: $\mu_1 = \mu_2$ H_a: $\mu_1 \neq \mu_2$
b. $t = 1.13$ (rounded)
c. $t_{cv} = \pm 2.365$
d. Retain H_0; $p > .05$.
e. $CI_{95} = (-1.375, 3.875)$

Chapter 12

12.1 a. H_0: $P = .07$ H_a: $P < .07$
b. $s_p = 0.0196$, $z = -1.53$
c. $z_{cv} = -2.326$
d. Retain H_0: $p > .05$.
e. $CI_{99} = (-.0105, .0905)$
12.2 a. H_0: $P_1 = P_2$ H_a: $P_1 > P_2$
b. $s_{P_1 - P_2} = 0.050$ $z = 1.90$
c. $z_{cv} = +1.645$
d. Reject H_0: $p < .05$.
12.3 H_0: $\rho = 0$ H_a: $\rho \neq 0$
a1. $r_{cv} = .316$ Reject H_0: $p < .05$.
a2. $r_{cv} = .403$ Retain H_0: $p > .01$.
b1. $r_{cv} = .274$
b2. $r_{cv} = .303$
12.4 a. H_0: $\rho = -0.21$ H_a: $\rho \neq -0.21$
b. $s_{z_r} = 0.192$ $z = 0.587$
c. $z_{cv} = \pm 1.96$
d. Reject H_0: $p < .05$
e. $CI_{95} = (-0.476, 0.276)$ for z_r
       $(-0.443, 0.269)$ for r (interpolated)
12.5 a. H_0: $\rho_1 = \rho_2$ H_a: $\rho_1 \neq \rho_2$
b. $s_{z_{r_1} - z_{r_2}} = 0.115$ $z = -0.521$
c. $z_{cv} = \pm 1.96$
d. Retain H_0: $p > .05$.
12.6 a. H_0: $\sigma^2 = 144$ H_a: $\sigma^2 \neq 144$
b. $\chi^2 = 39.06$
c. $\chi^2_{cv} \leq 14.611$ or ≥ 37.652
d. Reject H_0: $p < .05$.
12.7 a. H_0: $\sigma_1^2 = \sigma_2^2$ H_a: $\sigma_1^2 \neq \sigma_2^2$
b. $F = 1.34$
c. $F_{cv} = 2.17$
d. Retain H_0: $p > .10$

12.8 a. H_0: $\rho = 0$ H_a: $\rho \neq 0$
    b. $r = .20$
    c. $r_{cv} = .361$
    d. Retain H_0: $p > .05$.
    e. $CI_{95} = (-0.173, 0.579)$ for z_r
        $= (-0.171, 0.522)$ for r (interpolated)

12.9 a. H_0: $\rho = 0.35$ H_a: $\rho \neq 0.35$
    b. $s_{z_r} = 0.192$ $z = -0.844$
    c. $z_{cv} = \pm 1.96$
    d. Retain H_0: $p > .05$.
    e. Yes, the interval contains both 0.00 and 0.35.

12.10 a. H_0: P $= .60$ H_a: P $> .60$
    b. $s_p = .034$ $z = 1.471$
    c. $z_{cv} = +1.645$
    d. Retain H_0: $p > .05$; vote against bill.

12.11 a. H_0: P $= .60$ H_a: P $> .60$
    b. $s_p = .034$ $z = 1.818$
    c. $z_{cv} = +1.645$
    d. Reject H_0: $p < .05$; vote for bill.

12.12 a. H_0: $\sigma_1^2 = \sigma_2^2$ H_a: $\sigma_1^2 \neq \sigma_2^2$
    b. $F = 2.18$
    c. $F_{cv} = 2.35$
    d. Retain H_0: $p > .10$; there is no violation of the assumption of homogeneity of variance.

Chapter 13

13.1 a. H_0: $\mu_1 = \mu_2 = \mu_3 = \mu_4$ H_a: $\mu_i \neq \mu_j$ for some i,j
    b.

Summary ANOVA

Source	SS	df	MS	F	F_{cv}
Between	3.06	3	1.02	5.231	2.95
Within	5.46	28	0.195		
Total	8.52	31			

    c. Reject H_0: $p < .05$.
    d. Scheffé tests
        $F_{1-2} = 5.6481$ $F_{2-3} = 0.0128$
        $F_{1-3} = 6.1988$ $F_{2-4} = 2.3341$
        $F_{1-4} = 15.2440^*$ $F_{3-4} = 2.0012$
        Critical value $= 3(2.95) = 8.85$

        Tukey tests
        $Q_{1-2} = -3.3632$ $Q_{2-3} = -0.1602$
        $Q_{1-3} = -3.5234$ $Q_{2-4} = -2.1621$
        $Q_{1-4} = -5.5253^*$ $Q_{3-4} = -2.0019$
        Critical value $Q_{cv} = 3.87$

13.2 a. H_0: $\mu_1 = \mu_2 = \mu_3$ H_a: $\mu_i \neq \mu_j$ for some i,j
b.

Summary ANOVA

Source	SS	df	MS	F	F_{cv}
Between	11.112	2	5.556	1.01	3.68
Within	82.499	15	5.500		
Total	93.611	17			

c. Retain H_0: $p > .05$.

13.3 a. H_0: $\mu_1 = \mu_2 = \mu_3 = \mu_4 = \mu_5$ H_a: $\mu_i \neq \mu_j$ for some i,j
b.

Summary ANOVA

Source	SS	df	MS	F
Between	58	4	14.50	3.06
Within	166	35	4.74	
Total	224	39		

If $\alpha = .01$, retain H_0.
If $\alpha = .05$, reject H_0.

13.4 a. H_0: $\mu_1 = \mu_2 = \mu_3 = \mu_4 = \mu_5$ H_a: $\mu_i \neq \mu_j$ for some i,j
b.

Summary ANOVA

Source	SS	df	MS	F	F_{cv}
Occasions	77.668	4	19.417	13.703	4.43
Individuals	72.167	5	14.433		
Residual	28.332	20	1.417		
Total	178.167	29			

c. Reject H_0: $p < .01$.
d. Tukey tests

$Q_{1-2} = -0.700$ $Q_{2-4} = -5.823*$
$Q_{1-3} = -3.087$ $Q_{2-5} = -7.881*$
$Q_{1-4} = -6.523*$ $Q_{3-4} = -3.436$
$Q_{1-5} = -8.581*$ $Q_{3-5} = -5.494*$
$Q_{2-3} = -2.387$ $Q_{4-5} = -2.058$

13.5 a. H_0: $\mu_1 = \mu_2 = \mu_3$ H_a: $\mu_i \neq \mu_j$ for some i,j
b.

Summary ANOVA

Source	SS	df	MS	F	F_{cv}
Occasions	12.086	2	6.043	92.97	2.73
Individuals	5.665	7	0.809		
Residual	0.909	14	0.065		
Total	18.660	23			

c. Reject H_0: $p < .10$.

d. Scheffé tests
$Q_{1-2} = 41.885*$ Critical value $= 2(2.73) = 5.46$
$Q_{1-3} = 185.886*$
$Q_{2-3} = 51.297*$

13.6 a. H_0: $\mu_1 = \mu_2 = \mu_3 = \mu_4 = \mu_5$ H_a: $\mu_i \neq \mu_j$ for some i,j
b.

Summary ANOVA

Source	SS	df	MS	F	F_{cv}
Test Occasion	152	4	38	1.490	2.87
Persons	260	5	52		
Residual	510	20	25.5		
Total	922	29			

c. Retain H_0: $p > .05$.

13.7 a. H_0: $\mu_1 = \mu_2 = \cdots = \mu_8$ H_a: $\mu_i \neq \mu_j$ for some i,j
b.

Summary ANOVA

Source	SS	df	MS	F	F_{cv}
Between	3527.25	7	503.89	3.56	2.25
Within	5666.00	40	141.65		
Total	9193.25	47			

c. Reject H_0: $p < .05$.
d. Scheffé tests
$F_{4-6} = 1.715$ Critical value $= 7(2.25) = 15.75$
$F_{2-7} = 6.121$
$F_{4-7} = 12.199$

Tukey tests
$Q_{4-6} = 1.852$
$Q_{2-7} = 3.499$ Critical value $= Q_{cv} = 4.39$
$Q_{4-7} = 4.939*$

Chapter 14

14.1 a. H_0: $\mu_{1.} = \mu_{2.}$ H_a: $\mu_{1.} \neq \mu_{2.}$ (party main effect)
H_0: $\mu_{.1} = \mu_{.2}$ H_a: $\mu_{.1} \neq \mu_{.2}$ (age main effect)
H_0: All $(\mu_{rc} - \mu_{r.} - \mu_{.c} + \mu) = 0$
H_a: $(\mu_{rc} - \mu_{r.} - \mu_{.c} + \mu) \neq 0$ for some r,c (interaction)
b.

Summary ANOVA

Source	SS	df	MS	F	F_{cv}
Rows	360	1	360	81.82	4.12
Columns	490	1	490	111.36	4.12
Interaction	10	1	10	2.27	4.12
Within Cell	158.40	36	4.40		
Total	1018.40	39			

14.2 a. H_0: $\mu_{1.} = \mu_{2.} = \mu_{3.} = \mu_{4.}$ H_a: $\mu_{i.} \neq \mu_{j.}$ for some i,j (time)
   H_0: $\mu_{.1} = \mu_{.2} = \mu_{.3}$ H_a: $\mu_{.i} \neq \mu_{.j}$ for some i,j (drug)
   H_0: All $(\mu_{rc} - \mu_{r.} - \mu_{.c} + \mu) = 0$
   H_a: $(\mu_{rc} - \mu_{r.} - \mu_{.c} + \mu) \neq 0$ for some r,c (interaction)
   b.

Summary ANOVA

Source	SS	df	MS	F	F_{cv}
Time	150	3	50	14.62	4.22
Drug	40	2	20	5.85	5.08
Interaction	20	6	3.33	0.97	3.20
Within Cell	164	48	3.42		
Total	374	59			

14.3 a. H_0: $\mu_{1.} = \mu_{2.}$ H_a: $\mu_{1.} \neq \mu_{2.}$ (sex)
   H_0: $\mu_{.1} = \mu_{.2} = \mu_{.3}$ H_a: $\mu_{.i} \neq \mu_{.j}$ for some i,j (stressor)
   H_0: All $(\mu_{rc} - \mu_{r.} - \mu_{.c} + \mu) = 0$
   H_a: $\mu_{rc} - \mu_{r.} - \mu_{.c} + \mu \neq 0$ for some r,c (interaction)
   b.

Summary ANOVA

Source	SS	df	MS	F	F_{cv}
Rows	3.34	1	3.34	1.06	7.82
Columns	188.07	2	94.04	29.85	5.61
Interaction	146.46	2	73.23	23.25	5.61
Within Cell	75.60	24	3.15		
Total	413.47	29			

Chapter 15

15.1 a. H_0: Independence H_a: Dependence
   b. $\chi^2 = 4.0776$
   c. df $= 2$; $\chi^2_{cv} = 5.991$
   d. Retain H_0; $p > .05$.
15.2 a. $\chi^2 = 87.5951$
   b. df $= 4$; $\chi^2_{cv} = 13.277$
   c. Reject the hypothesis of goodness of fit; $p < .01$.
15.3 a. H_o: Independence H_a: Dependence
   Ib. $\chi^2 = 20.5249$
   c. df $= 2$; $\chi^2 = 5.991$
   d. Reject H_0; $p < .05$.
   IIb. $\chi^2 = 49.6933$
   c. df $= 4$; $\chi^2_{cv} = 9.488$
   d. Reject H_0; $p < .05$.
   IIIb. $\chi^2 = 11.4829$
   c. df $= 6$; $\chi^2_{cv} = 12.592$
   d. Retain H_0; $p > .05$.

15.4   Ia. $C = 0.237$
     b. $C_{max} = 0.707$
     IIa. $C = .354$
     b. $C_{max} = 0.817$
     IIIb. $C = 0.179$
         $C_{max} = 0.817$
15.5 a. $H_0: A = D$ $H_a: A \neq D$
     b. $\chi^2 = 4.55$
     c. $\chi^2_{cv} = 3.841$
     d. Reject H_0; $p < .05$.
15.6   *Median test*
     a. $H_0: \text{Mdn}_I = \text{Mdn}_{II}$
     b. Common median $= \dfrac{16 + 17}{2} = 16.5$
        $\chi^2 = 0.80$
     c. $\chi^2_{cv} = 3.841$
     d. Retain H_0; $p > .05$.
        Mann–Whitney U test
     a. H_0: Correct Responses$_I$ = Correct Responses$_{II}$
     b. $U_1 = 25$
     c. $U_{cv} = 27$
     d. Reject H_0; $p < .05$.
15.7 a. H_0: No differences between absentee rates
        H_a: Differences between absentee rates
     b. $H = 5.46$
     c. $H_{cv} = 7.82$
     d. Retain H_0; $p > .05$.
15.8 a. H_0: No differences in number of programming steps
     b. $T = 7$
     c. $T_{cv} = 10$
     d. Reject H_0; $p < .05$.

Index

Number	Page	Formula

Number	Page	Formula
12.2	229	$z = \dfrac{\text{statistic} - \text{hypothesized parameter}}{\text{standard error of the statistic}}$
12.4	231	$s_{p1-p2} = \sqrt{pq\left(\dfrac{1}{n_1} + \dfrac{1}{n_2}\right)}$
12.6	234	$z_r = \dfrac{1}{2}\log_e (1+r) - \dfrac{1}{2}\log_e (1-r)$
12.7	234	$s_{z_r} = \sqrt{\dfrac{1}{n-3}}$
12.10	236	$t = r\sqrt{\dfrac{n-2}{1-r^2}}$
12.11	237	$s_{z_{r1}-z_{r2}} = \sqrt{\dfrac{1}{n_1-3} + \dfrac{1}{n_2-3}}$
12.13	240	$\chi^2 = \dfrac{(n-1)s^2}{a}$
12.14	242	$F = \dfrac{s_1^2}{s_2^2}$
13.11	262	$SS_B = \text{sum of squares } \textit{between groups}$
		$= \displaystyle\sum_{j=1}^{k} n_j(\bar{X}_j - \bar{X})^2$
		$= \displaystyle\sum_{j=1}^{k} \left(\dfrac{T_j^2}{n_j}\right) - \dfrac{T^2}{N}$
3.12	262	$SS_W = \text{sum of squares } \textit{within groups}$
		$= \displaystyle\sum_{j=1}^{k}\sum_{i=1}^{n_j} (X_{ij} - \bar{X}_j)^2$
		$= \displaystyle\sum_{j=1}^{k}\sum_{i=1}^{n_j} X_{ij}^2 - \sum_{j=1}^{k}\left(\dfrac{T_j^2}{n_j}\right)$

Number	Page	Formula
13.13	262	$SS_T = \text{sum of squares } \textit{total}$
		$= \displaystyle\sum_{j=1}^{k}\sum_{i=1}^{n_j} (X_{ij} - \bar{X})^2$
		$= \displaystyle\sum_{j=1}^{k}\sum_{i=1}^{n_j} X_{ij}^2 - \dfrac{T^2}{N}$
13.14	266	$F = \dfrac{(\bar{X}_i - \bar{X}_j)^2}{MS_W\left(\dfrac{1}{n_i} + \dfrac{1}{n_j}\right)}$
13.15	268	$Q = \dfrac{\bar{X}_i - \bar{X}_j}{\sqrt{MS_W/n_j}}$
13.17	270	$SS_I = \displaystyle\sum_{i=1}^{n_j} \dfrac{T_i^2}{k} - \dfrac{T^2}{N}$
13.18	270	$SS_O = \displaystyle\sum_{j=1}^{k} \left(\dfrac{T_j^2}{n_j}\right) - \dfrac{T^2}{N}$
14.4	288	$SS_R = \text{Sum of squares for rows}$
		$= \dfrac{1}{nC}\displaystyle\sum_{r=1}^{R} T_r^2 - \dfrac{T^2}{N}$
14.5	288	$SS_C = \text{Sum of squares for columns}$
		$= \dfrac{1}{nR}\displaystyle\sum_{c=1}^{C} T_c^2 - \dfrac{T^2}{N}$
14.6	288	$SS_{RC} = \text{Sum of squares for interaction}$
		$= \dfrac{1}{n}\displaystyle\sum_{c=1}^{C}\sum_{r=1}^{R} T_{rc}^2 - \dfrac{1}{nC}\sum_{r=1}^{R} T_r^2$
		$- \dfrac{1}{nR}\displaystyle\sum_{c=1}^{C} T_c^2 + \dfrac{T^2}{N}$